Progress in
PHYSICAL ORGANIC CHEMISTRY

VOLUME 7

Progress in

PHYSICAL
ORGANIC
CHEMISTRY

VOLUME 7

Editors

ANDREW STREITWIESER, JR. *Department of Chemistry*
University of California, Berkeley, California

ROBERT W. TAFT, *Department of Chemistry*
University of California, Irvine, California

1970

INTERSCIENCE PUBLISHERS

a division of John Wiley & Sons · New York · London · Sydney · Toronto

The paper used in this book has pH of 6.5
or higher. It has been used because the
best information now available indicates
that this will contribute to its longevity.

2   3   4   5   6   7   8   9   10

Copyright © 1970 by John Wiley and Sons, Inc.

Library of Congress Catalog Card Number 63-19364

SBN-471   83353   3

PRINTED IN THE UNITED STATES OF AMERICA

Introduction to the Series

Physical organic chemistry is a relatively modern field with deep roots in chemistry. The subject is concerned with investigations of organic chemistry by quantitative and mathematical methods. The wedding of physical and organic chemistry has provided a remarkable source of inspiration for both of these classical areas of chemical endeavor. Further, the potential for new developments resulting from this union appears to be still greater. A closening of ties with all aspects of molecular structure and spectroscopy is clearly anticipated. The field provides the proving ground for the development of basic tools for investigations in the areas of molecular biology and biophysics. The subject has an inherent association with phenomena in the condensed phase and thereby with the theories of this state of matter.

The chief directions of the field are: (a) the effects of structure and environment on reaction rates and equilibria; (b) mechanism of reactions; and (c) applications of statistical and quantum mechanics to organic compounds and reactions. Taken broadly, of course, much of chemistry lies within these confines. The dominant theme that characterizes this field is the emphasis on interpretation and understanding which permits the effective practice of organic chemistry. The field gains its momentum from the application of basic theories and methods of physical chemistry to the broad areas of knowledge of organic reactions and organic structural theory. The nearly inexhaustible diversity of organic structures permits detailed and systematic investigations which have no peer. The reactions of complex natural products have contributed to the development of theories of physical organic chemistry, and, in turn, these theories have ultimately provided great aid to the elucidation of structures of natural products.

Fundamental advances are offered by the knowledge of energy states and their electronic distributions in organic compounds and the relationship of these to reaction mechanisms. The development, for example, of even an empirical and approximate general scheme for the estimation of activation energies would indeed be most notable.

The complexity of even the simplest organic compounds in terms of physical theory well endows the field of physical organic chemistry with

v

the frustrations of approximations. The quantitative correlations employed in this field vary from purely empirical operational formulations to the approach of applying physical principles to a workable model. The most common procedures have involved the application of approximate theories to approximate models. Critical assessment of the scope and limitations of these approximate applications of theory leads to further development and understanding.

Although he may wish to be a disclaimer, the physical organic chemist attempts to compensate his lack of physical rigor by the vigor of his efforts. There has indeed been recently a great outpouring of work in this field. We believe that a forum for exchange of views and for critical and authoritative reviews of topics is an essential need of this field. It is our hope that the projected periodical series of volumes under this title will help serve this need. The general organization and character of the scholarly presentations of our series will correspond to that of the several prototypes, e.g., *Advances in Enzymology*, *Advances in Chemical Physics*, and *Progress in Inorganic Chemistry*.

We have encouraged the authors to review topics in a style that is not only somewhat more speculative in character but which is also more detailed than presentations normally found in textbooks. Appropriate to this quantitative aspect of organic chemistry, authors have also been encouraged in the citation of numerical data. It is intended that these volumes will find wide use among graduate students as well as practicing organic chemists who are not necessarily expert in the field of these special topics. Aside from these rather obvious considerations, the emphasis in each chapter is the personal ideas of the author. We wish to express our gratitude to the authors for the excellence of their individual presentations.

We greatly welcome comments and suggestions on any aspect of these volumes.

ANDREW STREITWIESER, JR.
ROBERT W. TAFT

Contents

The Influence of Geometry on the Electronic Structure and Spectra of Planar Polyenes

By Howard E. Simmons

Central Research Department, Experimental Station,

E. I. du Pont de Nemours & Company, Wilmington, Delaware

CONTENTS

I. INTRODUCTION

The electronic structure of the polyenes has been studied extensively because of the importance of such molecules in synthetic, theoretical, and biochemistry. The mobile π-electrons are responsible for most of their interesting chemical and physical properties, and their ground and lower excited states are qualitatively understood from molecular orbital, valence-bond, free-electron, and exciton descriptions (1).

1

This chapter discusses some properties of the planar, linear di-, tri-, and tetraenes as a function of their geometry by calculation of their lower singlet and triplet states. The calculations are extended to a few long polyenes, cross-conjugated polyenes, and the cyclic radialenes, and they also allow a quantitative comparison to be made of the Hückel, Longuet-Higgins and Salem, and self-consistent field wavefunctions as bases for a configuration interaction description of the polyenes.

II. THEORY AND CALCULATIONS

One-dimensional, one-electron wavefunctions ψ_i for a system of N carbon atoms, first given by Hückel (2), are only partially determined by the translational symmetry of the linear polyenes,

$$\psi_i^H = \left\{ \frac{2}{N+1} \right\}^{1/2} \sum_{p=1}^{N} \sin\left(\frac{ip\pi}{N+1}\right) \cdot \chi_p \tag{1}$$

where χ_p is a localized $2p\pi$ atomic function, $i, j, \ldots$ label molecular orbitals, and $p, q, \ldots$ label atomic orbitals. Whereas the molecular orbitals of the cyclic polyenes (annulenes) are fully determined by their symmetry and correspond to traveling waves, those of the linear polyenes correspond to standing waves and their form is governed by the fixed nodes at the ends of the chain, i.e., by the precise formulation of the potential energy. The coefficients in eq. (1) are those prescribed by the simplest Hückel description which assumes an effective Hamiltonian whose matrix elements are constants, such that diagonal elements roughly give the energy of a $2p\pi$ electron on an isolated carbon atom p (α_p) and the off-diagonal elements are the binding energy between adjacent carbon $2p\pi$ atomic orbitals (β_{pq}). If the geometry of an annulene is not perfectly polygonal, then the molecular orbitals are influenced by the potential field of the nuclei in a manner that is not fully symmetry determined. Similarly, the specific geometry of a linear polyene of given length influences the molecular orbital coefficients. Hückel molecular orbitals do not take into account the geometry of the linear and cyclic polyenes and so do not distinguish differently shaped polyenes of the same size. It is only when the coulombic repulsion between electrons is introduced into π-electron theory that the appropriately modified orbitals are generated.

Hückel molecular orbitals are qualitatively useful but are a poor basis for detailed calculations of the properties of ground and excited states; for instance, they imply that as the chain becomes very long the central regions of polyenes undergo smoothing until all bonds are of equal length (3), and that the energy gap between the highest occupied and lowest vacant orbitals vanishes. This can be readily seen by considering the energy of the ith Hückel molecular orbital $\varepsilon_i = 2\beta \cos{(i\pi/N + 1)}$ where $i = 1, 2, \ldots, N$. The highest occupied and lowest vacant molecular orbitals in a linear polyene occur with $i = N/2$ and $(N/2) + 1$, respectively, and the argument of the cosine thus reduces to $\pi/2$ when N becomes large in both cases. This suggests that the longest wavelength transition in a polyene, which corresponds to absorption of a photon of energy equal to this gap, would occur at infinite wavelength. Since bond alternation and the existence of a large spectral limit are experimentally observed in the polyenes, the Hückel wavefunctions are clearly inadequate. These deficiencies are due in large part to neglect of the σ-electrons, and Lennard-Jones and Turkevich (4) showed how compression of the framework σ-bonds can be taken into account; Longuet-Higgins and Salem (5) developed this idea and gave a scheme for calculations. Dewar and Schmeising (6) have finally made compelling arguments for the role of hybridization and the small role of ground-state conjugation in polyenes, a view long held by Simpson (7).

The Longuet-Higgins and Salem theory shows how the π-electron energy (E^π) and σ-electron energy (E^σ) can be simultaneously taken into account under the framework of Hückel theory. It is only by the proper balancing of E^π and E^σ that bond distances can be adequately treated in conjugated olefins, where the total energy (E) of a conjugated hydrocarbon is $E = E^\pi + E^\sigma$. The π-electron energy is a function of the resonance integrals of *all* the (C—C) bonds, and the σ-electron energy is the sum of independent contributions from these bonds, i.e., $E^\pi = E^\pi(\beta_1, \beta_2, \ldots, \beta_i, \ldots)$ and $\beta_i = \beta(r_i)$, where r_i is the length of bond i, and $E^\sigma = \sum_i u(r_i)$. An expression for the independent contributions $u(r_i)$ can be derived from the vibrational force constants of, e.g., benzene, and from the linear relationship of bond length and bond order (see below). The form of the functions employed in the present work differs slightly from that in ref. 5 because of the adoption of more recent values of bond lengths and spectral parameters; these are r (in Å) $= 1.513 - 0.175P$, $u(r_i) = 12.270\,(r_i - 1.513 + a)\beta(r_i)$, where $\beta(r_i) = \beta_0\,\exp[-(r_i - 1.397)/a]$ and $a = 0.3624$ Å, $\beta_0 = 34.79$ kcal/mole. π-Wavefunctions can then be calculated by simply employing the appropriate $\beta(r_i)$.

The Longuet-Higgins and Salem wavefunctions resemble the self-consistent field wavefunctions to a considerable degree as shown in this study. Polyene molecular orbitals were calculated by the Hückel (HMO), Longuet-Higgins and Salem (L-H/S), and self-consistent field (SCF) methods and compared in terms of their ground-state properties. All three types of wavefunctions were used as bases for antisymmetrized molecular orbital calculations including configuration interaction (ASMO-CI) of excited states.

The SCFMO, ψ_i, were generated by solving the Hartree–Fock equations, eq. (2),

$$\sum_q F_{pq}C_{iq} = \varepsilon_i \sum_q S_{pq}C_{iq} \tag{2}$$

where the matrix elements F_{pq} are given by Pople's theory (8) as eqs. (3)–(7). U_{pp} is close to the valence-state energy of an electron in χ_p, β_{pq} is

$$F_{pp} = U_{pp} + \tfrac{1}{2} P_{pp}\gamma_{pp} + \sum_{r \neq p} (P_{rr} - 1)\gamma_{pr} \tag{3}$$

$$F_{pq} = \beta_{pq} - \tfrac{1}{2} P_{pq}\gamma_{pq}(p \neq q) \tag{4}$$

where

$$P_{pq} = 2 \sum_i^{occ} C_{ip}C_{iq} \tag{5}$$

$$U_{pp} = \langle \chi_p| - \tfrac{1}{2}\nabla^2 + V_p |\chi_p\rangle \tag{6}$$

$$\beta_{pq} = \langle \chi_p| - \tfrac{1}{2}\nabla^2 + V_q |\chi_q\rangle \tag{7}$$

$$\gamma_{pq} = \left\langle \chi_p(1)\chi_q(2) \left| \frac{1}{r_{12}} \right| \chi_p(1)\chi_q(2) \right\rangle \tag{8}$$

the resonance integral, γ_{pq} is the coulomb repulsion integral, P_{pq} is twice the corresponding element of the density matrix, and the other symbols have their usual meanings. The β_{pq} were computed theoretically from eq. (7), and the Goeppert–Mayer and Sklar form of the core potentials V_p was used throughout (9). Coulomb repulsion integrals for $r \geq 2.40$ Å were calculated from the usual theoretical expression, whereas for $r < 2.40$ Å these values were taken from an empirical curve which was determined from the spectrum of benzene (9): $\gamma_{pq} = 9.717 - 5.070r + 2.674r^2 - 0.526r^3$ (γ_{pq} in eV, r in Å). The one-center repulsion integral determined by this method is 9.717 eV and was used throughout, rather than the familiar Pariser value $\gamma_{11} = I - A = 10.53$ eV.

From an assumed geometry and starting set of HMO, the SCFMO ψ_i which correspond *to that geometry* can be calculated. Subsequent minimization of the energy with respect to the atomic coordinates is too heavy a computation, and to overcome this difficulty, a procedure used by Dewar (10) was employed in the following manner. The bond orders of ethylene, benzene, and graphite are fully determined by symmetry, and a plot of bond order P_{pq} vs. r_{pq} for these molecules closely determines a straight

$$r_{pq}(\text{in Å}) = 1.513 - 0.175 P_{pq} \tag{9}$$

line, eq. (9).* At the completion of an SCF calculation, new coordinates were computed by eq. (9), all integrals were reevaluated, and the SCF calculation was then cycled iteratively. The calculations converged rapidly in all cases to a final geometry which was self-consistent with respect to further variations in β_{pq} and γ_{pq}. All SCFMO were calculated in this manner and represent "best" wavefunctions of this type according to the usual π-electron theory.

The ψ_i^H and ψ_i serve as bases to describe electron configurations; e.g., a closed-shell determinantal wavefunction which closely resembles the ground state is given by eq. (10). Excited configurations arise when an electron is promoted from an orbital i, occupied in V_0, to some vacant

$$V_0 = (N!)^{-1/2}|\psi_1(1)\alpha(1)\psi_1(2)\beta(2)\cdots\psi_{N/2}(N-1)\alpha(N-1)\psi_{N/2}(N)\beta(N)|$$
$$= (1\ \bar{1}\ \cdots N/2\ \overline{N/2}) \tag{10}$$

orbital, k'; this gives a singlet and a triplet configuration, eq. (11), where only the $S_z = 0$ component of the triplet is considered and $V_{ik'}$ (upper sign) refers to the singlet and $T_{ik'}$ (lower sign) to the triplet. The energy of a, e.g., singlet configuration, $E(V_{ik'}) = \langle V_{ik'} \mid \mathbf{H} \mid V_{ik'} \rangle$,

$$\begin{cases} V_{ik'} \\ T_{ik'} \end{cases} = 2^{-1/2}[(1\ \bar{1}\cdots i\bar{k}'\cdots N/2\ \overline{N/2}) \mp (1\ \bar{1}\cdots ik'\cdots N/2\ \overline{N/2})] \tag{11}$$

is calculated using the full π-electron Hamiltonian $\mathbf{H}$,

$$\mathbf{H}(1,2,\cdots,N) = \sum_{\mu}^{N} \mathbf{H}_{\text{core}}(\mu) + \sum_{\mu > \nu}^{N} r_{\mu\nu}^{-1} \tag{12}$$

$$\mathbf{H}_{\text{core}}(\mu) = -\tfrac{1}{2}\mathbf{\nabla}_{\mu}^2 + \sum_{p} \mathbf{V}_p(\mu) + \sum_{r} \mathbf{V}_r^*(\mu) \tag{13}$$

where the sums μ,ν are over π-electrons, the $\mathbf{V}_p$ are potential energy operators, the sum p is over core nuclei, and the sum r is over atoms uncharged in the core whose potential energy operators are $\mathbf{V}_r^*$.

The theory of Pariser and Parr (11) was used to compute all CI matrix elements, and the matrices were factored by symmetry and solved. Transition moments, oscillator strengths, and density matrices of the final CI wavefunctions were computed for all states. The quite general formulas needed for the CI matrix elements have been given in detail by Pariser (13) and are not reproduced here. In all calculations (except that of β_{pq}), the overlap conditions $S_{pq} = \delta_{pq}$ was invoked in the usual spirit (8,11) and further details of the computations are given in the Appendix.

* This equation was used to calculate L-H/S and SCFMO and is similar to others in the literature (5,10).

The theory of the electronic spectra of alternant hydrocarbons has been developed largely by Moffitt (12), Pariser and Parr (11,13), Pople (14), Platt (15), and Dewar and Longuet-Higgins (16). Publications by Dewar et al. (10,17) on ground states and by Pople (18) and Allinger et al. (19,20) on excited states have recently emphasized important aspects of the molecular orbital theory of polyenes. Pople et al. (21) and Dewar (22) have developed theories that allow all valence electrons in hydrocarbons to be accounted for by SCFMO methods, but these have not yet been used extensively to study excited states.

The Pariser calculations on the polyacenes (13) were the first extensive study of configuration interaction in large aromatic hydrocarbons. They have withstood the test of time and have provided inspiration to experimentalists and theoreticians alike. Many of the predictions made have been verified experimentally, and the Pariser polyacene wavefunctions have been widely used in theoretical studies. The energies of the various configurations, eq. (11), were computed from the π-electron Hamiltonian, eq. (12), as were the off-diagonal matrix elements between different configurations, and the resulting matrix was finally diagonalized. There are $(N/2)^2 + 1$ singly excited configurations which contribute to each singlet state of an alternant hydrocarbon and $(N/2)^2$ such configurations which contribute to each triplet state, when N is the number of MO's. Because of computer limitations, Pariser used roughly $2N$ configurations and estimated that the electronic states were 0.1–0.5 eV too high when compared with calculations taking into account all singly excited configurations. The basis sets for these calculations were HMO's and the final states were in good agreement with the experimental spectra. Furthermore, configuration interaction made hardly any effect in the energies of the closed-shell ground states; e.g., the ground state of naphthalene was lowered by 0.0246 eV and that of pentacene by only 0.0552 eV. This means that the HMO are remarkably good representations for the polyacenes. As shown in the present study, however, HMO give poor descriptions of the polyenes.

In alternant hydrocarbons such as the polyenes, the configurational wavefunctions $V_{ij'}$ and $V_{ji'}$ or $T_{ij'}$ and $T_{ji'}$ are degenerate because of the perfect pairing of bonding and antibonding MO's. This means that the combinations $^1\Omega_{ij}^{\pm} = 2^{-1/2}[V_{ij'} \pm V_{ji'}]$ and $^3\Omega_{ij}^{\pm} = 2^{-1/2}[T_{ij'} \pm T_{ji'}]$ may be formed at once and used to factor the CI matrix. Therefore, the electronic states of the polyenes will be either plus-states (Ω^+) or minus-states (Ω^-). This basic symmetry property requires that plus-states do not interact with minus-states, and that the closed-shell, ground-state configuration V_0 behaves like a minus-state. In the polyene calculations presented here, the ($\pm$) notation introduced by Pariser (13) has been used to identify excited states when the polyene belongs to a nontrivial symmetry point group.

The calculations are summarized in Tables I–IX. All planar, linear polyenes either belong to point groups C_{2v} or C_{2h} or have no symmetry element other than the molecular plane (x,y); the C_2 axis is taken to be the y-axis under C_{2v} and the z-axis under C_{2h}. The axes are oriented with respect to the page as

The polyenes are named according to their stereochemistry at each single (lower case) and double bond (upper case), e.g., s-trans-1,3-butadiene is **t** and *s-trans-trans-s-cis*-1,3,5-hexatriene is **tTc** ($\equiv$ **cTt**).

III. COMPARISON OF WAVEFUNCTIONS

The HMO, L-H/S MO, and SCFMO were compared as basis sets for CI calculations. The SCF wavefunctions are a benchmark because the energy $H^0 = \langle V_0 |H| V_0 \rangle$ of a closed-shell configuration built from SCFMO's is invariant to further first-order changes in the orbitals, and by definition all matrix elements between V_0 and singly excited configurations $(V_{ik'})$ vanish. When V_0 is built from other than SCFMO's, the ground-state lowering (GSL) is a measure of how closely these orbitals resemble the self-consistent ones.

The data in Table I for the all-*trans* di-, tri-, and tetraenes show that HMO's rapidly deteriorate in their ability to represent polyenes as molecular size increases, unlike the polyacenes as discussed above. The GSL is 0.2 eV in butadiene, and in the tetraene the lowest excited singlet is 1.0 eV in "error" compared to the SCF value which is close to experiment. The energies of the triplets vary less with the source of the starting MO's, since the triplet wavefunction, eq. (11), insures strong spatial correlation even when the basis MO's are poor. The final bond lengths in the ground states computed from the CI wavefunctions differ considerably on comparing the L-H/S and SCFMO's, although the relationship between bond order and bond length employed was the same in both calculations. Apparently,

TABLE I

Comparison of Hückel, Longuet-Higgins/Salem, and Self-Consistent
Field Molecular Orbitals as Basis Functions for Excited States[a]

	(t)		
	HMO	L-H/S MO	SCFMO
H^0	-88.990	-88.778	-89.123
GSL	-0.202	-0.006	0
E^π	-54.606	-54.412	-54.688
E_b	-8.430	-8.236	-8.512
I_p	—	—	9.07
V_1	6.095 (0.972)	5.495 (0.955)	^1B_u, 5.573 (0.930)
V_2	6.885 (0)	6.683 (0)	$^1A_g{}^-$, 6.833 (0)
V_3	7.071 (0)	6.709 (0)	$^1A_g{}^+$, 6.862 (0)
V_4	9.775 (0.357)	9.585 (0.382)	^1B_u, 9.690 (0.396)
V_5	—	—	—
V_6	—	—	—
V_7	—	—	—
T_1	2.884 (ref.)	2.593 (ref.)	^3B_u, 2.780 (ref.)
T_2	4.691 (0)	4.289 (0)	$^3A_g{}^+$, 4.387 (0)
T_3	6.869 (0.487)	6.676 (0.525)	$^3A_g{}^-$, 6.833 (0.518)
T_4	8.127 (0)	8.043 (0)	^3B_u, 8.179 (0)
T_5	—	—	—
T_6	—	—	—
T_7	—	—	—
r_{12}	1.338	1.360	1.347
r_{23}	1.483	1.455	1.459
r_{34}	—	—	—
r_{45}	—	—	—

	(tTt)		
	HMO	L-H/S MO	SCFMO
H^0	-156.238	-156.028	-156.621
GSL	-0.454	$-0·022$	0
E^π	-81.858	-81.874	-82.312
E_b	-12.594	-12.610	-13.048
I_p	—	—	8.45
V_1	5.571 (1.407)	4.674 (1.401)	$^1B_u{}^+$, 4.736 (1.338)
V_2	6.519 (0)	5.803 (0)	$^1A_g{}^-$, 6.014 (0)
V_3	6.678 (0)	6.140 (0)	$^1A_g{}^+$, 6.238 (0)
V_4	6.900 (0.120)	6.478 (0.082)	$^1B_u{}^+$, 6.643 (0.085)
V_5	7.171 (0)	6.669 (0)	$^1B_u{}^-$, 6.821 (0)
V_6	9.101 (0)	8.708 (0)	$^1A_g{}^+$, 8.819 (0)
V_7	9.307 (0.305)	8.949 (0.365)	$^1B_u{}^+$, 9.091 (0.393)
T_1	2.430 (ref.)	1.951 (ref.)	$^3B_u{}^+$, 2.196 (ref.)

TABLE 1—*continued*

(tTt)			
	HMO	L-H/S MO	SCFMO
T_2	4.050 (0)	3.452 (0)	$^3A_g{}^+$, 3.569 (0)
T_3	5.117 (0)	4.461 (0)	$^3B_u{}^+$, 4.529 (0)
T_4	6.072 (0.835)	5.781 (0.926)	$^3A_g{}^-$, 6.014 (0.900)
T_5	7.171 (0)	6.669 (0)	$^3B_u{}^-$, 6.821 (0)
T_6	7.656 (0)	7.410 (0)	$^3B_u{}^+$, 7.593 (0)
T_7	7.815 (0)	7.589 (0)	$^3A_g{}^+$, 7.758 (0)
r_{12}	1.338	1.362	1.348
r_{23}	1.483	1.450	1.456
r_{34}	1.338	1.372	1.356
r_{45}	—	—	—

(tTtTt)			
	HMO	L-H/S MO	SCFMO
H^0	−231.338	−231.171	−232.041
GSL	−0.705	−0.042	0
E^π	−109.501	−109.326	−109.939
E_b	−17.149	−16.974	−17.587
I_p	—	—	8.08
V_1	5.270 (1.934)	4.139 (1.868)	$^1B_u{}^+$, 4.209 (1.745)
V_2	6.173 (0)	5.164 (0)	$^1A_g{}^-$, 5.448 (0)
V_3	6.445 (0)	5.616 (0)	$^1A_g{}^+$, 5.679 (0)
V_4	6.795 (0)	6.146 (0)	$^1B_u{}^-$, 6.348 (0)
V_5	6.917 (0.134)	6.253 (0.110)	$^1B_u{}^+$, 6.375 (0.133)
V_6	7.083 (0)	6.398 (0)	$^1A_g{}^+$, 6.552 (0)
V_7	7.444 (0)	6.658 (0)	$^1A_g{}^-$, 6.805 (0)
T_1	2.250 (ref.)	1.583 (ref.)	$^3B_u{}^+$, 1.883 (ref.)
T_2	3.640 (0)	2.831 (0)	$^3A_g{}^+$, 2.962 (0)
T_3	4.733 (0)	3.887 (0)	$^3B_u{}^+$, 3.965 (0)
T_4	5.467 (0)	4.565 (0)	$^3A_g{}^+$, 4.600 (0)
T_5	5.526 (1.106)	5.124 (1.248)	$^3A_g{}^-$, 5.448 (1.182)
T_6	6.795 (0)	6.146 (0)	$^3B_u{}^-$, 6.348 (0)
T_7	7.106 (0)[b]	6.658 (0.012)	$^3A_g{}^-$, 6.805 (0.015)
r_{12}	1.338	1.363	1.348
r_{23}	1.483	1.448	1.455
r_{34}	1.338	1.375	1.358
r_{45}	1.483	1.443	1.452

[a] All energies are given in eV. Oscillator strengths of the various transitions are given in parentheses. Bond lengths are given in Angstrom units. Calculated ionization potentials (I_p) have been scaled down by 2.45 eV uniformly.

[b] Computed to be ^3B_u.

the wavefunctions are very similar with respect to energy but differ considerably with respect to geometry. In similar studies (23) of cyclic polyenes and aromatic structures, L-H/S MO's were found to be close to the "best" SCFMO's when configuration energies are used as a criterion.

Some energy quantities of interest in the discussion of ground states are also given in Table I. The effective π-energy of the conjugated system, E^π (eq. 14), is obtained by adding to H^0 the energy of the sigma framework, E^{f+} (eq. 15), which allows for the repulsion of the atoms in the positively

$$E^\pi = H^0 + E^{f+} \tag{14}$$

$$E^{f+} = \sum_{p<q} (\gamma_{pq} - [q:pp] - [p:qq]) \tag{15}$$

charged core. The neutral atom penetration integrals ([q:pp]) were neglected in this work, since pilot calculations showed that their inclusion had little effect on the final results. E_b (eq. 16) is the effective bonding energy of the π-electrons relative to the appropriate valence state of carbon (I_c) (9,24).

$$E_b = E^\pi + NI_c \tag{16}$$

The SCFMO's obtained by iteration to stable geometry always have lowest H^0 when compared with other basis sets, such as HMO's. After configuration interaction is taken into account, however, the energy of the ground state, H^0 + GSL, may be lower for other basis sets. This is because simple MO theories do not account properly for framework energy. Bond energy is primarily a function of the core resonance integrals, which are assumed to increase negatively as $r_{pq} \to 0$, and a structure with short bonds can usually be found whose energy is less than that computed by SCF methods. In this study, the lengths of pure double and single (sp^2–sp^2) bonds were assumed to be 1.338 and 1.483 Å, respectively, and all calculations were begun with this geometry. Since the iterative SCF method finds stable double bond lengths greater than 1.338 Å, it is reasonable that the energy H^0 + GSL may be below the "best" SCF value for other basis sets when double bond lengths are fixed at 1.338 Å. For semiquantitative purposes, all reasonable basis sets give H^0, E^π, and E^b that are very close in magnitude.

IV. THE GROUND STATES OF THE LINEAR POLYENES

Since Dewar (10,17) and Adams and Miller (25) have considered the ground state of the all-*trans* polyenes in detail, the general conclusions will not be discussed here. In the present work, the influence of geometry was studied by calculating the energies and bond lengths of the planar conformations of the di-, tri-, and tetraenes (Tables II, III, and IV). It was found that the introduction of *s-cis-*, but not *cis-*, linkages into the all-*trans* polyenes alters bond lengths characteristically, lowers the π-bonding energy, and increases ionization potentials.

A. Bond Lengths

Longuet-Higgins and Salem (5) showed that an infinite polyene, all of whose bonds are equal, is unstable and that bond alternation, even in the middle of very long chains, is pronounced; they did not determine the chain size where no further changes in bond length occur. (In $(4N + 2)$-cyclic polyenes, they concluded that configurational instability should arise in rings of about 34 carbon atoms but might occur in rings with as few as 18 carbon atoms.) In the present work, direct calculation of the geometry of long polyenes shows that bond alternation is strong and that

TABLE II

Ground and Excited States of Linear Dienes[a,b]

	t	c
H^0	-89.123	-90.264
E^{π}	-54.688	-54.726
E_b	-8.512	-8.550
I_p	9.07	9.16
V_1	^1B_u, 5.573 (0.930)	$^1B_2$, 5.119 (0.432)
V_2	$^1A_g{}^-$, 6.833 (0)	$^1A_1{}^-$, 6.613 (0)
V_3	$^1A_g{}^+$, 6.862 (0)	$^1A_1{}^+$, 7.368 (0.779)
V_4	^1B_u, 9.690 (0.396)	$^1B_2$, 9.380 (0.001)
T_1	^3B_u, 2.780 (ref.)	$^3B_2$, 2.764 (ref.)
T_2	$^3A_g{}^+$, 4.387 (0)	$^3A_1{}^+$, 4.460 (0)
T_3	$^3A_g{}^-$, 6.833 (0.518)	$^3A_1{}^-$, 6.613 (0.395)
T_4	^3B_u, 8.179 (0)	$^3B_2$, 7.939 (0)

[a] All energies are given in eV. Oscillator strengths of the various transitions are given in parentheses. Bond lengths are given in Angstrom units. Calculated ionization potentials (I_p) have been scaled down by 2.45 eV uniformly.

[b] Explanation of the transition moment diagrams is given in the text.

TABLE III
Ground and Excited States of Linear Trienes[a,b]

trans

	$\alpha = 16°14'$ tTt	$\alpha = 41°31'$ tTc	$\alpha = 4°24'$ cTc
H^0	-156.621	-158.618	-160.166
E^π	-82.312	-82.351	-82.388
E_b	-13.048	-13.087	-13.124
I_p	8.45	8.51	8.56
V_1	$^1B_u{}^+$, 4.736 (1.338)	4.474 (0.777)	$^1B_u{}^+$, 4.349 (0.713)
V_2	$^1A_g{}^-$, 6.014 (0)	5.937 (0)	$^1A_g{}^+$, 5.921 (0)
V_3	$^1A_g{}^+$, 6.238 (0)	6.145 (0.449)	$^1A_g{}^-$, 5.952 (0)
V_4	$^1B_u{}^+$, 6.643 (0.085)	6.712 (0)	$^1B_u{}^-$, 6.564 (0)
V_5	$^1B_u{}^-$, 6.821 (0)	6.979 (0.238)	$^1B_u{}^+$, 7.389 (1.123)
T_1	$^3B_u{}^+$, 2.196 (ref.)	2.195 (ref.)	$^3B_u{}^+$, 2.203 (ref.)
T_2	$^3A_g{}^+$, 3.569 (0)	3.584 (0)	$^3A_g{}^+$, 3.597 (0)
T_3	$^3B_u{}^+$, 4.529 (0)	4.577 (0)	$^3B_u{}^+$, 4.614 (0)
T_4	$^3A_g{}^-$, 6.014 (0.900)	5.937 (0.704)	$^3A_g{}^-$, 5.952 (0.708)
T_5	$^3B_u{}^-$, 6.821 (0)	6.712 (0.071)	$^3B_u{}^-$, 6.564 (0)

cis

	tCt	$\alpha = 25°10'$ cCt	cCc
H^0	-159.015	-163.652	-169.617
E^π	-82.313	-82.355	-82.406
E_b	-13.049	-13.091	-13.142
I_p	8.45	8.53	8.65
V_1	$^1B_2{}^+$, 4.690 (1.214)	4.294 (0.468)	$^1B_2{}^+$, 3.660 (0.101)
V_2	$^1A_1{}^-$, 5.889 (0)	5.524 (0)	$^1A_1{}^-$, 4.856 (0)
V_3	$^1A_1{}^+$, 6.264 (0.090)	5.980 (0.664)	$^1A_1{}^+$, 6.261 (1.089)
V_4	$^1B_2{}^+$, 6.516 (0.092)	6.680 (0)	$^1B_2{}^-$, 6.650 (0)
V_5	$^1B_2{}^-$, 6.807 (0)	7.122 (0.413)	$^1B_2{}^+$, 7.025 (0.242)
T_1	$^3B_2{}^+$, 2.179 (ref.)	2.147 (ref.)	$^3B_2{}^+$, 2.092 (ref.)
T_2	$^3A_1{}^+$, 3.569 (0)	3.571 (0)	$^3A_1{}^+$, 3.596 (0)
T_3	$^3B_2{}^+$, 4.527 (0)	4.582 (0)	$^3B_2{}^+$, 4.670 (0)
T_4	$^3A_1{}^-$, 5.889 (0.667)	5.524 (0.345)	$^3A_1{}^-$, 4.856 (0.101)
T_5	$^3B_2{}^-$, 6.807 (0.198)	6.680 (0.348)	$^3B_2{}^+$, 6.166 (0)

[a] All energies are given in eV. Oscillator strengths of the various transitions are given in parentheses. Bond lengths are given in Angstrom units. Calculated ionization potentials (I_p) have been scaled down by 2.45 eV uniformly.

[b] Explanation of the transition moment diagrams is given in the text.

TABLE IV
Ground and Excited States of Linear Tetraenes[a,b]

trans–trans

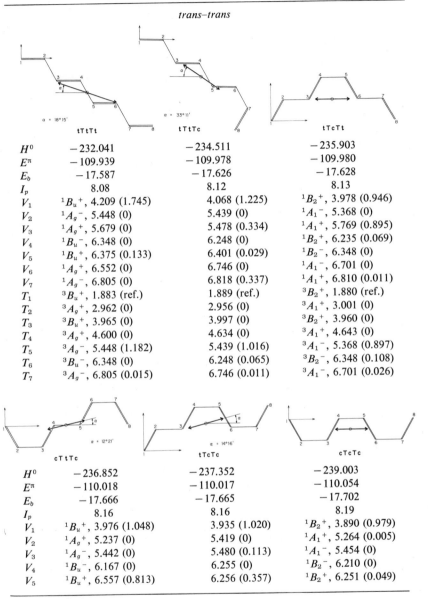

	tTtTt	tTtTc	tTcTt
H^0	-232.041	-234.511	-235.903
E^π	-109.939	-109.978	-109.980
E_b	-17.587	-17.626	-17.628
I_p	8.08	8.12	8.13
V_1	$^1B_u{}^+$, 4.209 (1.745)	4.068 (1.225)	$^1B_2{}^+$, 3.978 (0.946)
V_2	$^1A_g{}^-$, 5.448 (0)	5.439 (0)	$^1A_1{}^-$, 5.368 (0)
V_3	$^1A_g{}^+$, 5.679 (0)	5.478 (0.334)	$^1A_1{}^+$, 5.769 (0.895)
V_4	$^1B_u{}^-$, 6.348 (0)	6.248 (0)	$^1B_2{}^+$, 6.235 (0.069)
V_5	$^1B_u{}^+$, 6.375 (0.133)	6.401 (0.029)	$^1B_2{}^-$, 6.348 (0)
V_6	$^1A_g{}^+$, 6.552 (0)	6.746 (0)	$^1A_1{}^-$, 6.701 (0)
V_7	$^1A_g{}^-$, 6.805 (0)	6.818 (0.337)	$^1A_1{}^+$, 6.810 (0.011)
T_1	$^3B_u{}^+$, 1.883 (ref.)	1.889 (ref.)	$^3B_2{}^+$, 1.880 (ref.)
T_2	$^3A_g{}^+$, 2.962 (0)	2.956 (0)	$^3A_1{}^+$, 3.001 (0)
T_3	$^3B_u{}^+$, 3.965 (0)	3.997 (0)	$^3B_2{}^+$, 3.960 (0)
T_4	$^3A_g{}^+$, 4.600 (0)	4.634 (0)	$^3A_1{}^+$, 4.643 (0)
T_5	$^3A_g{}^-$, 5.448 (1.182)	5.439 (1.016)	$^3A_1{}^-$, 5.368 (0.897)
T_6	$^3B_u{}^-$, 6.348 (0)	6.248 (0.065)	$^3B_2{}^-$, 6.348 (0.108)
T_7	$^3A_g{}^-$, 6.805 (0.015)	6.746 (0.011)	$^3A_1{}^-$, 6.701 (0.026)

	cTtTc	tTcTc	cTcTc
H^0	-236.852	-237.352	-239.003
E^π	-110.018	-110.017	-110.054
E_b	-17.666	-17.665	-17.702
I_p	8.16	8.16	8.19
V_1	$^1B_u{}^+$, 3.976 (1.048)	3.935 (1.020)	$^1B_2{}^+$, 3.890 (0.979)
V_2	$^1A_g{}^+$, 5.237 (0)	5.419 (0)	$^1A_1{}^+$, 5.264 (0.005)
V_3	$^1A_g{}^-$, 5.442 (0)	5.480 (0.113)	$^1A_1{}^-$, 5.454 (0)
V_4	$^1B_u{}^-$, 6.167 (0)	6.255 (0)	$^1B_2{}^-$, 6.210 (0)
V_5	$^1B_u{}^+$, 6.557 (0.813)	6.256 (0.357)	$^1B_2{}^+$, 6.251 (0.049)

[a] All energies are given in eV. Oscillator strengths of the various transitions are given in parentheses. Bond lengths are given in Angstrom units. Calculated ionization potentials (I_p) have been scaled down by 2.45 eV uniformly.

[b] Explanation of the transition moment diagrams is given in the text.

TABLE IV—*continued*

V_6	$^1A_g^-$, 6.662 (0)	6.640 (0)	$^1A_1^-$, 6.533 (0)
V_7	$^1A_g^+$, 6.954 (0)	7.114 (0.508)	$^1A_1^+$, 7.413 (1.404)
T_1	$^3B_u^+$, 1.896 (ref.)	1.892 (ref.)	$^3B_2^+$, 1.903 (ref.)
T_2	$^3A_g^+$, 2.950 (0)	2.993 (0)	$^3A_1^+$, 2.984 (0)
T_3	$^3B_u^+$, 4.032 (0)	3.988 (0)	$^3B_2^+$, 4.020 (0)
T_4	$^3A_g^+$, 4.663 (0)	4.667 (0)	$^3A_1^+$, 4.689 (0)
T_5	$^3A_g^-$, 5.442 (0.942)	5.419 (0.942)	$^3A_1^-$, 5.454 (0.920)
T_6	$^3B_u^-$, 6.167 (0)	6.255 (0.005)	$^3B_2^-$, 6.210 (0.015)
T_7	$^3A_g^-$, 6.662 (0.010)	6.640 (0.050)	$^3A_1^-$, 6.533 (0.014)

trans–cis

tCtTt $\qquad$ $\alpha = 7°35'$

tCtTc $\qquad$ $\alpha = 22°27'$

cCtTt $\qquad$ $\alpha = 6°6'$

H^0	-235.666	-238.833	-242.105
E^π	-109.940	-109.980	-109.983
E_b	-17.588	-17.628	-17.631
I_p	8.07	8.12	8.13
V_1	4.170 (1.594)	4.017 (1.015)	3.956 (0.758)
V_2	5.341 (0)	5.308 (0)	5.044 (0)
V_3	5.665 (0.107)	5.472 (0.535)	5.234 (0.760)
V_4	6.303 (0)	6.201 (0)	6.183 (0)
V_5	6.335 (0.153)	6.341 (0.032)	6.392 (0.055)
V_6	6.489 (0.014)	6.730 (0)	6.710 (0)
V_7	6.791 (0)	6.758 (0.299)	6.988 (0.473)
T_1	1.865 (ref.)	1.870 (ref.)	1.831 (ref.)
T_2	2.961 (0)	2.955 (0)	2.956 (0)
T_3	3.965 (0)	3.997 (0)	3.993 (0)
T_4	4.599 (0)	4.633 (0)	4.635 (0)
T_5	5.341 (0.938)	5.308 (0.728)	5.044 (0.532)
T_6	6.303 (0.162)	6.201 (0.265)	6.183 (0.396)
T_7	6.791 (0.052)	6.730 (0.046)	6.441 (0)

tCcTt $\qquad$ $\alpha = 10°$

tCcTc $\qquad$ $\alpha = 24°38'$

cCtTc $\qquad$ $\alpha = 10°2'$

H^0	-244.890	-245.540	-245.972
E^π	-109.986	-110.023	-110.023
E_b	-17.634	-17.671	-17.671
I_p	8.14	8.17	8.18
V_1	3.816 (0.572)	3.817 (0.714)	3.783 (0.442)
V_2	5.031 (0)	5.128 (0)	4.960 (0)
V_3	5.739 (1.293)	5.414 (0.270)	5.124 (0.651)
V_4	6.031 (0.025)	6.060 (0.468)	6.095 (0)

TABLE IV—*continued*

V_5	6.189 (0)	6.117 (0)	6.465 (0.454)
V_6	6.681 (0)	6.609 (0)	6.643 (0)
V_7	6.912 (0.036)	7.242 (0.733)	7.043 (0.352)
T_1	1.840 (ref.)	1.856 (ref.)	1.833 (ref.)
T_2	2.982 (0)	2.974 (0)	2.950 (0)
T_3	3.952 (0)	3.981 (0)	4.032 (0)
T_4	4.650 (0)	4.671 (0)	4.667 (0)
T_5	5.031 (0.459)	5.128 (0.551)	4.960 (0.384)
T_6	6.189 (0.456)	6.117 (0.228)	6.095 (0.396)
T_7	6.412 (0)	6.521 (0)	6.330 (0)

$\alpha = 17°3'$

cCcTt

$\alpha = 35°36'$

cCcTc

H^0	−255.438	−255.459
E^π	−110.039	−110.076
E_b	−17.687	−17.724
I_p	8.25	8.27
V_1	3.260 (0.138)	3.361 (0.197)
V_2	4.387 (0)	4.545 (0)
V_3	5.542 (1.391)	5.046 (0.675)
V_4	6.022 (0)	5.959 (0)
V_5	6.384 (0.177)	6.588 (0)
V_6	6.611 (0.025)	6.646 (0.371)
V_7	6.644 (0)	6.812 (0.203)
T_1	1.779 (ref.)	1.803 (ref.)
T_2	2.988 (0)	2.979 (0)
T_3	3.991 (0)	4.019 (0)
T_4	4.387 (0.148)	4.545 (0.192)
T_5	4.724 (0)	4.730 (0)
T_6	5.640 (0)	5.823 (0)
T_7	6.022 (0.554)	5.959 (0.397)

cis–cis

$\alpha = 3°12'$

tCtCt

$\alpha = 16°47'$

cCtCt

H^0	−238.049	−243.844
E^π	−109.941	−109.985
E_b	−17.589	−17.633
I_p	8.07	8.12
V_1	$^1B_u{}^+$, 4.141 (1.571)	3.947 (0.850)
V_2	$^1A_g{}^-$, 5.266 (0)	5.014 (0)

TABLE IV—*continued*

V_3	$^1A_g{}^+$, 5.647 (0)	5.195 (0.425)
V_4	$^1B_u{}^-$, 6.253 (0)	6.119 (0)
V_5	$^1B_u{}^+$, 6.325 (0.244)	6.300 (0.155)
V_6	$^1A_g{}^+$, 6.388 (0)	6.700 (0)
V_7	$^1A_g{}^-$, 6.779 (0)	6.986 (0.582)
T_1	$^3B_u{}^+$, 1.849 (ref.)	1.818 (ref.)
T_2	$^3A_g{}^+$, 2.960 (0)	2.956 (0)
T_3	$^3B_u{}^+$, 3.966 (0)	3.994 (0)
T_4	$^3A_g{}^+$, 4.597 (0)	4.632 (0)
T_5	$^3A_g{}^-$, 5.266 (0.887)	5.014 (0.566)
T_6	$^3B_u{}^-$, 6.253 (0)	6.119 (0.108)
T_7	$^3B_u{}^+$, 6.734 (0)	6.421 (0)

$\alpha = 88°15'$

cCtCc

tCcCt

H^0	-249.616	-260.162
E^π	-110.029	-109.991
E_b	-17.677	-17.639
I_p	8.18	8.17
V_1	$^1B_u{}^+$, 3.850 (0.619)	$^1B_2{}^+$, 3.479 (0.266)
V_2	$^1A_g{}^+$, 4.687 (0)	$^1A_1{}^-$, 4.373 (0)
V_3	$^1A_g{}^-$, 4.823 (0)	$^1B_2{}^+$, 5.704 (0.029)
V_4	$^1B_u{}^-$, 5.941 (0)	$^1A_1{}^+$, 5.757 (1.567)
V_5	$^1A_g{}^-$, 6.620 (0)	$^1B_2{}^-$, 5.933 (0)
V_6	$^1B_u{}^+$, 6.773 (1.527)	$^1B_2{}^+$, 6.311 (0.017)
V_7	$^1A_g{}^+$, 7.041 (0)	$^1A_1{}^-$, 6.545 (0)
T_1	$^3B_u{}^+$, 1.788 (ref.)	$^3B_2{}^+$, 1.771 (ref.)
T_2	$^3A_g{}^+$, 2.951 (0)	$^3A_1{}^+$, 2.951 (0)
T_3	$^3B_u{}^+$, 4.027 (0)	$^3B_2{}^+$, 3.930 (0)
T_4	$^3A_g{}^+$, 4.658 (0)	$^3A_1{}^-$, 4.373 (0.162)
T_5	$^3A_g{}^-$, 4.823 (0.421)	$^3A_1{}^+$, 4.668 (0)
T_6	$^3B_u{}^-$, 5.941 (0)	$^3B_2{}^+$, 5.425 (0)
T_7	$^3B_u{}^+$, 6.229 (0)	$^3B_2{}^-$, 5.933 (0.523)

differential changes in bond length cease after about the C_{16}-hydrocarbon. In the center of very long chains, the "double" and "single" bonds become fixed with lengths of 1.36 and 1.45 Å, respectively.

The tables show that the introduction of *s-cis* linkages in linear polyenes tends to shorten those single bonds and lengthen attached double bonds (e.g., compare r_{45} in **tTt** and **tTc**), whereas the introduction of *cis* linkages has a much smaller effect; similar behavior is found in cross-conjugated polyenes. There is no experimental evidence bearing on this point. In the cyclic radialenes, there is virtually no change in bond length with ring size, the double and single bonds being near 1.35 and 1.46 Å, respectively.

B. Ground State and Conjugation Energies

Dewar and Gleicher (10) found that the bonds in all-*trans* polyenes and radialenes were strongly localized, and the present calculations support this conclusion. If this is indeed so, then there should be an approximately constant contribution to the π-electron energy from the "single" and "double" bond regions of a polyene. The additivity of bond contributions in the all-*trans* polyenes was shown by examining eq. (17),

$$E_b = (k + 1)E_d^\pi + kE_s^\pi = k(E_s^\pi + E_d^\pi) + E_d^\pi \qquad (17)$$

where k is the number of single bonds. A plot of k vs. E_b gave a straight line whose intercept is E_d^π, and E_s^π and E_d^π are constants equal to the π-energies of "single" and "double" bonds, respectively. A similar plot for the radialenes was also a straight line through the origin, eq. (18).

$$E_b = k(E_s^\pi + E_d^\pi) \qquad (18)$$

The data in Tables VII and IX give accurate straight lines (Fig. 1)

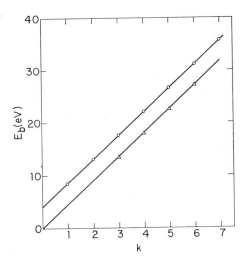

Fig. 1. Plot of $-E_b$ vs. k for all-*trans* polyenes ($\bigcirc$) and radialenes ($\triangle$) where k is the number of single bonds. The slope of both lines is $-(E_s^\pi + E_d^\pi) = 4.5392$ and the intercepts are 3.9707 and 0, respectively.

from which $E_s^\pi = -0.5685$ eV and $E_d^\pi = -3.9707$. By a different parametrization Dewar and Gleicher (10) found $E_s^\pi = -0.3142$ and $E_d^\pi = -2.1146$, and the difference, which has no effect on the calculated spectra, is attributed primarily to the way in which the core integrals are evaluated. Although additivity of single and double-bond π-energies holds accurately in the all-*trans* polyenes, the tables show that E_b fluctuates with changes in polyene geometry. E_s^π and E_d^π can be used, nevertheless, to estimate E_b in other systems; e.g., for the cross-conjugate trienes, $E_b = -13.049$ eV (eq. 17), whereas the average of the correct values is -13.045 eV (Table V).

TABLE V

Ground and Excited States of Cross-Conjugated Trienes[a,b]

	cross–cc	cross–ct	cross–tt
H^0	-162.767	-163.157	-165.189
E^π	-82.344	-82.309	-82.273
E_b	-13.080	-13.045	-13.009
I_p	8.84	8.76	8.69
V_1	$^1A_1{}^+$, 4.638 (0.158)	4.726 (0.349)	$^1A_1{}^+$, 4.881 (0.598)
V_2	$^1B_2{}^+$, 5.613 (0.960)	5.611 (0)	$^1B_2{}^+$, 5.246 (0.065)
V_3	$^1B_2{}^-$, 5.679 (0)	5.769 (0.795)	$^1B_2{}^-$, 5.226 (0)
V_4	$^1A_1{}^-$, 7.002 (0)	7.063 (0)	$^1A_1{}^-$, 7.095 (0)
V_5	$^1A_1{}^+$, 7.578 (0.466)	7.142 (0.248)	$^1A_1{}^+$, 7.174 (0.039)
T_1	$^3A_1{}^+$, 2.454 (ref.)	2.423 (ref.)	$^3A_1{}^+$, 2.364 (ref.)
T_2	$^3B_2{}^+$, 3.460 (0)	3.456 (0)	$^3B_2{}^+$, 3.450 (0)
T_3	$^3A_1{}^+$, 4.514 (0)	4.484 (0)	$^3A_1{}^+$, 4.427 (0)
T_4	$^3B_2{}^-$, 5.679 (0.405)	5.611 (0.303)	$^3B_2{}^-$, 5.266 (0.165)
T_5	$^3A_1{}^-$, 7.002 (0.097)	6.937 (0)	$^3A_1{}^+$, 6.426 (0)

[a] All energies are given in eV. Oscillator strengths of the various transitions are given in parentheses. Bond lengths are given in Angstrom units. Calculated ionization potentials (I_p) have been scaled down by 2.45 eV uniformly.

[b] Explanation of the transition moment diagrams is given in the text.

Nonempirical and empirical SCFMO methods usually predict that *s-cis*-butadiene is ca. 0.1 eV more stable than the *s-trans* isomer when only the π-electron energy is considered (8,9,26). Experimentally the reverse is true, and it has been suggested (9) that nonbonded hydrogen repulsions can account for this discrepancy. In the present work, this difference is reduced to ca. 0.04 eV (cf. E_b in Table II), which is more in accord with this suggestion, since the s-cis and s-trans conformers differ by ≤ 5 kcal/mole and the nonbonded repulsion energy is probably in this range.

As s-cis linkages are introduced in the trienes (Table III), the π-bonding energy decreases, but the effect of cis linkages is less important (cf., **tTt** → **tTc** → **cTc** and **tTt** → **tCt**). The difference between the two extreme isomers **tTt** and **cCc** is 0.09 eV. Similarly, the tetraenes (Table IV) become more stable on introduction of s-cis linkages and the difference between the extremes, **tTtTt** and **cCcTc**, is 0.14 eV. It is apparent, however, that this effect contributes only in a minor way to molecular stability. In very large polyenes, e.g., the β-carotenes (Table VIII), the effect is negligible when only one or two cis linkages occur in the long chain.

The effective π-bonding energies of the all-trans polyenes are given in Table VII. The average π-bonding energies per π-electron of **t**, **tTt**, and **tTtTt** are -2.128, -2.175, and -2.198 eV, respectively, and this quantity decreases smoothly to a sensibly constant value of -2.24 eV for the C_{20} polyene. Polyenes longer than the decaene are not appreciably stabilized by increased conjugation and show little differential changes in bond lengths.

C. Ionization Potentials

Ionization potentials (I_p) are often overestimated (by 1–2 eV) in MO theories, although refinements can remove this difficulty. In the present work, the values in the tables are scaled by a constant that brings 1,3-butadiene into agreement with experiment. Adams and Miller (25) computed the ionization potentials of the all-trans polyenes by a modified SCFMO method designed to treat ionization potentials more properly, and our results are in good agreement with theirs (Table VII; lit. values (25) in parentheses).

The geometry of the linear polyenes influences the ionization potentials in a characteristic fashion: (1) the introduction of rigid cis linkages causes virtually no effect, compare **tTt** (8.45 eV) with **tCt** (8.45 eV), and **tTtTt** (8.08 eV) with **tCtTt** (8.07 eV) and **tCtCt** (8.07 eV); and (2) the introduction of s-cis linkages increases the molecular ionization potential, compare **t** (9.07 eV) with **c** (9.16 eV), **tTt** (8.45 eV) with **tTc** (8.51 eV) and **cTc** (8.56 eV), and **tTtTt** (8.08 eV) with **tTtTc** (8.12 eV), **cTtTc** (8.16 eV) and **cTcTc** (8.19 eV). These small effects are consistently observed and are apparently associated with a general energy lowering in structures where the density of double bonds is largest.

TABLE VI

SCF Bond Orders and Bond Lengths of Polyenes in States V_0, V_1, and T_1[a]

Polyene		$p_{12}(r_{12})$	$p_{23}(r_{23})$	$p_{34}(r_{34})$	$p_{45}(r_{45})$	$p_{56}(r_{56})$	$p_{67}(r_{67})$	$p_{78}(r_{78})$
t	V_0:	0.9511 (1.347)	0.3088 (1.459)					
	V_1:	0.4755	0.6443					
	T_1:	0.4755	0.6002					
c	V_0:	0.9420 (1.348)	0.3358 (1.454)					
	V_1:	0.4710	0.6638					
	T_1:	0.4710	0.6226					
tTt	V_0:	0.9446 (1.348)	0.3265 (1.456)	0.8970 (1.356)				
	V_1:	0.7005	0.5718	0.4553				
	T_1:	0.7195	0.5336	0.4250				
tTc	V_0:	0.9444 (1.348)	0.3268 (1.456)	0.8879 (1.358)	0.3536 (1.451)	0.9348 (1.349)		
	V_1:	0.7664	0.5352	0.4483	0.6212	0.6309		
	T_1:	0.7246	0.5308	0.4251	0.5593	0.7018		
cTc	V_0:	0.9359 (1.349)	0.3509 (1.452)	0.8806 (1.359)				
	V_1:	0.7060	0.5876	0.4243				
	T_1:	0.7084	0.5540	0.4243				
tCt	V_0:	0.9439 (1.348)	0.3280 (1.456)	0.8964 (1.356)				
	V_1:	0.6939	0.5743	0.4669				
	T_1:	0.7163	0.5369	0.4300				
cCt	V_0:	0.9318 (1.350)	0.3600 (1.450)	0.8849 (1.358)	0.3334 (1.455)	0.9416 (1.348)		
	V_1:	0.6160	0.6274	0.4679	0.5335	0.7563		
	T_1:	0.7009	0.5649	0.4385	0.5471	0.7058		
cCc	V_0:	0.9223 (1.352)	0.3796 (1.447)	0.8658 (1.361)				
	V_1:	0.6408	0.6106	0.5227				
	T_1:	0.6809	0.5860	0.4536				
tTtTt	V_0:	0.9429 (1.348)	0.3305 (1.455)	0.8885 (1.358)	0.3482 (1.452)			
	V_1:	0.7962	0.4939	0.5736	0.6019			
	T_1:	0.8234	0.4586	0.5539	0.5727			
tTtTc	V_0:	0.9430 (1.348)	0.3304 (1.455)	0.8884 (1.358)	0.3484 (1.452)	0.8796 (1.359)	0.3573 (1.450)	0.9331 (1.350)
	V_1:	0.8277	0.4629	0.6261	0.5914	0.5363	0.5511	0.7346
	T_1:	0.8242	0.4575	0.5568	0.5714	0.5506	0.4848	0.8082
	V_0:	0.9426 (1.348)	0.3313 (1.455)	0.8785 (1.359)	0.3757 (1.447)			

Conformer	Quantity							
tTcTt	V_1:	0.8189	0.4846	0.5469	0.6335	0.8716 (1.361)	0.3548 (1.451)	0.9342 (1.350)
	T_1:	0.8237	0.4597	0.5468	0.5951	0.5110	0.5261	0.7836
	V_0:	0.9333 (1.350)	0.3570 (1.451)	0.8797 (1.359)	0.3483 (1.452)	0.5449	0.4819	0.8114
cTtTc	V_1:	0.7741	0.5191	0.5803	0.5920	0.8876 (1.358)	0.3308 (1.455)	0.9428 (1.348)
	T_1:	0.8091	0.4835	0.5534	0.5700	0.5740	0.4909	0.8009
	V_0:	0.9428 (1.348)	0.3308 (1.455)	0.8799 (1.359)	0.3726 (1.448)	0.5538	0.4579	0.8251
tTcTc	V_1:	0.8355	0.4649	0.5894	0.6258	0.8785 (1.359)	0.3580 (1.450)	0.9328 (1.350)
	T_1:	0.8237	0.4589	0.5492	0.5913	0.5414	0.5481	0.7389
	V_0:	0.9343 (1.350)	0.3546 (1.451)	0.8727 (1.360)	0.3698 (1.448)	0.5510	0.4843	0.8098
cTcTc	V_1:	0.7999	0.5078	0.5529	0.6220	0.8852 (1.358)	0.3313 (1.455)	0.9426 (1.348)
	T_1:	0.8111	0.4815	0.5473	0.5876	0.6371	0.4416	0.8544
	V_0:	0.9421 (1.348)	0.3323 (1.455)	0.8879 (1.358)	0.3500 (1.452)	0.5472	0.4583	0.8266
tCtTt	V_1:	0.7819	0.5038	0.5803	0.6018	0.8749 (1.360)	0.3325 (1.455)	0.9421 (1.348)
	T_1:	0.8183	0.4639	0.5557	0.5753	0.5493	0.4771	0.8292
	V_0:	0.9420 (1.348)	0.3326 (1.455)	0.8876 (1.358)	0.3505 (1.452)	0.5528	0.4560	0.8296
tCtTc	V_1:	0.8122	0.4752	0.6286	0.5921	0.8685 (1.361)	0.3551 (1.451)	0.9340 (1.350)
	T_1:	0.8183	0.4637	0.5584	0.5744	0.5138	0.5102	0.8059
	V_0:	0.9296 (1.350)	0.3647 (1.449)	0.8760 (1.360)	0.3558 (1.451)	0.5502	0.4773	0.8180
cCtTt	V_1:	0.8002	0.5775	0.5269	0.5933	0.8754 (1.360)	0.3598 (1.450)	0.9320 (1.350)
	T_1:	0.7003	0.4963	0.5581	0.5839	0.6090	0.5071	0.7849
	V_0:	0.9391 (1.349)	0.3392 (1.454)	0.8748 (1.360)	0.3833 (1.446)	0.5445	0.4861	0.8103
tCtTt	V_1:	0.7921	0.5030	0.5531	0.6361	0.8630 (1.362)	0.3365 (1.454)	0.9404 (1.348)
	T_1:	0.8033	0.4803	0.5471	0.6015	0.5783	0.4461	0.8636
	V_0:	0.9397 (1.349)	0.3378 (1.454)	0.8766 (1.360)	0.3794 (1.447)	0.5436	0.4603	0.8297
tCcTc	V_1:	0.8114	0.4853	0.5822	0.6337	0.8587 (1.363)	0.3561 (1.451)	0.9336 (1.350)
	T_1:	0.8049	0.4779	0.5488	0.5974	0.5567	0.4473	0.8720
	V_0:	0.9288 (1.350)	0.3662 (1.449)	0.8751 (1.360)	0.3576 (1.450)	0.5417	0.4775	0.8209
cCtTc	V_1:	0.7265	0.5540	0.5747	0.5945			
	T_1:	0.7991	0.4974	0.5610	0.5840			
	V_0:	0.9169 (1.353)	0.3895 (1.445)	0.8529 (1.364)	0.4069 (1.442)			
cCcTt	V_1:	0.6750	0.5903	0.5642	0.6317			
	T_1:	0.7692	0.5300	0.5500	0.6218			
	V_0:	0.9194 (1.352)	0.3851 (1.446)	0.8567 (1.363)	0.3999 (1.443)			
cCcTc	V_1:	0.6881	0.5840	0.5610	0.6364			
	T_1:	0.7731	0.5253	0.5502	0.6161			
	V_0:	0.9420 (1.348)	0.3324 (1.455)	0.8870 (1.358)	0.3516 (1.451)			

TABLE VI—continued

Polyene		$p_{12}(r_{12})$	$p_{23}(r_{23})$	$p_{34}(r_{34})$	$p_{45}(r_{45})$	$p_{56}(r_{56})$	$p_{67}(r_{67})$	$p_{78}(r_{78})$
tCtCt	V_1:	0.7880	0.5000	0.5797	0.6017	0.6416	0.4505	0.8427
	T_1:	0.8204	0.4627	0.5553	0.5777	0.5487	0.4621	0.8229
	V_0:	0.9298 (1.350)	0.3644 (1.449)	0.8755 (1.360)	0.3569 (1.451)	0.8849 (1.358)	0.3325 (1.455)	0.9420 (1.348)
cCtCt	V_1:	0.7081	0.5736	0.5248	0.5923			
	T_1:	0.8024	0.4948	0.5572	0.5858			
	V_0:	0.9299 (1.350)	0.3642 (1.449)	0.8735 (1.360)	0.3620 (1.450)			
cCtCc	V_1:	0.7739	0.5266	0.5738	0.6066			
	T_1:	0.8049	0.4941	0.5508	0.5934			
	V_0:	0.9361 (1.349)	0.3453 (1.453)	0.8684 (1.361)	0.3958 (1.444)			
tCcCt	V_1:	0.7668	0.5118	0.5841	0.6225			
	T_1:	0.7996	0.4857	0.5582	0.6095			
	V_0:	0.9541 (1.346)	0.2960 (1.461)	0.9082 (1.354)				
cross-tt	V_1:	0.7166	0.4797	0.4663				
	T_1:	0.7578	0.4719	0.3878				
	V_0:	0.9552 (1.346)	0.2934 (1.462)	0.9007 (1.355)				
cross-tc	V_1:	0.8596	0.4011	0.4348	0.5817	0.6014		
	T_1:	0.7783	0.4510	0.3856	0.4977	0.7297		
	V_0:	0.9471 (1.347)	0.3166 (1.458)	0.8942 (1.356)	0.3205 (1.457)[b]	0.9454 (1.348)		
cross-cc	V_1:	0.7442	0.5071	0.4055				
	T_1:	0.7510	0.4792	0.3838				

[a] Bond lengths (r) are given in Angstrom units.

[b] $p_{35}(r_{35})$.

V. SPECTRA OF LINEAR POLYENES

The energies of the lower singlet and triplet states of all configurations of the di-, tri-, and tetraenes are given in Tables II, III, and IV. Two tetraene configurations, cCcCt and cCcCc, are incapable of existence with planar geometry, either as the unsubstituted hydrocarbon or when incorporated in cyclic structures, and were omitted from the study. The various configurations occur in substituted molecules, and corrections for alkyl substituent effects on the spectra were made by applying the Woodward rules (19,20). Furthermore, the geometries of most molecules of interest are seldom known, so the spectral comparisons must be viewed qualitatively. There is also considerable uncertainty in locating Franck–Condon transitions and determining the number of absorptions under an envelope. In the following, absorptions are given for the envelope maximum in hydrocarbon solution when possible. Experimental oscillator strengths are not generally known for desired model compounds. Theoretical oscillator strengths were compared with molar extinction coefficients, to which they are proportional provided the absorption bands from

TABLE VII

Ground and Excited States of All-*Trans* Polyenes[a]

Polyene	E_b	I_p	$V_1 (f)$	T_1	r_{12}
t	−8.512	9.07 (9.23)[b]	5.573 (0.930)	2.780	1.347
tTt	−13.048	8.45 (8.52)	4.736 (1.338)	2.196	1.348
tTtTt	−17.587	8.08 (8.10)	4.209 (1.745)	1.883	1.348
tTtTtTt	−22.127	7.83 (7.82)	3.846 (2.121)	1.699	1.348
tTtTtTtTt	−26.666	7.66 (7.62)	3.583 (2.474)	1.581	1.348
tTtTtTtTtTt	−31.206	7.54 (7.47)	3.388 (2.810)	1.502	1.348
tTtTtTtTtTtTt	−35.746	7.45 (7.36)	3.240 (3.134)	1.447	1.348

Polyene	r_{23}	r_{34}	r_{45}	r_{56}	r_{67}	r_{78}	r_{89}
t	1.459						
tTt	1.456	1.356					
tTtTt	1.455	1.358	1.452				
tTtTtTt	1.455	1.358	1.451	1.359			
tTtTtTtTt	1.455	1.358	1.451	1.360	1.450		
tTtTtTtTtTt	1.455	1.358	1.451	1.360	1.450	1.360	
tTtTtTtTtTtTt	1.455	1.358	1.451	1.360	1.450	1.360	1.450

[a] All energies are given in eV. Oscillator strengths of the various transitions are given in parentheses. Bond lengths are given in Angstrom units. Calculated ionization potentials (I_p) have been scaled down by 2.45 eV uniformly.

[b] Values in parentheses taken from ref. 25.

molecule to molecular are of similar shape. Singlet–triplet absorption data is sparse, and 0–0 absorption values have been quoted consistently.

In Tables II–V, scale drawings of the isomers are given with the numbering referred to in Tables VI and VII. In each diagram, the transition moment vector of $V_1 \leftarrow N$ is shown by a two-headed arrow which passes through the electrical center of gravity and whose length is drawn to a relative scale. The angles α refer to the orientation with respect to the coordinate axes of calculation.

A. Dienes (Fig. 2 and Table II)

The lowest excited singlets are ordered properly with respect to

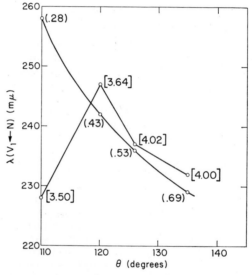

Fig. 2. Calculated (smooth curve) and observed (broken curve) wavelengths of the $V_1 \leftarrow N$ transition of *cisoid*-1,3-dienes. The points refer to cyclopentadiene (110°), 1,3-cyclohexadiene (120°), 1,2-dimethylenecyclopentane (126°), and 1,2-dimethylenecyclobutane (135°), where θ measures the interior CCC angle. In all cases the experimental values of λ were corrected for alkyl substitution (see text). The values in parentheses are theoretical oscillator strengths and those in brackets are experimental log ε.

energy and oscillator strength with V_1 at 5.57 eV ($f = 0.93$) in **t** and at 5.12 eV ($f = 0.43$) in **c**; experimentally, these values are 217 mμ (log $\varepsilon = 4.32$, hexane) (27) or 5.72 eV in 1,3-butadiene and 260 mμ (log $\varepsilon = 3.90$, cyclohexane) (28) or 5.00 eV (corrected) in 1,3-cyclohexadiene. 1,3-Cyclohexadiene is not planar, but microwave (29) and theoretical (19) studies suggests that the deviation is small (17.5 ± 2°).

The lowest triplets are ordered properly with T_1 at 2.78 eV in **t** and 2.76 eV in **c**. Experimentally, these values are 2.60 eV in **t** and 2.52 eV (cyclopentadiene) and 2.28 eV (1,3-cyclohexadiene) in **c** (30), or when corrected, 2.57 and 2.32 eV, respectively (31). The T_2 state in **t** is predicted at 3.56 eV and found at 3.84 eV (30). Previously in Pariser–Parr–Pople calculations the *s-cis* triplet T_1 has been found at higher energy than the *s-trans* (9,11,14). Correct ordering of the triplets has been obtained by including resonance integrals between nonbonded carbon atoms (20), but such procedures are inconsistent with the simultaneous neglect of differential overlap and cannot be easily assessed theoretically. The present results emerge naturally from the use of energy-minimized SCFMO.

The transition energies in butadiene have been studied by Allinger and Miller (19) as a function of the dihedral angle between the two double bonds. They found that small deviations from planarity caused only a small effect on the $V_1 \leftarrow N$ transition energy and virtually no effect on $V_2 \leftarrow N$. For a dihedral angle of 30°, the transition energy to V_1 and V_2 increased by 0.51 and 0.00 eV, respectively, whereas the oscillator strengths were practically invariant. In larger molecules the effects are considerably smaller, and it seems likely that in trienes, tetraenes, etc., deviations from planarity by contiguous double bonds can be neglected for dihedral angles $\leq 20°$.

In this study, another kind of deformation was investigated, i.e., the in-plane bending of the σ-bond framework. It was found that the lowest excited singlets of *s-cis* dienes are ordered properly with respect to this deformation. Calculations were performed for cyclopentadiene (35), 1,3-cyclohexadiene (28), 1,2-dimethylenecyclopentane (36) and 1,2-dimethylenecyclobutane (37) assuming that the interior angle θ is 110°, 120°, 126°, and 135°, respectively. None of these geometries is known accurately, but the assumed models are ordered properly in direction and the true values are probably close to these. The experimental value for cyclopentadiene is 110° (38) and 120° 10' for 1,3-cyclohexadiene (29). The results are plotted in Figure 2, and both energy and oscillator strength parallel

experiment satisfactorily for $\theta \geq 120°$. The absorption spectra of cyclopentadiene (λ_{max} 233 mμ; log ε 3.5) (39) and 1,3-cyclohexadiene (λ_{max} ca. 246 mμ; log ε 3.8) (40) have been determined in the vapor phase, where it is still apparent that the Franck–Condon transition is at higher energy in cyclopentadiene. 1,3-Cyclohexadiene has been considered to have an anomalous ultraviolet spectrum (31,41), but the present results suggest that cyclopentadiene ($\theta = 110°$) behaves exceptionally. This might be accounted for by the larger 1,3 and 1,4 interactions in this diene. At the calculated 1,4 distance (2.37 Å), $\beta_{14} = -0.265$ eV and inclusion of this term in the SCF calculation lowered V_1, however, when all 1,3 and 1,4 interactions were included, V_1 was raised to about the proper value. Since this procedure introduced elements outside the theory used for the calculation (neglect of differential overlap), it is difficult to interpret this result, but it is clearly in the proper direction.

B. Trienes (Fig. 3 and Table III)

Of the six possible triene configurations, five are known: **cTc** in cholestane-5,7,14-trienes (42,43), **tTc** in cholestane-2,4,6-trienes (42,43), **tTt** in *trans*-1,3,5-hexatriene (44), **cCt** in androsta-5,7,9(11)-trienes (42,43), and **tCt** in *cis*-1,3,5-hexatriene (45) and in ergosta-4,6,8(14)-trienes (44).

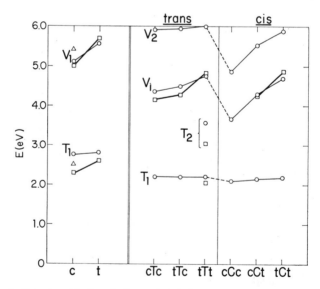

Fig. 3. Calculated ($\bigcirc$) and observed ($\square$) singlet (V_1, V_2) and triplet (T_1) energy levels of the linear dienes and trienes. Values given by $\triangle$ are for cyclopentadiene. The model compounds are identified in the text. The configurations are ordered for each structural type in terms of decreasing energy of the ground state (H^0).

The calculated $V_1 \leftarrow N$ transitions reproduce the observed spectra satisfactorily as the geometry of the triene changes. Apparently, small variations in planarity in the polyenes do not greatly affect the spectra. In both *cis* and *trans* trienes, the $V_1 \leftarrow N$ and $V_2 \leftarrow N$ transition energies decrease as the number of *s-cis* configurations increases. The core potential in going from *s-trans* to *s-cis* configurations ("long field" to "round field") (13,46) acts to lower both the ground and excited singlet states, the latter more than the former. In all cases, V_1 is a plus-state (B_u^+ or B_2^+) and V_2 is minus (A_g^- or A_1^-). Plus-states are expected (13,46) to be of lowest energy in "long-field" molecules, but minus-states will be favored by "round fields." As the configurations become more "round," the $V_2 - V_1$ separation was found to decrease, but in no case does a minus-state become the lowest excited singlet; in fact, in very large polyenes the V_1 state is always plus (cf., discussion in ref. 13).

Comparison of the calculated oscillator strengths with experimental molar extinction coefficients shows agreement in the qualitative ordering of the configurations. The approximate molar extinction coefficients are average values for several members of a class in some cases.

Configuration	f, calc.	ε, obs.
cCc	0.101	—
cCt	0.468	~12,000
cTc	0.713	~16,000
tTc	0.777	~17,000
tCt	1.214	~30,000
tTt	1.338	~40,000

C. Tetraenes (Fig. 4 and Table IV)

Of the 18 configurations considered, experimental data are available for only five: **tTtTt** in *trans,trans*-1,3,5,7-octatetraene (44); **tCcTc** in cholesta-4,6,8(9),11-tetraenes (42,43); **tCcTt** in cholesta-3,5,7,9(11)-tetraenes (42,43); and **tCtTt** in *cis,trans*-1,3,5,7-octatetraene (47). *cis-cis*-1,3,5,7-Octatetraene, which presumably exists in the **tCtCt** configuration and readily undergoes valence isomerization to *cis,cis,cis*-1,3,5-cyclo-octatriene, has a spectrum very similar to that of the *trans,trans* isomer (48).

The calculated $V_1 \leftarrow N$ transition energies agree well with experiment. For a given geometrical isomer (e.g., *trans–cis*), the $V_1 \leftarrow N$ transitions move to progressively longer wavelengths as more *s-cis* linkages are introduced. This behavior, like that of the trienes, seems general, and V_1 is a plus-state in all cases.

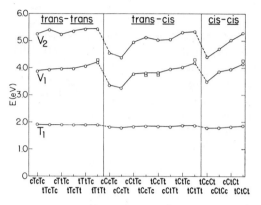

Fig. 4. Calculated (○) and observed (□) singlet (V_1, V_2) and triplet (T_1) energy levels of the linear tetraenes. The model compounds are identified in the text. The configurations are ordered for each structural type in terms of decreasing energy of the ground state (H^0).

Oscillator strengths and molar extinction coefficients agree satisfactorily considering the complicated structures of the model compounds.

Configuration	f, calc.	ε, obs.
tCcTt	0.572	~20,000
tCcTc	0.714	~14,000
tCtTt	1.597	~58,000
tTtTt	1.745	~70,000

D. All-*trans* Polyenes (Fig. 5 and Table V)

The experimental spectra, taken from the recent work of Sondheimer, Ben-Efraim, and Wolovsky (44), show well-developed vibrational progressions in the V_1 absorptions with the second band most intense. This band was chosen for comparison with the theoretical spectra, and the agreement is very good for both singlet and triplets. (The value observed

all-*trans* $CH_2=CH-(CH=CH)_n-CH=CH_2$	f, calc.	f, obs.	ε, obs.
0	0.930	0.74	23,000
1	1.338	1.11	42,700
2	1.745	1.66	70,000
3	2.121	2.17	115,000
4	2.474	—	217,000
5	2.810	—	—
6	3.134	—	—

for $n = 6$ (Fig. 5) suggests that the hexadecaoctaene may be contaminated with a homolog absorbing at longer wavelengths.)

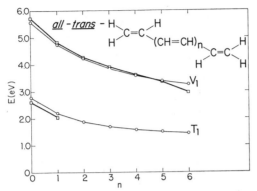

Fig. 5. Calculated (○) and observed (□) singlet (V_1) and triplet (T_1) energy levels of the all-*trans* polyenes. The V_1 experimental values are taken from ref. 44 and the T_1 values from refs. 30 and 31.

The oscillator strengths computed for this series agree well with those found for the all-*trans* α,ω-dimethylpolyenes (49) and with the molar extinction coefficients of the parent polyenes (44).

E. β-Carotenes (Table VIII)

The studies by Zechmeister (50–53) of the carotenoid pigments developed some of the early concepts of the influence of polyene geometry

all-*trans*-β-carotene: 450 mμ (ε 150,000)

9-*cis*-β-carotene: 445 mμ (ε 135,000)
340 mμ (ε 15,000)

TABLE VIII

Ground and Excited States of β-Carotenes[a]

	all-*trans* β-Carotene	9-*cis*-β-Carotene	15-*cis*-β-Carotene	11,11'-di-*cis*-β-Carotene
H^0	-879.729	-890.408	-896.757	-906.505
E^π	-303.331	-303.335	-303.334	-303.336
E_b	-49.363	-49.367	-49.366	-49.368
I_p	7.27	7.27	7.27	7.27
V_1	^1B_u, 2.963 (4.078)	2.945 (3.854)	$^1B_2$, 2.914 (3.505)	^1B_u, 2.910 (3.551)
V_2	^1A_g, 3.811 (0)	3.780 (0.124)	$^1A_1$, 3.842 (0.602)	^1A_g, 3.780 (0)
V_3	^1A_g, 4.273 (0)	4.245 (0)	$^1A_1$, 4.176 (0)	^1A_g, 4.174 (0)
V_4	^1B_u, 4.558 (0.423)	4.555 (0.389)	$^1B_2$, 4.513 (0.413)	^1B_u, 4.537 (0)[b]
T_1	^3B_u, 1.356 (ref.)	1.348 (ref.)	$^3B_2$, 1.338 (ref.)	^3B_u, 1.341 (ref.)
T_2	^3A_g, 1.638 (0)	1.626 (0)	$^3A_1$, 1.637 (0)	^3A_g, 1.624 (0)
T_3	^3B_u, 2.019 (0)	2.014 (0)	$^3B_2$, 2.011 (0)	^3B_u, 2.023 (0)
T_4	^3A_g, 2.449 (0)	2.449 (0)	$^3A_1$, 2.449 (0)	^3A_g, 2.449 (0)

[a] All energies are given in eV. Oscillator strengths of the various transitions are given in parentheses. Bond lengths are given in Angstrom units. Calculated ionization potentials (I_p) have been scaled down by 2.45 eV uniformly.
[b] The values for V_5 are ^1B_u, 4.617 (1.029).

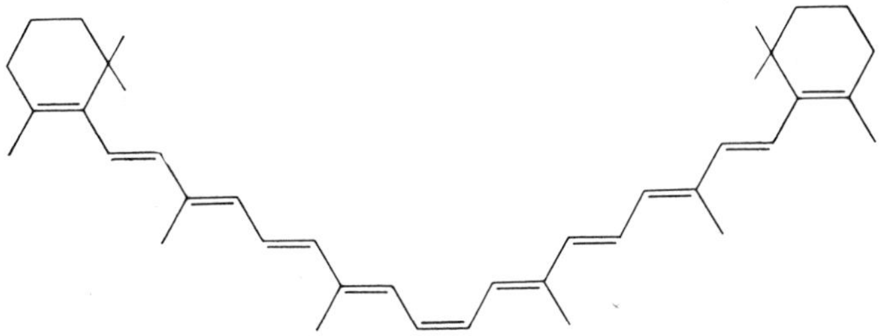

15-*cis*-β-carotene: 445 mμ (ε 90,000)
340 mμ (ε 50,000)

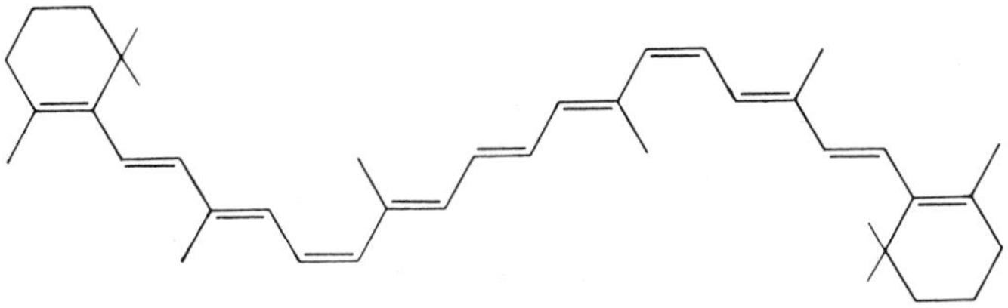

11,11′-di-*cis*-β-carotene: 455 mμ (ε 125,000)

on the location of excited singlet states. The lowest energy transition in an all-*trans* polyene is allowed and intense, whereas transition to the second singlet is forbidden and not usually observed. The introduction of a *cis* linkage is predicted by simple molecular orbital theory to decrease the intensity of $V_1 \leftarrow N$ and make $V_2 \leftarrow N$ allowed; the latter band gains intensity as the *cis* linkage moves toward the center of the chain.

Computations for all-*trans* β-carotene, 9-*cis*-β-carotene, 15-*cis*-β-carotene, and 11,11′-di-*cis*-β-carotene (54) were carried out including determination of the minimum energy geometries (Table VIII). Comparison with experiment shows the following:

β-carotene	V_1, eV	f_1	V_2, eV	f_2
all-*trans*	calc. 2.96	4.08	3.81	0
	obs. 3.10	(ε 150,000)	3.65	~0
9-*cis*	calc. 2.95	3.85	3.78	0.12
	obs. 3.14	(ε 135,000)	3.65	(ε 15,000)
15-*cis*	calc. 2.91	3.51	3.84	0.60
	obs. 3.14	(ε 90,000)	3.65	(ε 50,000)
11,11′-di-*cis*	calc. 2.91	3.55	3.78	0
	obs. 3.06	(ε 125,000)	—	—

Qualitatively, the results reflect the experimental observations, which unfortunately may not refer to pure isomers because of difficulties in purification and of poor stability (e.g., the spectrum of the 11,11'-di-*cis*-isomer must be measured at $-185°$ (54)). The experimental V_2 values were not corrected for substituent effects, since there is no firm basis for doing this in Woodward's rules.

F. Cross-Conjugated Polyenes (Table V)

Three configurations of 2-vinyl-1,3-butadiene, the smallest cross-conjugated polyene, were studied. As more *s-cis* linkages are introduced (cross-**tt** → cross-**ct** → cross-**cc**), H^0 increases but E^π and E_b decrease. In the linear dienes, trienes, and tetraenes, H^0, E^π, and E_b change differentially in the same sense for these structural changes. In any event, E_b decreases in both systems as the number of *s-cis* linkages increases.

An accumulation of *s-cis* linkages lowers V_1 but raises V_2, whereas in linear trienes both are lowered. The linear systems absorb at longer wavelengths than their cross-conjugated counterparts, and the lowest singlet is again a plus-state. The cross-**tt** configuration occurs in 1,2,4,5,6,6-hexamethyl-3-methylene-1,4-hexadiene (55) which absorbs at 256 mμ (ε 21,400) or 5.21 eV (corrected). Cross-**tc** occurs in cholesta-6,8(14)9(11)-triene-3β-acetate (42,43) which absorbs at 285 mμ (ε 9,100) or 4.77 eV (corrected); 2-vinyl-1,3-butadiene, which is presumably not planar and cross-**tc** in configuration, absorbs at 231 mμ (ε 20,500) (56) or 5.37 eV. The calculated values for cross-**tt** and cross-**tc** are 4.88 ($f = 0.60$) and 4.73 eV ($f = 0.35$), respectively. A second band is observed in the cholestatriene at 243 mμ (ε 12,100) or 5.11 eV, and the calculated value of V_2 is 5.61 eV ($f = 0$). Although data are scarce, the computed spectra agree only moderately well in this series, presumably because of significant deformation of the planar geometry.

G. Radialenes (Fig. 6 and Table IX)

Calculations were carried out for the radialenes $\left(\begin{array}{c} \diagdown \\ \diagup \end{array} C{=}CH_2 \right)_n$ with $n = 3, 4, 5$, and 6, which are alternant and nonalternant hydrocarbons when n is even and odd, respectively. Strong bond alternation was found, and when $n > 6$ the double and single bonds are fixed at 1.355 and 1.460 Å, respectively, regardless of alternant character. These results closely parallel those of Dewar and Gleicher (10). The ionization potentials show maximum values for the nonalternant members of the series.

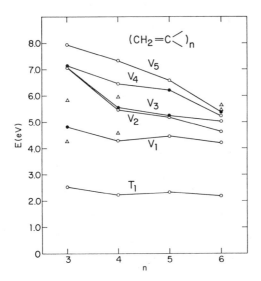

Fig. 6. Calculated (○) and observed (△) spectra of the radialenes. The solid circles (●) represent levels to which transition is allowed. The model compounds are discussed in the text.

The energies of the lowest singlet and triplet states are predicted to alternate in magnitude as n increases, the alternant hydrocarbons generally absorbing at longer wavelengths. For V_2 and higher states, the wavelength increases smoothly with molecular size. The radialenes belong to the point group D_{nh} and symmetry considerations are thus important in determining the allowedness of their low-lying transitions. Trimethylenecyclopropane is unique in that transition is allowed to its lowest singlet ($^1E'$) but is forbidden to the two next lowest singlets, $^1A_2'$ and $^1A_1'$. Transition to the lowest singlet of all larger radialenes is symmetry forbidden. In tetramethylenecylobutane, the two lowest singlet transitions are forbidden but the third (^1E_u) is strongly allowed; in fact, this is the only allowed transition among the first seven calculated states. In pentamethylenecyclopentane, the two lowest singlets are similarly forbidden, while the third ($^1E_1'$) is allowed. In hexamethylene–cyclohexane, the four lowest singlets are forbidden, the first allowed transition being to $^1E_{1u}$.

Trimethylenecyclopropane absorbs at 295 mμ (envelope maximum) or 4.21 eV in isopentane (57,58) and a second maximum has been reported at 213 mμ (ε 19,400) (59); it is likely, however, that the latter band is spurious (58). In the vapor phase, the long wavelength maximum absorption occurs at 288 mμ or 4.31 eV. Trisisopropylidenecyclopropane has a a single maximum at 309.5 mμ (ε 18,000) or 4.54 eV (corrected) (60,61).

The $^1E'$ state of trimethylenecyclopropane is calculated to lie at 4.80 eV.

Tetramethylenecyclobutane has two maxima at 271 mμ (ε ~ 7,600) or 4.58 eV and at 208 mμ (ε 49,500) or 5.96 eV in hexane solution (58,62). The intense absorption at 5.96 eV must be identified with the strongly allowed $V_3(^1E_u) \leftarrow N$ transition calculated to lie at 5.55 eV. The weak absorption at 4.58 eV most likely corresponds to the calculated $^1B_{2g}$ level

TABLE IX

Ground and Excited States of the Radialenes[a,b]

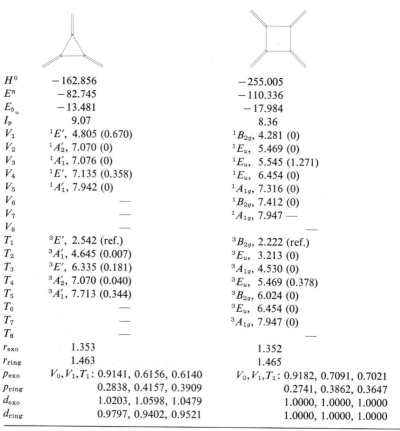

H^0	-162.856	-255.005
E^π	-82.745	-110.336
E_b	-13.481	-17.984
I_p	9.07	8.36
V_1	$^1E'$, 4.805 (0.670)	$^1B_{2g}$, 4.281 (0)
V_2	$^1A_2'$, 7.070 (0)	^1E_u, 5.469 (0)
V_3	$^1A_1'$, 7.076 (0)	^1E_u, 5.545 (1.271)
V_4	$^1E'$, 7.135 (0.358)	^1E_u, 6.454 (0)
V_5	$^1A_1'$, 7.942 (0)	$^1A_{1g}$, 7.316 (0)
V_6	—	$^1B_{2g}$, 7.412 (0)
V_7	—	$^1A_{1g}$, 7.947 —
V_8	—	—
T_1	$^3E'$, 2.542 (ref.)	$^3B_{2g}$, 2.222 (ref.)
T_2	$^3A_1'$, 4.645 (0.007)	^3E_u, 3.213 (0)
T_3	$^3E'$, 6.335 (0.181)	$^3A_{1g}$, 4.530 (0)
T_4	$^3A_2'$, 7.070 (0.040)	^3E_u, 5.469 (0.378)
T_5	$^3A_1'$, 7.713 (0.344)	$^3B_{2g}$, 6.024 (0)
T_6	—	^3E_u, 6.454 (0)
T_7	—	$^3A_{1g}$, 7.947 (0)
T_8	—	—
r_{exo}	1.353	1.352
r_{ring}	1.463	1.465
p_{exo}	V_0,V_1,T_1: 0.9141, 0.6156, 0.6140	V_0,V_1,T_1: 0.9182, 0.7091, 0.7021
p_{ring}	0.2838, 0.4157, 0.3909	0.2741, 0.3862, 0.3647
d_{oxo}	1.0203, 1.0598, 1.0479	1.0000, 1.0000, 1.0000
d_{ring}	0.9797, 0.9402, 0.9521	1.0000, 1.0000, 1.0000

[a] All energies are given in eV. Oscillator strengths of the various transitions are given in parentheses. Bond lengths are given in Angstrom units. Calculated ionization potentials (I_p) have been scaled down by 2.45 eV uniformly.

[b] Charge densities (d) and bond orders (p) are given for the ground and lowest singlet and triplet states.

TABLE IX—*continued*
Ground and Excited States of the Radialenes[a,b]

H^0	-361.137	-476.582
E^π	-138.033	-165.711
E_b	-22.593	-27.183
I_p	8.61	8.52
V_1	$^1E_2'$, 4.468 (0)	$^1B_{2u}$, 4.205 (0)
V_2	$^1E_2'$, 5.171 (0)	$^1E_{2g}$, 4.630 (0)
V_3	$^1E_1'$, 5.233 (1.152)	$^1B_{1u}$, 5.033 (0)
V_4	$^1E_1'$, 6.214 (0.249)	$^1E_{2g}$, 5.226 (0)
V_5	$^1A_2'$, 6.578 (0)	$^1E_{1u}$, 5.368 (1.438)
V_6	$^1E_2'$, 7.182 —	$^1B_{2u}$, 6.019 (0)
V_7	$^1A_1'$, 7.220 —	$^1A_{2g}$, 6.568 —
V_8	$^1E_1'$, 7.784 —	$^1E_{1u}$, 6.617 —
T_1	$^3E_2'$, 2.323 (ref.)	$^3B_{2u}$, 2.177 (ref.)
T_2	$^3E_1'$, 3.393 (0.0002)	$^3E_{2g}$, 2.567 (0)
T_3	$^3A_1'$, 4.687 (0)	$^3E_{1u}$, 3.619 (0)
T_4	$^3E_2'$, 5.077 (0.126)	$^3A_{1g}$, 4.736 (0)
T_5	$^3E_1'$, 6.048 (0.100)	$^3B_{1u}$, 5.033 (0)
T_6	$^3A_2'$, 6.578 (0)	$^3E_{2g}$, 5.226 (0.191)
T_7	$^3E_2'$, 6.595 —	$^3B_{2u}$, 5.672 (0)
T_8	$^3A_1'$, 7.272 —	$^3A_{2g}$, 6.568 —
r_{exo}	1.355	1.355
r_{ring}	1.461	1.460
p_{exo}	V_0, V_1, T_1: 0.9057, 0.7347, 0.7292	V_0, V_1, T_1: 0.9007, 0.7607, 0.7548
p_{ring}	0.2962, 0.3785, 0.3634	0.3035, 0.3694, 0.3615
d_{exo}	0.9985, 1.0040, 0.9974	1.0000, 1.0000, 1.0000
d_{ring}	1.0015, 0.9960, 1.0026	1.0000, 1.0000, 1.0000

[a] All energies are given in eV. Oscillator strengths of the various transitions are given in parentheses. Bond lengths are given in Angstrom units. Calculated ionization potentials (I_p) have been scaled down by 2.45 eV uniformly.

[b] Charge densities (d) and bond orders (p) are given for the ground and lowest singlet and triplet states.

at 4.28 eV. Transition to the lowest $^1B_{2g}$ state is formally forbidden by symmetry, but here there are appropriate vibrations to mix the closely spaced $^1B_{2g}$ and ^1E_u levels. In fact, another ^1E_u level lies between these two states at 5.47 eV, and this level happens to be nonessentially forbidden but is well suited for vibronic mixing. The intense band observed probably contains both the allowed (5.55 eV) and the forbidden (5.47 eV) ^1E_u transitions.

Hexapropylidenecyclohexane has a maximum at 220 mμ ($\varepsilon \geq 31,700$) with a shoulder at 263 mμ ($\varepsilon \sim 7,800$) in cyclohexane, and hexaethylidenecyclohexane shows this shoulder at 266 mμ ($\varepsilon \sim 5,700$) (63,64). If the shoulder represents an independent band, then there are two levels in the hexaethyl derivative at 5.64 and 5.47 eV (corrected). The strong absorption at 5.64 eV must be identified with the allowed $^1E_{1u}$ absorption calculated to lie at 5.37 eV. The shoulder then might correspond to one (or a combination) of the four predicted lower-lying states (4.2–5.2 eV). Vibronic mixing is possibly important, but without analysis no further conclusions can be drawn concerning the lower-lying state(s).

The larger radialenes are probably not planar, but the strong band in each spectrum can be identified with the theoretically allowed transition, which is clearly predicted in each case because of the high symmetry. When these correlations are inspected, it is seen that the agreement is moderately satisfactory.

	Strong transition	
n-Radialene	obs., eV (ε)	calc., eV (f)
3	4.2 (3,750)	4.8 (0.67), $^1E'$
4	6.0 (49,500)	5.5 (1.27), ^1E_u
5	—	5.2 (1.15), $^1E_1$
6	5.6 (31,700)	5.4 (1.44), $^1E_{1u}$

The nonalternant radialenes have nonuniform charge distributions. In trimethylenecyclopropane, the exocyclic methylene groups acquire excess negative charge in the ground state, and this polarization increases in both the V_1 and T_1 states. In the ground state of pentamethylenecyclopentane, however, a small net positive charge occurs on the exocyclic methylene groups. In the V_1 state the polarity is reversed, the exocyclic carbons becoming faintly negatively charged, but in the T_1 state these carbons become slightly more positively charged than in the ground state. The differing polarizations of the ground states of these two molecules are not unexpected, and experimental information on this point will be interesting.

VI. TRIPLET STATES

The triplet levels of the polyenes are given in the tables, but there is little experimental data available for comparison. s-trans-1,3-Butadiene

(30,31) (see discussion above) and all-*trans* hexatriene (30) have been studied, and Tables II and III and Figure 3 show that satisfactory agreement was obtained.

The T_1 state has been measured (30,31) in cyclopentadiene (2.6 eV) and cyclohexadiene (2.3 eV); the calculated values are 2.735 and 2.764 eV, respectively. The quantitative agreement is not very good and the ordering is not correct; a similar difficulty was found in the ordering of the V_1 states. This may be due to geometry changes in excited states of these small molecules, although the ordering of the triplets of *s-trans*-1,3-butadiene (2.78 eV) and either of the *s-cis* dienes is predicted correctly. Strain in the sigma framework of the ground and excited states is probably of comparable importance to the changes in π-energy, and at present such subtleties cannot be adequately predicted by π-electron theories.

Interestingly, the calculations predict that the configuration of a given polyene has virtually no effect on the location of triplet levels (see Figs. 3–5). This is particularly striking in the tetraenes where the V_1 states range over 1 eV, whereas the T_1 states vary over 0.1 eV. Antisymmetrized MO's strongly correlate the motions of unpaired electrons, and introduction of electron repulsion disturbs both the ground state and low triplet to roughly equal extents with such wavefunctions.

The triplet energies T_1 of the all-*trans* polyenes fall off smoothly as the chain length increases, but not so rapidly as for the singlets V_1 (Fig. 5).

Lewis and Kasha (34) observed that all-*trans* lycopene exhibited weak phosphorescence that was enhanced on irradiation. It was thought that a *cis* linkage was introduced by photochemical isomerization and that the *cis*-lycopene was the source of the enhanced triplet emission, but current considerations suggest that no authentic cases of polyene phosphorescence are known (33). Evidence of triplet states of long polyenes has come, however, from absorption experiments. It was recently found that all-*trans*-β-carotene formed a transient, absorbing species when flash-illuminated in the presence of sensitizers, such as chlorophyll *a*, anthracene, tetracene, and benzanthracene (32). The carotene showed no transients when flashed alone in deoxygenated systems, and the transient appeared only when the actinic light was partially or entirely absorbed by the sensitizer, but not when it was absorbed by the carotene. The transient appeared as a broad, well-defined absorption with a maximum at 515 mμ or 2.41 eV, and the authors suggested that they were observing triplet absorption.

The triplet spectrum of all-*trans*-β-carotene was calculated, and the oscillator strengths of the transitions $T_j \leftarrow T_1$ were all zero for the first ten states, except $T_9 \leftarrow T_1$ ($f = 2.05$). The corresponding energies are T_1 (^3B_u) = 1.356 eV and $T_9(^3A_g)$ = 4.273 eV, which predict that the triplet spectrum of all-*trans* β-carotene should show only one intense absorption of energy 2.92 eV. This is in reasonable agreement with the experimental observation considering the high J value of the calculation and the unknown influence of alkyl groups on this triplet spectrum. T_8 is also a ^3A_g state which lies at 2.74 eV above T_1, and, although $T_8 \leftarrow T_1$ is accidentally forbidden, it is likely that the actual absorption can be assigned to both of these states. The detailed calculations thus provide support for the experimental assignment.

Predictions can be made of other triplet–triplet absorptions expected in the polyenes, although these have not yet been determined in small molecules. Some selected values are given below.

Polyene	$T_J \leftarrow T_1$	ΔE, eV (λ, mμ)	f
t	3	4.053 (306)	0.518
c	3	3.849 (322)	0.395
tTt	4	3.818 (325)	0.900
tTc	4	3.742 (332)	0.704
	5	4.517 (275)	0.071
cTc	4	3.749 (331)	0.708
tCt	4	3.710 (334)	0.667
	5	4.628 (268)	0.198
cCt	4	3.377 (367)	0.345
	5	4.533 (274)	0.348
cCc	4	2.764 (449)	0.101
tTtTt	5	3.565 (348)	1.182

VII. EXCITED-STATE STRUCTURES

It is of interest to investigate whether additivity of bond energies also holds in polyene excited states. In triplet states the electrons are highly correlated, the antisymmetrized wavefunctions keeping the unpaired electrons far apart, so that the usual chemist's pictures for T_1 states, such as for the all-*trans* polyenes, might be reasonable. Such structures might be

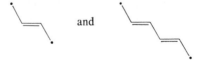

less reasonable, however, for singlet states. The expression corresponding to eq. (17) for the lowest triplet state of the all-*trans* polyenes now becomes

$$E_b^T = kE_s^T + (k-1)E_d^T = k(E_s^T + E_d^T) - E_d^T \qquad (19)$$

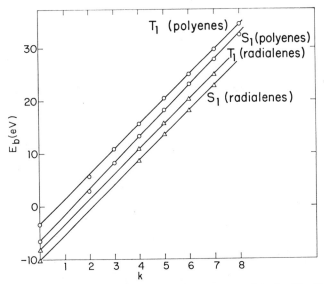

Fig. 7. Plot of $-E_b^{T,V}$ vs. k for the lowest triplet (T) and singlet $(S \equiv V)$ states of the all-*trans* polyenes ($\bigcirc$) and radialenes ($\triangle$) where k is the number of formal single bonds in the diradical excited-state structures. The slopes and intercepts are, respectively: polyene singlets, 4.9078 and -6.4937 eV; polyene triplets, 4.7425 and -3.4499 eV; radialene singlets, 4.7328 and -10.1599 eV; radialene triplets, 4.6709 and V -7.6957 eV.

A plot of E_b^T vs. k (Fig. 7) gives eq. (20)

$$E_b^T = -4.7425k + 3.4499 \qquad (20)$$

with $E_s^T = -1.2926$ and $E_d^T = -3.4499$, and similarly eq. (21) for the lowest singlet state

$$E_b^V = -4.9078k + 6.4937 \qquad (21)$$

with $E_s^V = +1.5859$ and $E_d^V = -6.4937$. In neither case are the plots accurate straight lines as in the ground state, but both are capable of predicting excited states to ca. 0.1 eV. For example, for cross-**tt** eqs. (20) and (21) give 10.778 and 8.230 eV, respectively, and the accurate values are 10.645 and 8.128 eV.

From these results, it might be said that if the lowest triplet state is formulated by structures such as $\cdot CH_2$—CH=CH—$CH_2\cdot$, then the central bond has roughly $(3.4499/3.9707) \times 100 = 87\%$ of the character of the double bonds of the ground state, whereas a single bond in this structure has over twice the π-character of the single bond in the ground state. In the lowest excited singlet state the double bonds are over one and a half times as strong as in the ground state, but the single bonds make a large repulsive contribution to the binding. Thus, the familiar "diradical" pictures often employed for triplet states are actually a rather good representation.

The plots (Fig. 7) for the π-bonding energies in the T_1 and $V_1 \equiv S_1$ states of the radialenes (eq. 22) correspond to eqs. (23) and (24) with

$$E_b^{T,V} = kE_s^{T,V} + (k - 2)E_d^{T,V} = k(E_s^{T,V} + E_d^{T,V}) - 2E_d^{T,V} \tag{22}$$

$$E_b^T = -4.6709k + 7.6957 \tag{23}$$

$$E_b^V = -4.7328k + 10.1599 \tag{24}$$

$E_s^T = -0.8230$, $E_d^T = -3.8479$, $E_s^V = +0.3472$, and $E_d^V = -5.0800$. In the T_1 state, the radialenes have bond energies similar to those in the ground state, but in the V_1 state the formal double bonds are considerably stronger than those in the ground state.

A. Excited-State Bond Orders

The bond orders of the polyenes in the ground and lowest excited singlet and triplet states are given in Table VI. It was shown above that in the ground state, *trans* and *cis* double bonds tend to lengthen when surrounded by adjacent *s-cis* linkages. Bond orders in the various excited states also show some interesting patterns, but these will not be specifically discussed. The table should be useful, however, for photochemical predictions; for instance, in the all-*trans* 1,3,5-heptatriene, the bond orders of the (34) and (56) bonds in V_1 are 0.455 and 0.701, re-

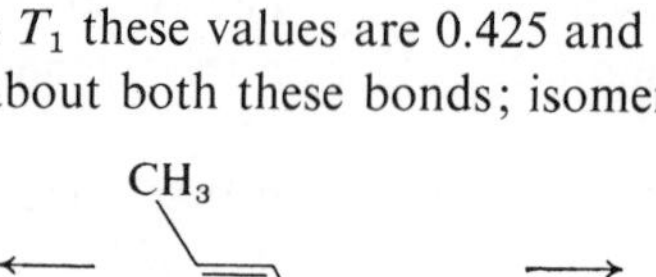

spectively, whereas in T_1 these values are 0.425 and 0.720. Irradiation can cause isomerization about both these bonds; isomerization at (56) would

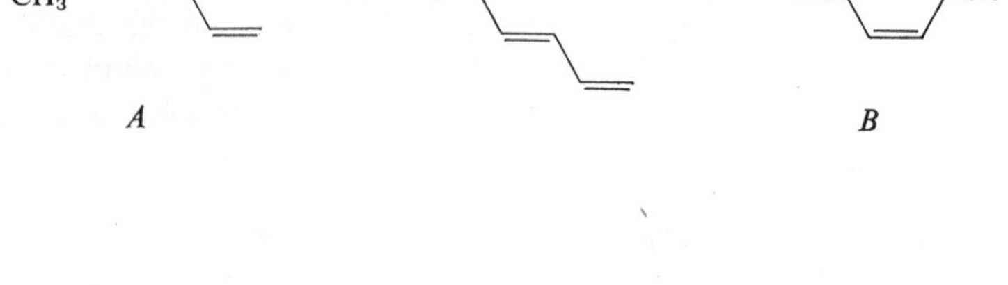

A B

be favored by direct irradiation (V_1) and at (34) by sensitized irradiation (T_1), provided that the planar excited states are at energy maxima. Thus, the initial ratio A/B should be larger in direct compared to sensitized irradiations. This conclusion has not yet been tested.

VIII. INTRAMOLECULAR RADIATIVE AND RADIATIONLESS PROCESSES

Two theories of intramolecular mixing of singlet and triplet states in radiationless processes have been suggested, which differ in the nature of the perturbation that causes mixing. Gouterman (65) views these processes as arising by interaction of the molecular system with the time-dependent, oscillatory phonon field, whereas Robinson and Frosch (66) stress the importance of time-independent perturbations, such as spin–orbit coupling, and of the crowded lattice vibronic levels which serve to increase the resonance between initial and final states. Some experimental evidence supports the latter view (66,67). The rate of such intersystem crossing depends on the vibrational overlap function of the initial and final states and on the energy gap separating these states. The probability P_{VT} for intersystem crossing between a singlet state and a triplet state is given by eq. (25),

$$P_{VT} \sim |\langle V_J| \mathbf{H}_{SO} |T_K\rangle\langle \chi_{V_J}|\chi_{T_K}\rangle|^2 \tag{25}$$

where $\mathbf{H}_{SO}$ is the perturbing spin–orbit Hamiltonian, V_J and T_K are purely electronic singlet and triplet wavefunctions, and χ are vibrational wavefunctions for the appropriate electronic states indicated by subscripts.

McClure (68) has given selection rules which govern spin–orbit interaction between singlet and triplet wavefunctions, and because this perturbation operator (69) involves individual angular momentum operators (which have the same symmetry properties as rotations), selection rules are easy to derive for π-electron systems.

The radiationless crossing of states in the same spin manifold, e.g., internal conversion, also depends on the energy gap separating the initial and final states and is governed by matrix elements which formally can be expressed as products of vibrational overlap integrals and an electronic part which is largely coulombic in nature. For some perturbation $\mathbf{P}$, the probability P_{VV} of radiationless transition between two, e.g., singlet, states is given by eq. (26).

$$P_{VV} \sim |\langle V_J |\mathbf{P}| V_K\rangle\langle \chi_{V_J}|\chi_{V_K}\rangle|^2 \tag{26}$$

The perturbation $\mathbf{P}$ is time dependent and presumably involves nuclear deformations, and so in large molecules with many vibrational modes it can have many symmetries. This means that the electronic part of eq. (26) will rarely vanish strongly, and the vibrational overlaps will be very important in determining P_{VV}. In the C_{2h} and C_{2v} polyenes there are always vibrations with the same symmetry as the dipole moment operators $\mathbf{M}$ for allowed, radiative transitions. Because of the similarity between the radiative $\langle V_J | \mathbf{M} | V_K \rangle$ and radiationless $\langle V_J | \mathbf{P} | V_K \rangle$ transition moment integrals, the latter may often be large whenever the former is large. This weak restriction suggests that whenever a radiative transition is strong, the corresponding radiationless process may also be facile, but more cannot be said.

A survey of the data in Tables II–IV along with the above considerations allows some conclusions to be drawn regarding the polyenes.

1. The radiative process connecting the lowest excited singlet state and the ground state ($V_1 \leftrightarrow V_0$) is allowed in all polyenes.

2. Oscillator strengths for the radiative processes $V_0 \leftrightarrow V_J$ are largest for $J = 1$, when there is a maximum accumulation of *trans* and *s-trans* linkages ("long-field" polyenes). The introduction of *cis* (but not *s-cis*, except in the case of **c**) linkages causes the major absorption to shift to high energies ("round-field" polyenes); e.g., the oscillator strength of $V_0 \leftrightarrow V_3$ is largest in **c, cCc, cCt, tCcTt, cCtTc, cCcTt,** and **cCcTc.**

3. When no elements of symmetry are present, all adjacent states are, in general, connected ($V_J \leftrightarrow V_{J+1}$, for all J) by the radiative process, but the probabilities of adjacent transitions decrease as J increases. When symmetry elements are present, transitions between adjacent states can become forbidden; as J increases, an increasing proportion of the symmetrical polyenes suffer forbidden adjacent transitions.

4. Radiationless processes connecting all adjacent singlet states are, in general, allowed ($V_J \longleftrightarrow V_{J+1}$) within the framework of the above discussion. The presence of symmetry elements is expected to make certain radiationless transitions less likely because of the electronic part of eq. (26).

5. The radiative process connecting adjacent lower excited triplet states is strongly forbidden in all polyenes ($T_J \leftrightarrow T_{J+1}$ for $J \leq 2$; even the oscillator strength of $T_4 \leftrightarrow T_3$ seldom exceeds 0.01).

6. Radiationless processes connecting the lower triplet states are expected to be less probable than those connecting the singlet states. Internal conversion is, therefore, particularly facile in the singlet manifold, but considerably less so in the triplet manifold. This is suggested both by the fact that radiative oscillator strengths are zero between all adjacent triplet states below V_1 in energy and that the average spacing of the lower triplets is larger than that of the lower singlets. Recently it has shown that some higher triplets live long enough to undergo chemical reactions. For example, the lifetime of T_2 in 9,10-dibromoanthracene (2×10^{-11} sec) is sufficiently long for external energy transfer, whereas typical excited singlet state lifetimes (other than V_1) are of the order 10^{-12} to 10^{-13} sec (70).

7. In polyenes of C_{2v} symmetry, the radiationless crossing $T_1 \longleftrightarrow V_0$ is allowed, whereas in those of C_{2h} symmetry this crossing is forbidden, with respect to the spin–orbit coupling operator.

8. Radiationless transitions from the lowest excited singlet state to the lower triplet states depend on molecular geometry. In the all-*trans* polyenes, radiationless transition is allowed to a triplet state very close in energy and just below V_1; incorporation of *s-cis* linkages either raises the energy of this triplet above V_1 or causes transition to this triplet to become forbidden. "Long-field" polyenes with *cis* linkages, such as **tCtCt**, similarly have a triplet just below V_1 to which radiationless transition is allowed, and the incorporation of *s-cis* linkages generally raises the energy of this triplet above V_1. A listing of the "spin–orbit" allowedness" of the $V_1 \longleftrightarrow T_J$ transitions in symmetrical polyenes follows, in which all triplets below V_1 are included.

		$V_1 \longleftrightarrow T_J$	
		Allowed	Forbidden
t	$J =$	1	2
c		2	1
tTt		1,3	2
cTc		1	2
tCt		2	1,3
cCc		2	1
tTtTt		1,3	2
tTcTt		2	1,3
cTtTc		1	2
cTcTc		2	1
tCtCt		1,3	2
tCcCt		2	1
cCtCc		1	2

These points can be amplified by considering some specific examples. Because of the interest in the dienes, the state diagrams for **t** and **c** are given below.

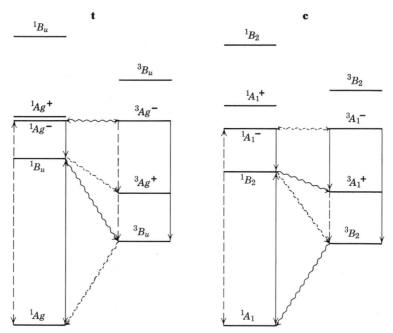

Full lines represent favorable processes and dotted, unfavorable; in a given manifold, this applies to both radiative and radiationless processes. The most important absorption process in **t** is to V_1 (^1B_u). Intersystem crossing to T_1 (^3B_u) is allowed by the spin–orbit selection rules, but not to the triplet closer in energy T_2 ($^3A_g{}^+$). Furthermore, the same consideration forbids the radiationless connection of T_1 to the ground state. The most important absorption process in **c** is also to V_1 ($^1B_2$), but now intersystem crossing to the triplet closest in energy T_2 ($^3A_1{}^+$) is allowed and furthermore the V_1–T_2 gap is smaller in **c** than in **t**. Once the lowest triplet T_1 is reached, however, there is an allowed radiationless crossing to the ground state. It is therefore concluded that intersystem crossing is more important in *cis* than in *trans* dienes, but that T_1 will be longer lived in *trans* than in *cis* dienes. In both **t** and **c**, V_2 and T_3 are degenerate, and crossing is allowed in **t** but not in **c**.

The results obtained above for **c** and **t** can be easily extended to the tri- and tetraenes. For instance, in **tTt**, V_1 ($^1B_u{}^+$) and T_3 ($^3B_u{}^+$) are separated by only 0.207 eV and the crossing is allowed, whereas the crossing from T_1 ($^3B_u{}^+$) to the ground state is forbidden. When a *cis*-linkage is introduced, as in **tCt**, the crossing from V_1 ($^1B_2{}^+$) to T_3 ($^3B_2{}^+$), which are separated by the smaller gap of 0.163 eV, is forbidden, but return to the ground state from T_1 ($^3B_2{}^+$) is allowed.

Radiative emission or fluorescence is observed only if the radiationless processes are not much faster than radiative emission. Furthermore, fluorescence generally occurs only from the V_1 state, since internal conversion from higher singlets to V_1 is normally very fast compared to emission from such states. The rate of spontaneous radiative decay from a higher to a lower state is proportional to the oscillator strength connecting the states and to the square of the energy gap. Factors influencing fluorescence ($V_1 \leftrightarrow V_0$) are different in the "long" and "round-field" polyenes. The radiative oscillator strength is high and the energy gap is large in the former, and both are small in the latter. For example, in **tTt**, $E(V_1)$ − $E(V_0)$ = 4.74 eV, f = 1.34, whereas in **cCc** these values are 3.66 eV and 0.10, respectively. It is therefore expected that "long-field" polyenes are more likely to exhibit fluorescence than "round-field" polyenes with a comparable number of double bonds. Another factor which must be considered is whether intersystem crossing can occur to some triplets near V_1; this would depopulate V_1 and so reduce fluorescence.

Fluorescence is rarely observed in small, nonrigid polyenes, but it has been detected (71) from the *s-cis* dienes ergosterol (E) and lumisterol$_2$ (L)

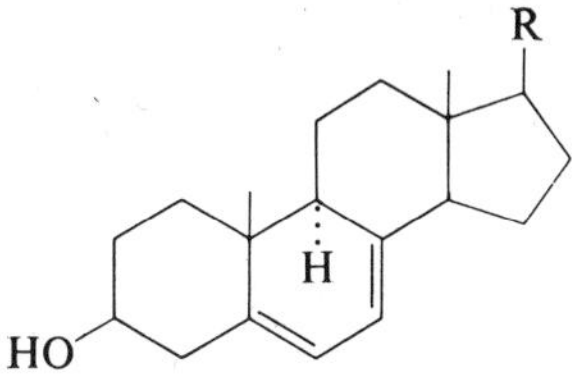

Ergosterol 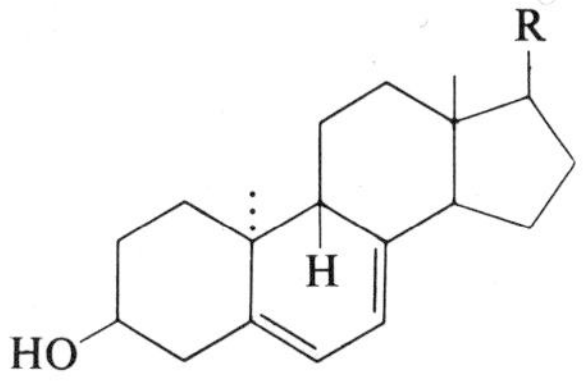   Lumisterol$_2$

R = C$_9$H$_{17}$

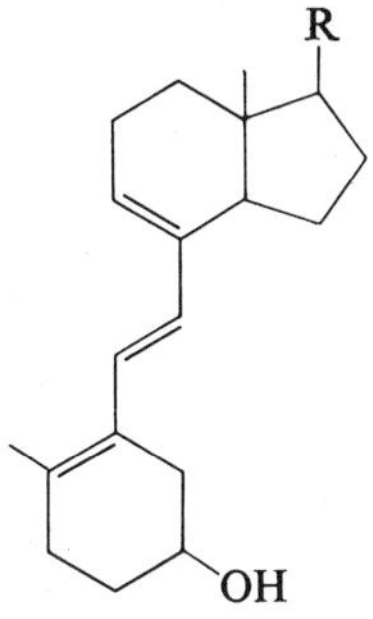

Tachysterol$_2$ 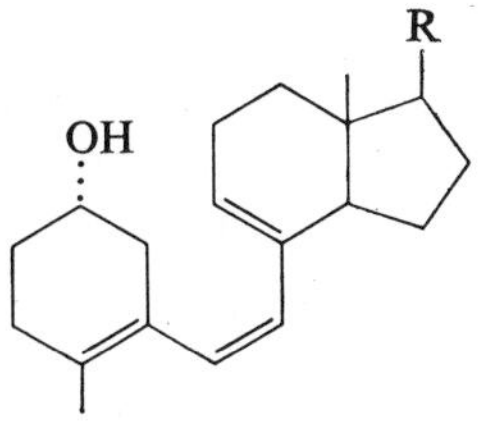   Preergocalciferol and from the *trans* triene tachysterol$_2$ (T) but not from the *cis* triene preergocalciferol (P). None of these compounds shows phosphorescence. The dienes E and L have rigid **c** conformations, and apparently the $V_1 - V_0$ energy gap is sufficiently large to overcome the small transition probability. Furthermore, the $V_1 - T_2$ gap is sufficiently large in these small polyenes so that V_1 may not be depopulated by intersystem crossing. For the same reason, E and L do not show phosphorescence.

The stereochemistry of T and P is not known, and examination of Dreiding molecular models suggests that the preferred structures will be close to those shown, i.e., T ≡ **tTc** and P ≡ **tCc**. Clearly, however, neither molecule is rigid and the π-electron portions of both are probably not planar; this is particularly so in P, which experiences considerable steric crowding. The fact that P does not fluoresce may simply be associated with its less rigid and less planar structure and with its "round-field" nature. As pointed out in conclusion (7) above, however, incorporation of *s-cis* linkages in the *trans* triene raises T_3 above V_1, so that intersystem crossing may become much less likely because of the large $V_1 - T_2$ gap in T. Furthermore, the $V_1 - V_0$ gap is larger in *trans* than in *cis* trienes, which facilitates fluorescence in T compared to P. In the *cis* trienes, T_3 lies very close and below V_1 in **tCt** and **cCc** and above V_1 in **tCc**; however, the $V_1 - T_2$ gap is smaller in the *cis* triene **tCc** than in the *trans* triene **tTc**. For these reasons, intersystem crossing is probably more favorable in planar P than in planar T, and the depopulation of V_1 in P may also help account for its failure to fluoresce. It will be interesting to test these views in some rigid polyenes.

Very long polyenes, such as the carotenes and lycopene (72), are reported not to fluoresce. This is readily understood in the dense packing of lower singlet and triplet states, where many crossings are to be expected and $V_1 - V_0$ gaps are small.

Appendix

A single program was written for the Univac 1108 to carry out the entire calculation. Input was an initial atomic geometry, effective nuclear charge, valence-state ionization potentials, one-center coulomb integrals, and singly excited configuration designations with their symmetries. Subroutines computed atomic coordinates and all integrals over atomic orbitals including β_{pq} which involves overlap, penetration, coulomb repulsion, and hybrid coulomb integrals (9). The following outline (Fig. 8) is a skeleton of the program. The superscript t on integrals means calculation

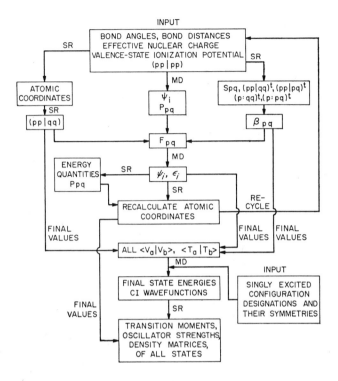

Fig. 8. Flow diagram of computer program.

according to ref. 9. SR means subroutine, and MD means matrix diagonalization. The SCF calculation was recycled to self-consistent geometry as explained in the text, and then the final values of all recomputed integrals, wavefunctions, etc., were carried forward to the configuration interaction calculation. An alternate subroutine allowed calculation of Longuet-Higgins and Salem wavefunctions (in place of ψ_i) and these were carried directly to the configuration interaction calculation. The various subroutines calculated automatically the following molecular properties of interest: SCF wavefunctions and energies, molecular ionization potential, total π-electron energy, framework repulsion energy, effective π-energy of the molecule, effective π-binding energy, all singlet and triplet state energies with their symmetries, x and y components of all transition moments, state dipole moments, oscillator strengths for all states to all states, and bond lengths, bond orders, and charge densities of ground and all singlet and triplet states.

Acknowledgment

The author expresses his deep appreciation of Mr. Lee A. Stone, Du Pont Engineering Department, for his invaluable assistance in programming and to Dr. R. Pariser, Du Pont Elastomers Department, for his long encouragement and valuable discussions. He is grateful also to Dr. R. S. H. Liu, Du Pont Central Research Department, for advice on experimental spectra.

References

1. J. N. Murrell, *The Theory of the Electronic Spectra of Organic Molecules*, John Wiley & Sons, New York, 1963; L. Salem, *The Molecular Orbital Theory of Conjugated Systems*, Benjamin, New York, 1966.
2. E. Hückel, *Z. Phys.*, *70*, 204 (1931); *76*, 628 (1932).
3. J. E. Lennard-Jones, *Proc. Roy. Soc. (London)*, Ser. A, *158*, 280 (1937).
4. J. E. Lennard-Jones and J. Turkevich, *Proc. Roy. Soc. (London)*, Ser. A, *158*, 297 (1937).
5. H. C. Longuet-Higgins and L. Salem, *Proc. Roy. Soc. (London)*, Ser. A, *251*, 172 (1959); *A257*, 445 (1960).
6. M. J. S. Dewar and H. W. Schmeising, *Tetrahedron*, *5*, 166 (1959).
7. W. T. Simpson, *Theories of Electrons in Molecules*, Prentice-Hall, Englewood Cliffs, N.J., 1962.
8. J. A. Pople, *Trans. Faraday Soc.*, *49*, 1375 (1953); A. Bristock and J. A. Pople, *ibid.*, *50*, 901 (1954).
9. H. E. Simmons, *J. Chem. Phys.*, *40*, 3554 (1964).
10. M. J. S. Dewar and G. J. Gleicher, *J. Am. Chem. Soc.*, *87*, 685, 692 (1965).
11. R. Pariser and R. G. Parr, *J. Chem. Phys.*, *21*, 466 (1953); *21*, 767 (1953); R. G. Parr and R. Pariser, *J. Chem. Phys.*, *23*, 711 (1955).
12. W. Moffitt, *J. Chem. Phys.*, *22*, 320 (1954).
13. R. Pariser, *J. Chem. Phys.*, *24*, 250 (1956).
14. J. A. Pople, *Proc. Phys. Soc. (London)*, *A68*, 81 (1955).
15. J. R. Platt, *J. Chem. Phys.*, *17*, 484 (1949).
16. M. J. S. Dewar and H. C. Longuet-Higgins, *Proc. Phys. Soc. (London)*, *A67*, 795 (1954).
17. A. L. H. Chung and M. J. S. Dewar, *J. Chem. Phys.*, *42*, 756 (1965).
18. J. A. Pople and S. H. Walmsley, *Trans. Faraday Soc.*, *58*, 441 (1962).
19. N. L. Allinger and M. A. Miller, *J. Am. Chem. Soc.*, *86*, 2811 (1964).
20. N. L. Allinger and J. C. Tai, *J. Am. Chem. Soc.*, *87*, 2081 (1965).
21. J. A. Pople, D. P. Santry, and G. A. Segal, *J. Chem. Phys.*, *43*, S129 (1965).
22. M. J. S. Dewar and G. Klopman, *J. Am. Chem. Soc.*, *89*, 3089 (1967).
23. H. E. Simmons, unpublished results.
24. K. Ruedenberg, *J. Chem. Phys.*, *34*, 1861 (1961).
25. O. W. Adams and R. L. Miller, *J. Am. Chem. Soc.*, *88*, 404 (1966).
26. R. G. Parr and R. S. Mulliken, *J. Chem. Phys.*, *18*, 1338 (1950).
27. R. W. Kierstead, R. P. Linstead, and B. C. L. Weedon, *J. Chem. Soc.*, *1953*, 1803.

28. E. Merkel, *Z. Elektrochem.*, *63*, 373 (1959).
29. S. S. Butcher, *J. Chem. Phys.*, *42*, 1830 (1965).
30. D. F. Evans, *J. Chem. Soc.*, *1960*, 1735.
31. R. E. Kellogg and W. T. Simpson, *J. Am. Chem. Soc.*, *87*, 4230 (1965).
32. M. Chessin, R. Livingston, and T. F. Truscott, *Trans. Faraday Soc.*, *62*, 1519 (1966).
33. R. E. Kellogg, private communication.
34. G. N. Lewis and M. Kasha, *J. Am. Chem. Soc.*, *66*, 2100 (1944).
35. S. L. Friess and V. Boekelheide, *J. Am. Chem. Soc.*, *71*, 4145 (1949), λ_{max} 241 mμ (ε 3160).
36. A. T. Blomquist, J. Wolinsky, Y. C. Meinwald, and D. T. Longone, *J. Am. Chem. Soc.*, *78*, 6057 (1956), λ_{max} 248 mμ (ε 10,500).
37. A. T. Blomquist and J. A. Verdol, *J. Am. Chem. Soc.*, *78*, 109 (1956), λ_{max} 242 mμ (ε 10,000).
38. V. Schomaker and L. Pauling, *J. Am. Chem. Soc.*, *61*, 1769 (1939).
39. L. W. Pickett, E. Paddock, and E. Sackter, *J. Am. Chem. Soc.*, *63*, 1073 (1941).
40. V. Henri and L. W. Pickett, *J. Chem. Phys.*, *7*, 439 (1939).
41. W. C. Price and A. D. Walsh, *Proc. Roy. Soc. (London), Ser. A*, *179*, 201 (1941).
42. L. Dorfman, *Chem. Rev.*, *53*, 47 (1953).
43. L. F. Fieser and M. Fieser, *Steroids*, Reinhold, New York, 1959.
44. F. Sondheimer, D. A. Ben-Efraim, and R. Wolovsky, *J. Am. Chem. Soc.*, *83* 1675 (1961).
45. C. W. Spangler and G. F. Woods, *J. Org. Chem.*, *30*, 2218 (1965).
46. J. R. Platt, *J. Chem. Phys.*, *18*, 1168 (1950).
47. W. Ziegenkein, *Ber.*, *98*, 1427 (1965).
48. T. D. Goldfarb and L. Lindqvist, *J. Am. Chem. Soc.*, *89*, 4588 (1967).
49. P. Naylor and M. C. Whiting, *J. Chem. Soc.*, *1955*, 3037.
50. L. Zechmeister and A. Polgar, *J. Am. Chem. Soc.*, *65*, 1522 (1943).
51. L. Zechmeister, *Chem. Rev.*, *34*, 267 (1944).
52. A. Sandoval and L. Zechmeister, *J. Am. Chem. Soc.*, *69*, 553 (1947).
53. J. H. Pinckard and L. Zechmeister, *J. Am. Chem. Soc.*, *70*, 1939 (1948).
54. L. Jurkowitz, J. N. Loeb, P. K. Brown, and G. Wald, *Nature*, *184*, 614 (1959).
55. H. Hart, P. M. Collins, and A. J. Waring, *J. Am. Chem. Soc.*, *88*, 1005 (1966).
56. W. J. Bailey and J. Economy, *J. Am. Chem. Soc.*, *77*, 1133 (1955).
57. P. A. Waitkus, L. I. Peterson, and G. W. Griffin, *J. Am. Chem. Soc.*, *88*, 181 (1966).
58. P. A. Waitkus, E. B. Sanders, L. I. Peterson, and G. W. Griffin, *J. Am. Chem. Soc.*, *89*, 6318 (1967).
59. E. A. Dorko, *J. Am. Chem. Soc.*, *87*, 5518 (1965).
60. G. Körbrich and H. Heinemann, *Angew. Chem.*, *77*, 590 (1965).
61. G. Köbrich, H. Heinemann, and W. Zündorf, *Tetrahedron*, *23*, 565 (1967).
62. G. W. Griffin and L. I. Peterson, *J. Am. Chem. Soc.*, *85*, 2268 (1963).
63. H. Hopff and A. Wick, *Helv. Chim. Acta*, *44*, 19 (1961).
64. H. Hopff and A. Gati, *Helv. Chim. Acta*, *48*, 1289 (1965).
65. M. Gouterman, *J. Chem. Phys.*, *36*, 2846 (1962).
66. G. Robinson and R. Frosch, *J. Chem. Phys.*, *37*, 1962 (1962); *38*, 1187 (1963).
67. R. Williams and G. Goldsmith, *J. Chem. Phys.*, *39*, 2008 (1963).
68. D. S. Clure, *J. Chem. Phys.*, *17*, 665 (1949); *20*, 682 (1952).

69. H. F. Hameka and L. J. Oosterhoff, *Mol. Phys.*, *1*, 358 (1958).
70. R. S. H. Liu and J. R. Edman, *J. Am. Chem. Soc.*, *90*, 213 (1968).
71. E. Havinga, R. J. de Kock, and M. P. Rappoldt, *Tetrahedron*, *11*, 276 (1960).
72. T. Förster, *Fluoreszenz Organischer Verbindungen*, Vandenhoeck and Ruprecht, Göttingen, Germany, 1951, p. 96.

Optically Active Deuterium Compounds

By Lawrence Verbit

*Department of Chemistry, State University of New York
at Binghamton, Binghamton, New York*

CONTENTS

I. INTRODUCTION

Substitution of deuterium for hydrogen in a molecule provides a useful tool for investigating the course of chemical reactions. Deuterated compounds differ from their hydrogen analogs in such properties as NMR coupling constants, infrared and ultraviolet absorption, dipole moments, nuclear quadrupole coupling constants, and by the various manifestations included in the term kinetic isotope effect.

Deuterium as a substituent is also capable of contributing to the dissymmetry of a molecule, i.e., a deuterated compound may exist in enantiomeric forms capable of rotating the plane of polarized radiation.

We shall define an optically active deuterium compound as a molecule which owes its chirality solely to the isotopic differences between substituent

H and D atoms.* Hence, the review includes compounds such as 1-butanol-1-d, (1), but does not consider those such as 2-butanol 2-d, (2), since 2 would still be optically active if the D were replaced by H.

$$CH_3CH_2CH_2-\overset{\overset{\displaystyle H}{|}}{\underset{\underset{\displaystyle D}{|}}{C}}-OH \qquad\qquad CH_3CH_2-\overset{\overset{\displaystyle CH_3}{|}}{\underset{\underset{\displaystyle D}{|}}{C}}-OH$$

$$(1)\qquad\qquad\qquad\qquad\qquad (2)$$

For a subject which -was virtually nonexistent two decades ago, the author has found himself forced to limit the scope of the chapter in order to focus on the stereochemistry and mechanism of reactions involving optically active deuterium compounds. Hence, many biochemical aspects are treated only as they pertain to the chemically synthesized deuterium compounds. In part, this was done to keep the review within the spirit of the series for which it was written and, in part, because several reviews on biochemical aspects of optically active deuterium compounds have recently appeared (1–6).

Several possible ways of organizing the material were considered. In the end it seemed most desirable to attempt to preserve the historical perspective so that developments are presented in chronological order, insofar as this is possible. As an aid in locating compounds discussed in this chapter a table is included at the end which lists by class all the compounds discussed, references, highest reported rotation and absolute configuration (if known), and where they may be found in this review.

II. EARLY ATTEMPTS TO SYNTHESIZE OPTICALLY ACTIVE DEUTERIUM COMPOUNDS

The discovery of the isotope of hydrogen of mass number 2 by Urey and co-workers (7) in 1932 was viewed with great interest by workers in the area of optical activity. The question immediately arose whether it would be possible to obtain an optically active compound whose chirality depended entirely upon the differences between a substituent D and H atom. The simplest case would be a molecule of structure 3. Another possibility would be a molecule which owes its optical activity to isotopically different substituents *not* attached to the chiral center, e.g., 4.

---

* It is recognized that the term "optically active" is a manifestation of the chirality of a molecule. However, optical rotation measurements have played such an important part in the history of the compounds discussed here that it was felt useful to retain the operational term.

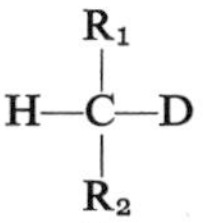

$$
\begin{array}{cc}
R_1 & R_1 \\
| & | \\
H{-}C{-}D & CH_3{-}C{-}CD_3 \\
| & | \\
R_2 & R_2 \\
(3) & (4)
\end{array}
$$

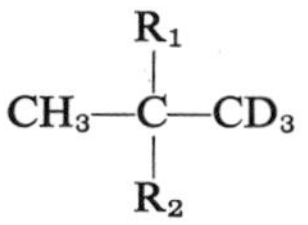

The years prior to 1948 saw many reports of attempts to prepare chiral deuterium compounds. The few claims of optical activity were soon shown to be due to the presence of optically active impurities so that as late as 1948 no deuterium compound had been reported which had demonstrable optical activity (however, see below). The many unsuccessful attempts in this area are cited by Buchanan (8), by Alexander and Pinkus (9), and by Arigoni and Eliel (2), and will not be repeated here. However, two attempts which undoubtably produced chiral deuterium compounds, although no rotation could be measured, are discussed.

Burwell and co-workers (10), Figure 1, and Brown and Groot (11), Figure 2, carried out reactions on optically active starting compounds which did not involve the breaking of bonds to the chiral center.* Hence, the products should be of the same enantiomeric purity as the starting materials. As we shall see, such compounds exhibit extremely small rotations in the region of the Na D line. Note that the products in Figures 1 and 2 both belong to the class of compounds represented by **4**. It would be of interest to investigate the optical activity of the above two deuterated

$$
C_6H_5\overset{*}{C}H(CH_3)CH_2Br \xrightarrow[\text{2. D}_2\text{O}]{\text{1. Mg}} C_6H_5\overset{*}{C}H(CH_3)CH_2D
$$

Configuration undetermined   No detectable optical rotation

Figure 1

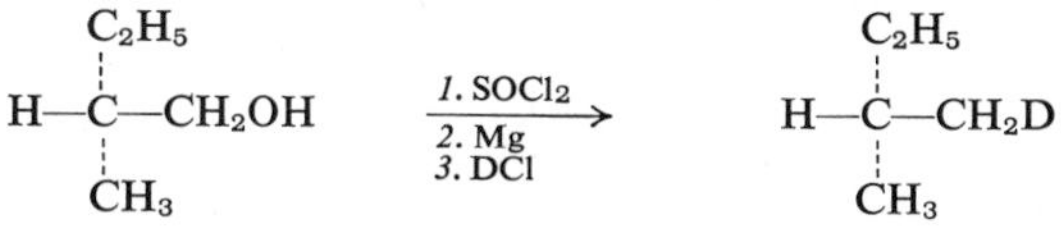

$$
\begin{array}{c}
C_2H_5 \\
| \\
H{-}C{-}CH_2OH \\
| \\
CH_3
\end{array}
\xrightarrow[\substack{2.\ \text{Mg} \\ 3.\ \text{DCl}}]{1.\ \text{SOCl}_2}
\begin{array}{c}
C_2H_5 \\
| \\
H{-}C{-}CH_2D \\
| \\
CH_3
\end{array}
$$

(R)-(+)-2-Methyl-1-butanol   No detectable optical rotation

Figure 2

compounds using a modern spectropolarimeter operating in the ultraviolet region where rotational values will be greater than in the vicinity of the Na D line.

* The Fischer representation of stereochemistry will generally be used: horizontal bonds to the chiral center project above the plane of the paper, and the (dashed) vertical bonds project behind this plane. Occasionally it will be necessary to depict a three-dimensional arrangement using a wedge-shaped bond to indicate projection above the plane of the paper.

III. FIRST DEMONSTRATIONS OF OPTICAL ACTIVITY

The first synthesis of a compound with demonstrable optical activity due to hydrogen–deuterium dissymmetry was reported in 1949. Alexander and Pinkus (9) reduced *trans*-2-menthene, $[\alpha]_D^{25}$ +132.0°, with D_2 and Raney nickel to give the dideutero-*trans*-menthane (**5**), $[\alpha]_D^{25}$ −0.09 ± 0.01° (neat). A control reaction using H_2 instead of D_2 yielded inactive *trans*-menthane.

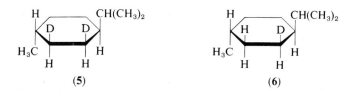

(5)   (6)

Note that the four chiral centers of **5** contain both types of dissymmetry shown in structures **3** and **4**. Although the 2,3-dideutero-*trans*-menthane obtained by Alexander and Pinkus may well be a mixture of diastereomers, the positions of the D atoms in the predominant stereoisomer is assigned tentatively by the present author as in structure **5**. The reasoning is partly based on Alexander's preparation the following year (12) of (−)-*trans*-menthane-2-*d* (**6**), $[\alpha]_D$ −0.09 ± 0.02° (neat) by $LiAlD_4$ displacement of the tosylate of (−)-menthol. The reaction should proceed predominantly with inversion of configuration (see below) to yield the stereoisomer shown in **6**, i.e., the configuration at C-2 is *S*.* The fact that both **5** and **6** are levorotatory together with the reasonable assumption that the addition of H_2 to the double bond occurred in the *cis*-manner (15) allows the tentative assignment of the deuterium atoms in the predominant stereoisomer, **5**, as *cis* to the isopropyl group.

The first unambiguous example of a molecule containing only one chiral center of the type R_1R_2CHD was provided by Eliel (16) with the synthesis of (*R*)-(−)-ethyl-benzene-α-*d* (**7**), shown in Figure 3.† Treatment of (*S*)-(−)-α-phenylthyl chloride, $[\alpha]_D^{25}$ −49.2°, with $LiAlD_4$–LiD in

---

* The *R*, *S* nomenclature system of Cahn, Ingold, and Prelog will be used (13). For further extensions of this system see the paper by Hanson (14). Although this method is unambiguous when considering an individual compound, one must be alert to the fact that upon comparing two compounds which clearly belong to the same configurational family, the priorities of the sequence rule *may* cause the assignment of the *S*-designation to one and the *R*-designation to the other compound.

† The symbol $\overset{\sigma}{\longrightarrow}$ indicates a reaction proceeding with inversion of configuration.

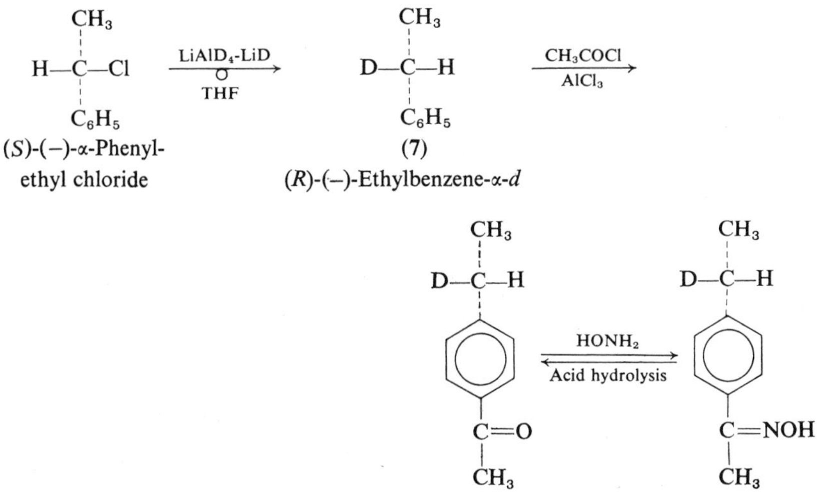

CH$_3$

H—C—Cl $\xrightarrow[\text{THF}]{\text{LiAlD}_4\text{-LiD}}$ D—C—H $\xrightarrow[\text{AlCl}_3]{\text{CH}_3\text{COCl}}$

C$_6$H$_5$

(S)-(−)-α-Phenyl-
ethyl chloride

CH$_3$

D—C—H

C$_6$H$_5$

(7)

(R)-(−)-Ethylbenzene-α-d

CH$_3$

D—C—H

$\underset{\text{Acid hydrolysis}}{\overset{\text{HONH}_2}{\rightleftarrows}}$

CH$_3$

D—C—H

C=O

CH$_3$

C=NOH

CH$_3$

Figure 3

tetrahydrofuran yielded the hydrocarbon (7) having $[\alpha]_D^{25}$ −0.30°. The R-configuration is assigned to this isomer on the basis (for which much independent evidence exists, see below) that the displacement of halide by deuteride proceeds chiefly with inversion of configuration. Based on the value of $[\alpha]_D$ 103.9° reported by Dauben and McCoy (17) for optically pure α-phenylethyl chloride,* and assuming no racemization in the deuteride reduction, ethylbenzene-α-d may be calculated to have a *minimum* $[\alpha]_D$ of 0.63°.†

Friedel–Crafts acylation of (−)-ethylbenzene-α-d, Figure 3, yielded (R)-(−)-para-acetylethylbenzene-α-d, $[\alpha]_D^{23}$ −0.27°. The crystalline oxime of this ketone had $[\alpha]_D^{28}$ −0.17°. Acid hydrolysis of the oxime regenerated the ketone of unchanged optical activity.

IV. THEORETICAL TREATMENTS OF OPTICALLY ACTIVE DEUTERIUM COMPOUNDS

In order to set the stage for future developments, it is appropriate to mention here some theoretical work on optically active deuterium compounds.

* Varying values have been reported in the literature for the specific rotation of optically pure α-phenylethyl chloride and a value as high as 125° has been calculated; ref. 18, p. 85.

† This value may be too low by a factor of 2; see p. 65.

Theoretical treatments of isotopic dissymmetry began shortly after the discovery of deuterium. Two early papers predicted only negligible optical rotation (19,20). However, in 1952 a paper by Fickett (21) appeared which predicted substantial optical activity and which stimulated further experimental work.

Assuming that the principal contributions to the optical activity of deuterium compounds arise from the C—H and C—D stretching vibrations, Fickett (21) applied Kirkwood's theory of optical rotatory power (22) to the case of isotopic dissymmetry. The theoretical results indicated that (S)-ethylbenzene-α-d (8), should have $[\alpha]_D^{25}$ +0.41° for the neat liquid, and that (R)-butane-2-d (9), an unknown compound at the time, should have $[\alpha]_D^{25}$ +1.1° (neat).

$$
\begin{array}{cc}
\text{CH}_3 & \text{CH}_3 \\
| & | \\
\text{H—C—D} & \text{D—C—H} \\
| & | \\
\text{C}_6\text{H}_5 & \text{C}_2\text{H}_5 \\
\textbf{(8)} & \textbf{(9)}
\end{array}
$$

$[\alpha]_D^{25}$, (neat)	$[\alpha]_D^{25}$, (neat)
predicted (21) + 0.41°	predicted (21) + 1.1°
found* (16) + 0.63°	found† (25) + 0.69°

It is encouraging that, for the two examples cited by Fickett, the absolute configurations are predicted correctly although the rotational values seem to be in somewhat unsatisfactory agreement with the experimentally determined ones (p. 55 and p. 59). Such results are perhaps not too surprising when one considers that Fickett's treatment of the optical activity of deuterium compounds in terms of molecular vibrations is based on polarizability formulas (22) which were themselves derived on the assumption that the nuclei were at rest in their equilibrium configuration!

A general theory for molecules which are optically active because of isotopic substitution has been derived by Cohan and Hameka (23). Their theory is also based on the premise that the optical activity of deuterium compounds is due to the effect of molecular vibrational differences, and derives in part from Hameka's theory of optical rotatory power (24). The reader is referred to the papers for details of the treatment; we shall mention here some of the results.

Because the vibrational rotatory strength calculations require an accurate set of force constants. Cohan and Hameka selected the com-

---

* Calculated minimum value (enantiomer actually prepared), see p. 65 and also p. 55.

† Calculated minimum value, see p. 58.

TABLE I
Predicted Absolute Configurations and Rotations (23)

		$[\alpha]_D$, calculated
(R)-(+)-Bromochlorodeuteromethane	Cl | D—C—H | Br	+ 0.0002°
(R)-(−)-Bromodeuterotritiomethane	T | D—C—H | Br	− 0.0003°

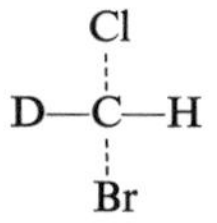

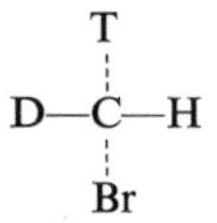

pounds BrClCHD and BrTCHD, for which good force constant data are available. The theoretical results are shown in Table I. Neither of these compounds has yet been prepared in chiral form, but the magnitudes predicted for the D-line rotations are very close to the limits of present-day spectropolarimeters. Once obtained in optically active forms, determination of the rotations of these compounds should not be a serious problem. However, the preparation of the tritiated compound in anything approaching optical purity will be complicated by the radioactive decay of the tritium. With full confidence in the abilities and resourcefulness of organic chemists, one may predict that it will not be long before these compounds are available so that the theoretical values may be tested against the experimental ones.

V. OPTICALLY ACTIVE BUTANE-2-d

The synthesis of optically active butane-2-d, whose rotation and configuration had been predicted theoretically by Fickett (p. 56), was reported by Helmkamp, Joel, and Sharman (25) in 1956. The synthetic route, Figure 4, started from $(2R,3R)$-(+)-2,3-epoxybutane having an estimated optical purity of 97%. Ring opening of the epoxide with LiAlD$_4$, eq. (1), gave 2-butanol-3-d, $[\alpha]_D^{25}$ − 13.59°. The nondeuterated analog, (R)-(−)-2-butanol is reported to have $[\alpha]_D$ − 13.8° for the optically pure material (26). If the assumption is made that the rotatory contribution of the deuterated carbon to the magnitude and sign of the D-line rotation of (−)-2-butanol-3-d is negligibly small, then by analogy with the non-deuterated alcohol, the R-configuration may be assigned to C-2. Since the

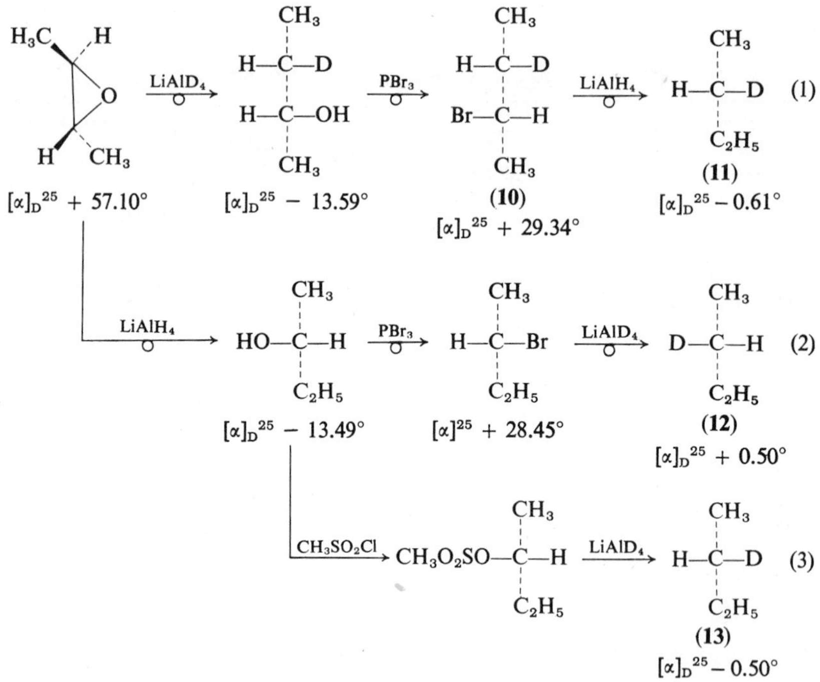

Figure 4

ring opening of epoxides with hydride has been shown to proceed with inversion (27), the configuration at C-3 may be designated as S. This assignment has been confirmed by Weber, Seibl, and Arigoni (28), p. 71, who obtained the same stereoisomer by the deuteroboration of cis-2-butene followed by oxidation. Hence, in eq. (1), the product of epoxide ring opening is $(2R,3S)$-$(-)$-2-butanol-3-d possessing an optical purity of ca. 97%.

Treatment of the alcohol with PBr$_3$ gave $(2S,3S)$-$(+)$-2-bromobutane-3-d, (10), $[\alpha]_D^{25}$ +29.34°, a value well below the highest reported rotation of 38.9° (29). Hence, it appears that approximately 25% racemization occurred at C-2 during conversion of the deuterated alcohol to the bromide (10). However, the bromination is not apt to cause racemization at C-3, the deuterium-bearing carbon.

Removal of the bromine atom by treatment of 10 with LiAlH$_4$ afforded butane-2-d (11), $[\alpha]_D^{25}$ −0.61° (the rotation was measured in a specially constructed polarimeter tube and is corrected for the presence of 8.4% of nondeuterated butane). Based on the method of preparation, the absolute configuration of $(-)$-butane-2-d is assigned as S, in agreement with the theoretical prediction.

The same starting epoxide was subsequently utilized by Helmkamp and Rickborn (30) for alternative syntheses of butane-2-d which also provided important information on the stereochemistry of LiAlH$_4$ reductions, Figure 4. Ring opening of the epoxide with LiAlH$_4$ gave (R)-$(-)$-2-butanol which was essentially optically pure. Conversion of the alcohol to the bromide having $[\alpha]_D^{25}$ $+28.45°$, eq. (2), was effected with PBr$_3$. Reduction of the bromide with LiAlD$_4$ gave $(+)$-butane-2-d (12), $[\alpha]_D^{25}$ $+0.50$, thus indicating that the PBr$_3$ bromination of 2-butanol and the deuteride reduction of 2-bromobutane had both proceeded with inversion of configuration (see also below). However, it is clear that substantial racemization occurred in the bromination step; the bromobutane of specific rotation $+ 28.45°$ is no more than 72% optically pure (29), p. 58. If one makes the assumption for the moment (however, see below) that the deuteride reduction steps involved no racemization, one may calculate that the minimum $[\alpha]_D$ of butane-2-d is 0.50 × (100/72) = 0.69°. If this minimum value is correct then the possibility of a racemization reaction at the deuterium-bearing carbon in the original synthesis (25), Figure 4, eq. (1), is indicated.

Helmkamp and Rickborn (30) demonstrated that repeated preparations of (S)-2-bromobutane from (R)-2-butanol using PBr$_3$ yielded bromide whose specific rotations varied as much as 5°. The HBr produced in this reaction could cause racemization of the bromobutane to a variable extent, depending upon the reaction conditions. Such racemization has been observed in other deuterated systems (31).

In another reaction sequence (30), Figure 4, eq. (3), the $(-)$-2-butanol obtained from the epoxide ring opening was converted to the methanesulfonate which was then reduced with LiAlD$_4$ to yield $(-)$-butane-2-d, (13), $[\alpha]_D^{25}$ $-0.50°$ (after correction for 5% nondeuterated butane), enantiomeric with that synthesized via eq. (2). Consideration of the reaction sequence confirms the stereochemistry of the hydride (and deuteride) displacement as proceeding with inversion of configuration. However, the butane-2-d (13) has an $[\alpha]_D$ which is only 82% of the highest value obtained (eq. (1)). Repeated preparations of the sulfonate under different reaction conditions yielded product of the same specific rotation, thus indicating that racemization does not occur during preparation and isolation of the sulfonate. The evidence then indicates that the deuteride displacement of the sulfonate involves about 20% racemization.

In order to further investigate the stereochemistry of hydride reductions, Helmkamp and Rickborn (30) prepared the levorotatory and *meso*

CH₃ —

$$
\begin{array}{c}
CH_3 \\
H\!-\!C\!-\!D \\
H\!-\!C\!-\!OH \\
CH_3 \\
(2R,3S)\text{-2-Butanol-3-}d \\
[\alpha]_D^{25} - 13.59°
\end{array}
\quad
\xrightarrow[\;2.\ LiAlD_4\;]{1.\ CH_3SO_2Cl}
\quad
\begin{array}{c}
CH_3 \\
H\!-\!C\!-\!D \\
D\!-\!C\!-\!H \\
CH_3 \\
(2S,3S)\text{-Dideuterobutane} \\
[\alpha]_D^{25} - 1.01°
\end{array}
$$

$$\Big\downarrow PBr_3$$

$$
\begin{array}{c}
CH_3 \\
H\!-\!C\!-\!D \\
Br\!-\!C\!-\!H \\
CH_3 \\
(10) \\
(2S,3S)\text{-2-Bromo-butane-3-}d \\
[\alpha]_D^{25} + 29.34° \\
ca.\ 75\%\ optically\ pure
\end{array}
\quad
\xrightarrow{LiAlD_4}
\quad
\begin{array}{c}
CH_3 \\
H\!-\!C\!-\!D \\
H\!-\!C\!-\!D \\
CH_3 \\
\\
meso\text{-2,3-Dideuterobutane} \\
[\alpha]_D^{25} - 0.03°
\end{array}
$$

Figure 5

diastereomers of 2,3-dideuterobutane, Figure 5. Reduction of the sulfonate ester of $(2R,3S)$-2-butanol-3-d with $LiAlD_4$ yielded $(2S,3S)$-dideuterobutane, $[\alpha]_D^{25} - 1.01°$, while corresponding reduction of the bromide (10) afforded the *meso*-isomer which exhibited $[\alpha]_D^{25} - 0.03°$.* The expected stereochemistry of the reactions, Figure 5, is in agreement with that observed and provides confirmatory evidence that the reduction of sulfonates and halides with $LiAlH_4$ occurs with inversion of configuration.

VI. OPTICALLY ACTIVE 1-BUTANOL-1-*d*, BENZYL-α-*d* ALCOHOL, AND ETHANOL-1-*d*: SYNTHESES, INTER-CONVERSIONS, AND CHEMICAL INTERMEDIATES

The chemical synthesis of optically active 1-butanol-1-*d* (14) by Streitwieser (31) in 1953 began a new chapter in the study of reaction mechanisms since now the stereochemistry of reactions at primary carbon

* By definition, the *meso*-2,3-dideuterobutane should be optically inactive. However, we have noted (p. 58) that the bromide (10) is only about 75% optically pure. Reduction should then yield about 12.5% of the $(2S,3S)$-dideuterobutane and the observed rotation should then be $0.125 \times -1.01° = -0.13°$. The lower value of $-0.03°$ may indicate the possibility of additional racemization occurring at the CHD carbon or, indeed, may be of no significance since the discrepancy involves a specific rotation of $0.1°$.

$$CH_3CH_2CH_2—\overset{\overset{\displaystyle D}{|}}{\underset{\underset{\displaystyle H}{|}}{C}}—OH$$

(14)

was opened to investigation. At the same time the enzymatic synthesis of chiral ethanol-1-*d* was reported (32), signaling the beginning of intensive studies of enzyme stereochemistry. The chemical and enzymatic syntheses were to provide an important complement to each other. This section will deal with the compounds initially synthesized by these methods and the configurational correlations which were established between them.

Streitwieser's initial preparation (31) of 1-butanol-1-*d* involved the partial asymmetric reduction of butyraldehyde with optically active 2-octyloxy-2-*d*-magnesium bromide prepared from 2-octanol-2-*d*. Use of (+)-octanol-2-*d* gave (−)-1-butanol-1-*d* whereas use of the dextrorotatory isomer afforded (+)-1-butanol-1-*d*. Although the amount of deuterium transfer and the magnitude of the optical activity of the 1-butanol-1-*d* were found to vary from run to run, Streitwieser was able to show that the small observed rotations (a few tenths of a degree in a four-decimeter polarimeter tube) were real and could only be due to hydrogen–deuterium dissymmetry in the molecule.

Levorotatory 1-butanol-1-*d* was later synthesized by the reduction of butyraldehyde-1-*d* with 2-octyloxymagnesium bromide from (−)-2-octanol (33).

In order to demonstrate that the optical activity of the deuterated butanol was an intrinsic property of the molecule, the (+)-alcohol was converted to (−)-1-bromobutane-1-*d* by treatment with PBr_3 and the kinetics of racemization of the bromide with LiBr in 90% aqueous acetone were measured (31). The polarimetric rate constant, 3×10^{-5} liter mole^{-1} sec^{-1}, was in good agreement with the value reported for the exchange of 1-bromobutane with radioactive bromide (34). Similar observations made in *secondary systems* that the *rate* of racemization is just twice the *rate* of incorporation of radioactive halide and that the two second-order *rate constants* are equal were an important part of the evidence put forth by Hughes and Ingold (35) for the nature of the S_N2 reaction. Hence, Streitwieser (31) was able to demonstrate that in the S_N2 reaction of *primary* halides, each displacement is accompanied by an inversion of configuration.

A more useful stereoselective reducing agent which allowed a prediction of the configuration of the deuterated alcohol formed in the reaction was utilized subsequently by Streitwieser and Wolfe (36). It had been shown by Vavon and Antonini (37) that the magnesium bromide salt of isoborneol, **16**, functions as an asymmetric reducing agent. Noyce and Denney (38) had demonstrated that optically pure isoborneol containing 10–15% of borneol is obtained from the $LiAlH_4$ reduction of natural gum

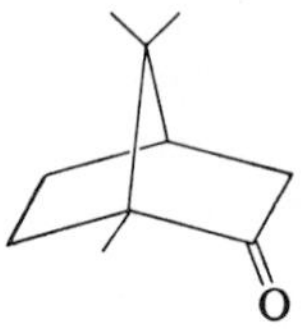

(15) 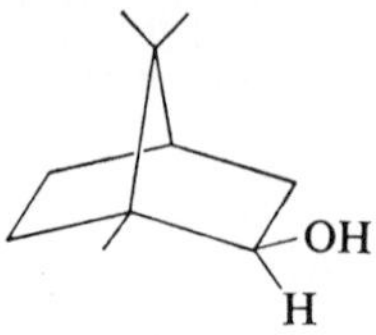   **(16)**

camphor. In addition, the bornylmagnesium bromide has been found to react much slower than the corresponding isobornyl reagent (37); hence, its presence in the reducing mixture is not considered objectionable.

The absolute configuration of naturally occurring (+)-camphor is **(15)** (39). Hence, (−)-isoborneol which is formed on reduction has the configuration **(16)**.

The reduction was first applied by Streitwieser and Wolfe (36) to benzaldehyde-1-*d* to yield (−)-benzyl-α-*d* alcohol, Figure 6, R = C_6H_5, $[\alpha]_D^{25}$ −0.645 ± 0.002°.*

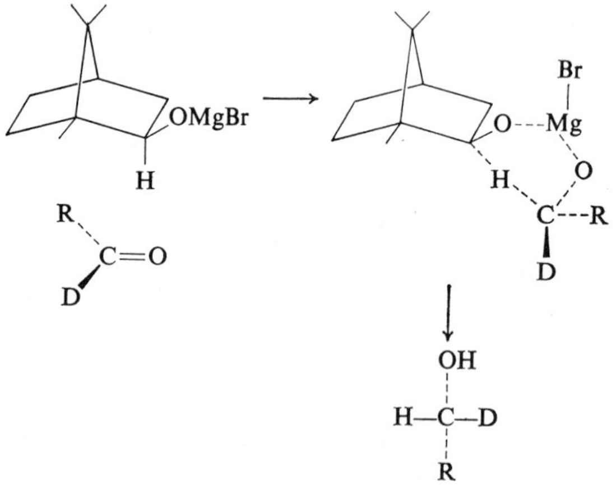

Figure 6

Consideration of the two possible product-determining transition states for reduction (41) leads to the prediction that the transition state depicted in Figure 6 will be favored. The resulting benzyl-α-*d* alcohol, **(17)**,

---

* A later preparation by the same method afforded benzyl-α-*d* alcohol having $[\alpha]_D$ − 0.715° (0.98 D/molecule) (40); see also footnote on p. 63.

R = C_6H_5, should then have the R-configuration. Since the benzyl-α-d alcohol obtained by this method was levorotatory, Streitwieser and Wolfe assigned the R-configuration to this enantiomer.

The same method used to establish the optical activity of 1-butanol-1-d, p. 61, was applied to the benzyl-α-d alcohol: conversion to the bromide and comparison of the rate constant for racemization with that for exchange with radioactive bromide. As in the butanol case, good agreement was found between the two rate constants.

Some derivatives of benzyl-α-d alcohol were also prepared and from the methods of synthesis Streitwieser and Wolfe (36) found that the (−)-alcohol, (−)-bromide, (−)-acetate, and (+)-hydrogen phthalate all possess the same configuration, (see also Table VIII at end of chapter).

Reduction of *nondeuterated* butyraldehyde with isobornyloxy-2-d-magnesium bromide yielded dextrorotatory 1-butanol-1-d, Figure 6 (enantiomer of structure shown), R = n-C_3H_7, $[\alpha]_D$ +0.172 ± 0.004° (0.55 D/molecule)* (40). Hence, by the same argument described above for benzyl-α-d alcohol, *levorotatory* 1-butanol-1-d is predicted to have the R-configuration (Fig. 6, (17), R = n-C_3H_7).

The prediction that 1-butanol-1-d and benzyl-α-d alcohol of the same sign of rotation have the same configuration was verified by relating each alcohol to n-butylbenzene-α-d, Figure 7 (40). The tosylate prepared from (−)-benzyl-α-d alcohol was treated with sodioacetoacetic ester in benzene to give (−)-ethyl benzyl-α-d-acetoacetate (18). This reaction is presumed to proceed with essentially complete inversion of configuration by analogy with the reaction of benzyl-α-d tosylate with ethoxide ion in ethanol, p. 94. Hydrolysis of 18 gave (−)-4-phenyl-2-butanone-4-d which, upon Clemmensen reduction afforded, (−)-n-butylbenzene-α-d (20), $[\alpha]_D$ −0.78 ± 0.04°. Since the starting benzyl-α-d alcohol was shown subsequently to be ca. 44% optically pure (p. 80), a minimum specific rotation of 1.77° may be calculated for optically pure 20 on the assumption that the only inversion step in the sequence, the acetoacetic ester alkylation, proceeds without racemization (however, see below).

1-Butanol-1-d, $[\alpha]_D^{25}$ −0.044°, was converted to (+)-1-chlorobutane-1-d (19) by thermal decomposition of the chlorosulfite, Figure 7, a reaction which Streitwieser and Schaeffer (44) previously had shown to proceed with 91 ± 2% inversion of configuration.

---

* Gerlach (42) (p. 86), has shown that a side reaction of this asymmetric Meerwein–Ponndorf–Verley-type reduction is the disproportionation of the starting aldehyde to the corresponding ester. The alcohol component of this ester is found not to be deuterated. Ester interchange with already formed primary deuterated alcohol also may occur, resulting in isolated alcohol which is only partially deuterated.

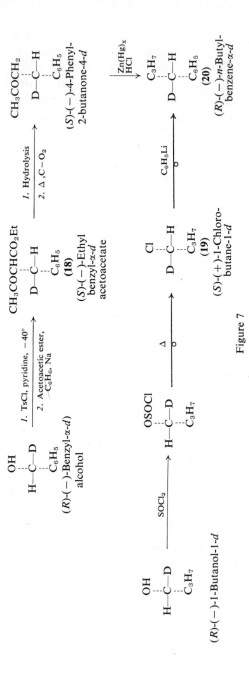

Figure 7

Treatment of **19** with phenyllithium in benzene yielded n-butyl-benzene-α-d (**20**), $[\alpha]_D^{25}$ $-0.18°$. This reaction has been shown (45) to follow second-order kinetics in dibutyl ether solution, and hence is very probably a direct displacement reaction with inversion of configuration. Correcting for the approximately 10% racemization occurring during the chlorosulfite decomposition (44) gives $[\alpha]_D^{25}$ $-0.20°$ for **20**. Since the 1-butanol-1-d starting material was shown later to be only about 9% optically pure (p. 80), a minimum value of $[\alpha]_D$ 2.2° may thus be calculated for optically pure n-butylbenzene-α-d.

Using this minimum figure of 2.2°, one may calculate that in the acetoacetic ester alkylation, Figure 7, top line, approximately 20% racemization took place since the resultant n-butylbenzene-α-d had a calculated minimum rotation of only 1.77°.

From the stereochemistry of the reactions used in the interconversions of Figure 7, Streitwieser and co-workers (40) concluded that 1-butanol-1-d and benzyl-α-d alcohol of the same sign of rotation possess the same configuration. Since the absolute configuration of the reference compound, $(-)$-n-butylbenzene-α-d, was not known at that time, the configurations of the levorotatory alcohols were inferred to be R based on their method of preparation, p. 62.

The correctness of this conclusion was subsequently confirmed by the direct synthesis of (S)-$(+)$-n-butylbenzene-α-d from (S)-$(-)$-1-phenyl-1-butanol. Figure 8 (46). Since both steps in the reaction sequence are

$$
\begin{array}{ccccc}
\text{C}_3\text{H}_7 & & \text{C}_3\text{H}_7 & & \text{C}_3\text{H}_7 \\
| & \text{POCl}_3 & | & \text{LiAlD}_4 & | \\
\text{H—C—OH} & \xrightarrow[\substack{\text{Pyridine} \\ \text{(mostly} \\ \text{inversion)}}]{} & \text{Cl—C—H} & \xrightarrow{} & \text{H—C—D} \\
| & & | & & | \\
\text{C}_6\text{H}_5 & & \text{C}_6\text{H}_5 & & \text{C}_6\text{H}_5 \\
(21) & & (22) & & (23) \\
(S)\text{-}(-)\text{-1-Phenyl-} & & (R)\text{-}(+)\text{-1-Phenyl-} & & (S)\text{-}(+)\text{-}n\text{-Butyl-} \\
\text{1-butanol} & & \text{1-chlorobutane} & & \text{benzene-}\alpha\text{-}d
\end{array}
$$

Figure 8

assumed to involve inversion of configuration, the S-configuration is assigned to the resultant n-butylbenzene-α-d, which, after correction for the 85% optical purity of the starting alcohol and 4% of nondeuterated material in the product had $[\alpha]_D^{25}$ $+0.84°$. The value is less than half that calculated from Streitwieser's results (p. 65) and indicates that considerable racemization has occurred in the reactions **21** → **23**. This racemization is most important since Eliel's synthesis of ethylbenzene-α-d (p. 55) utilized these same reactions starting from α-phenylethanol. Hence, the minimum calculated rotation of 0.63° (p. 55) for ethylbenzene-α-d may be too low by a factor of two.

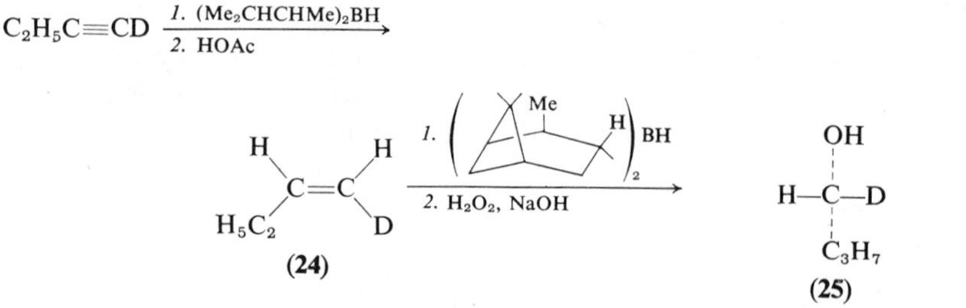
$C_2H_5C{\equiv}CD$
1. $(Me_2CHCHMe)_2BH$
2. HOAc
(24)
1.
Me
BH
2. H_2O_2, NaOH
(25)

(*R*)-(−)-1-Butanol-1-*d*
$[\alpha]_D{}^{21} = 0.262 \pm 0.007°$
(neat)

Figure 9

An alternative synthesis of 1-butanol-1-*d* was carried out by Streitwieser and co-workers (47) using the asymmetric hydroboration reaction (48). The reaction scheme is shown in Figure 9. Hydroboration of 1-butyne-1-*d* with the sterically hindered di(3-methyl-2-butyl)borane followed by protonolysis gave *cis*-1-butene-1-*d* (**24**). The geometry of **24** was characterized by NMR; $J_{(\text{vinyl H's})}$ 10.4 Hz, a coupling attributable to *cis*-olefinic protons in the butene series (49). Hydroboration of **24** with (−)-diisopinocampheylborane* (derived from reaction of diborane with (+)-α-pinene), followed by treatment of the intermediate organoborane with alkaline hydrogen peroxide afforded 1-butanol-1-*d* (**25**), $[\alpha]_D^{21}$ −0.262 ± 0.007° (after correction for the 80% optically pure α-pinene utilized in the reaction). The hydrogen phthalate derivative had $[\alpha]_D^{22}$ +0.55 ± 0.01° (c 24, acetone).

Based on the value of $[\alpha]_D$ 0.47° for optically pure 1-butanol-1-*d* (p. 80) the alcohol produced by the asymmetric hydroboration reaction is 56% optically pure. The highest specific rotation previously obtained in the isobornyloxymagnesium bromide synthesis of 1-butanol-1-*d* was 0.185° (40), corresponding to an optical purity of 39%.

The 56% optical purity found in the case of 1-butanol-1-*d* (47) is fairly typical for the asymmetric hydroboration of *cis*-olefins although some values have ranged as high as 95% (51). The reason for these relatively high optical yields has been ascribed to a particularly good steric fit of the olefin into the reagent. From an assumed lowest-energy conformation for diisopinocampheylborane, **26**, and a proposed four-center tran-

* The organoborane is discussed more conveniently in terms of the monomer although it is recognized (48) that the dimer is probably the actual hydroborating species. The dimer, *sym*-tetraisopinocampheyldiborane, has been used in a mechanistic discussion of the asymmetric hydroboration reaction (50). The steric requirements of the two species are not necessarily the same, in part due to possible differences in the hybridization of boron.

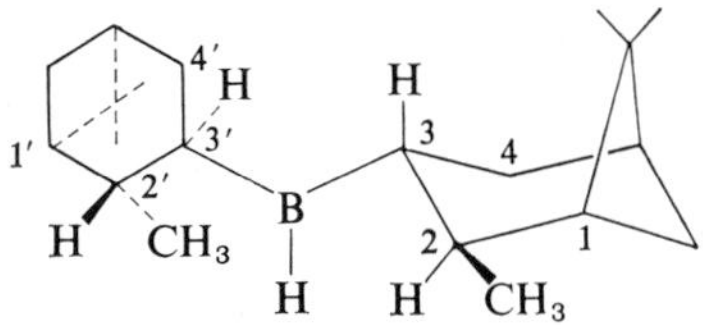

(26)

sition state (52) for the addition of the boron–hydrogen bond to a double bond, Brown (51) has devised a useful model for predicting the abolute configuration of the product. Application of Brown's model to the case of *cis*-1-butene-1-*d* leads to the prediction of the *S*-configuration for (−)-1-butanol-1-*d*. The prediction is not correct; the levorotatory alcohol is known to possess the *R*-configuration (pp. 62, 65, 80, and 104).

Streitwieser and co-workers (47) have suggested a small modification of Brown's model which preserves the predictions for *cis*-olefins but increases its applicability to the present case. Their proposal consists of a triangular π-complex for the borane and olefin, Figure 10. The product-determining transition state is then postulated to involve a relatively small perturbation from this arrangement.

Using this model, the two possible transition states for the hydro-boration of *cis*-1-butene-1-*d* are depicted by 27 and 28 (for details of the symbolism, see ref. 51). The important modification in this model is that the olefin is pushed further into the reagent so that the significant non-bonded interaction is between the ethyl group of the *cis*-1-butene-1-*d* and the M (medium) group in 27 or the S′ (small) group in 28.* On this basis

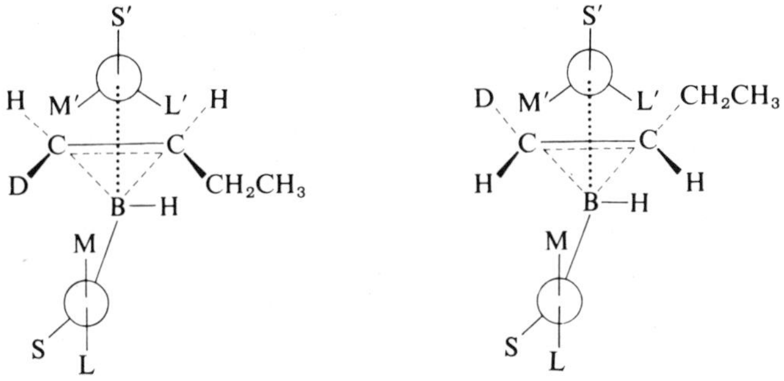

(27)   **(28)**

Figure 10

* Since the differences in nonbonded interaction of hydrogen and deuterium have been shown to be negligibly small in a sensitive Meerwein–Ponndorf–Verley reduction (53), only the relative steric interactions of the group on C-2 of the olefin are considered.

28 is preferred, and the correct *R*-configuration of the final hydroboration–oxidation product, $(-)$-1-butanol-1-*d* (**25**), is obtained.

The modifications required for the transition state model to achieve agreement with the experiment indicate that caution is necessary in the use of the model. The model was not found useful when applied to allenes (98) and has caused confusion when applied to the case of aldehydes (110, 134). Until the details of the asymmetric hydroboration reaction are better understood, it appears that configurational assignments based on transition state models can be equally well decided by the toss of a coin (135).

Arigoni and co-workers (54) have reported a similar use of the asymmetric hydroboration reaction to prepare 1-hexanol-1-*d*. However, no rotational values were reported for the alcohol. By analogy with the results obtained in the case of *cis*-1-butene-1-*d*, one would predict that hydroboration of *cis*-1-hexene-1-*d* with $(-)$-diisopinocampheylborane would give rise to (R)-$(-)$-1-hexanol-1-*d*.

One of the most important and well-studied enzymatic reactions is the oxidation–reduction reaction catalyzed by nicotinamide-nucleotide-linked dehydrogenases. The subject has been reviewed a number of times (4–6) and only the results pertinent to the discussion of optically active deuterium compounds will be treated here.

The coenzyme nicotinamide adenine dinucleotide (abbreviated NAD^+) (**29**), R = adenosinediphosphoribosyl, in conjunction with the enzyme alcohol dehydrogenase (**ADH**) oxidizes ethanol to acetaldehyde with the formation of the reduced coenzyme NAD-H (**30**), Figure 11. The reduction of acetaldehyde is the microscopic reverse of this reaction.

Westheimer, Vennesland, and co-workers (5,32,55) have shown that the hydrogen transfer is direct, reversible, and stereospecific. Alcohol dehydrogenase catalyzes the reversible addition of hydrogen to only one side of the plane of the nicotinamide ring of NAD^+. The *pro-R* hydrogen (13) of ethanol is always transferred in the presence of the enzyme and becomes the *pro-R* hydrogen at C-4 of reduced NAD, Figure 11. Isotopic substitution does not alter the stereospecificity, i.e., reduction of acetalde-

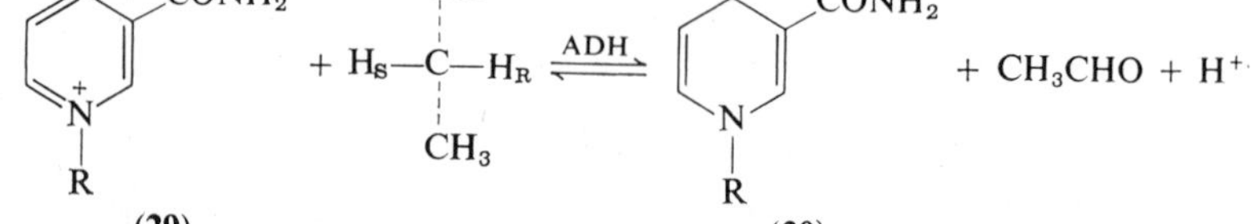

(**29**)
NAD^+

(**30**)
NAD—H

Figure 11

hyde-1-*d* with NAD-H in the presence of ADH produces ethanol-1-*d* which, upon enzymatic reoxidation yields only acetaldehyde-1-*d* with no loss of deuterium (32).

By carrying out the reduction of CH_3CDO with glucose in the presence of glucose dehydrogenase, ADH, and a catalytic amount of NAD^+ (a coupled redox system), Levy, Loewus, and Vennesland (55) were able to obtain enough ethanol-1-*d* to allow measurement of the optical rotation. The alcohol was purified by vapor phase chromatography and had $[\alpha]_D^{28}$ $-0.28 \pm 0.03°$.

The enzymatic synthesis of $(-)$-ethanol-1-*d* posed two important questions: (*1*) what was the absolute configuration of the alcohol? and (*2*) was the ethanol-1-*d* optically pure?

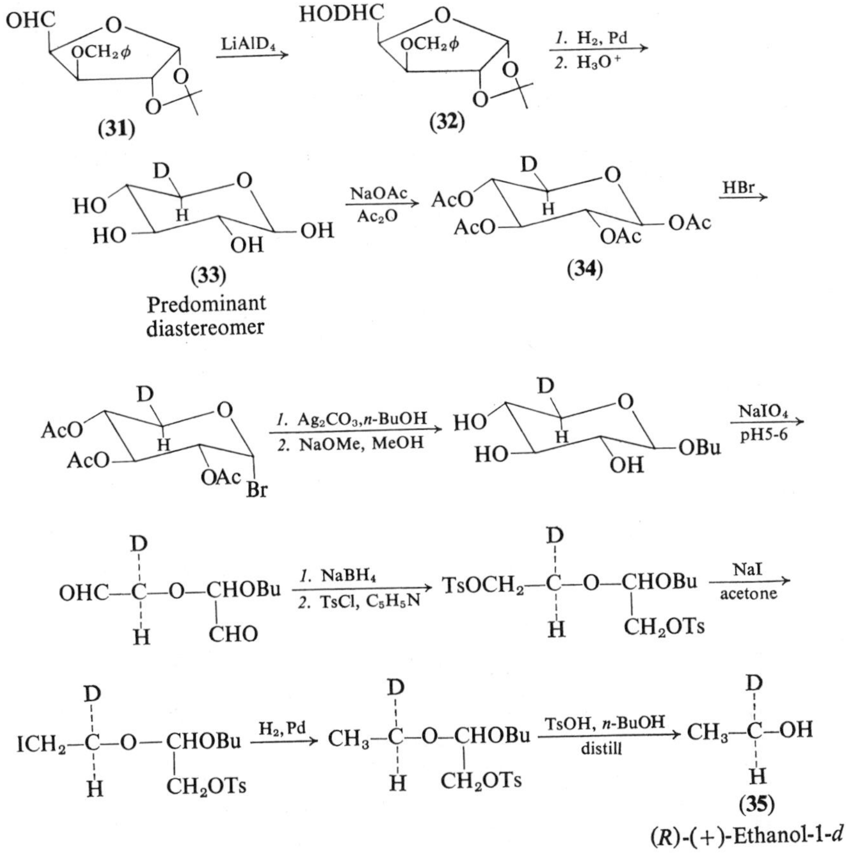

Figure 12

Both of these questions were answered by Lemieux and Howard (56) utilizing an elegant synthesis starting in the sugar series, Figure 12.

The xylofuranose (**31**) was reduced with $LiAlD_4$ to give **32**, followed by catalytic hydrogenolysis of the benzyl group and acid hydrolysis to yield D-xylose-5-*d* (**33**). A NMR investigation of the corresponding acetate (**34**) indicated a 30% excess of the β-deuterium isomer, with the C-5 hydrogen chiefly axial (hence, deuterium mainly equatorial) in the predominant diastereomer. The known absolute configuration of D-xylose and the fact that the hydroxy (acetoxyl) groups are equatorial in the predominant conformation allows the assignment of the *R*-configuration at C-5, with this center having an optical purity of 30%.

The *R*-C-5 carbon of the tetra-acetate (**34**) was then converted to ethanol-1-*d* (**35**) by the series of reactions shown in Figure 12, none of which involves breaking any of the bonds to the deuterated carbon. Comparison of the NMR spectrum of **35** with ethanol and with the ethanol–water azeotrope, Figure 13, indicated that the sample was essentially pure ethanol-1-*d*.

The observed rotation of the ethanol-1-*d*, α +0.066° (ℓ 1, neat), for material expected to be 30% optically pure, corresponds to a specific rotation of 0.22° for optically pure ethanol-1-*d*, in good agreement with that found by Levy, Loewus, and Vennesland, p. 68. From the method of synthesis, the *R*-configuration is assigned to (+)-ethanol-1-*d* (56).

Assignment of the *R*-configuration to (+)-ethanol-1-*d* in turn defines

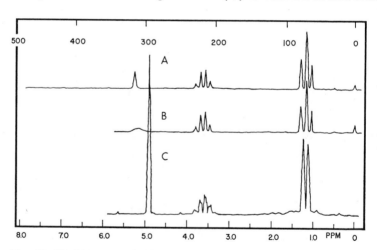

Fig. 13. Nuclear magnetic resonance spectra of ethanol (*A*), ethanol–water azeotrope (*B*), and (+)-ethanol-1-*d* (**35**) (*C*). Reproduced by permission of the National Research Council of Canada (56).

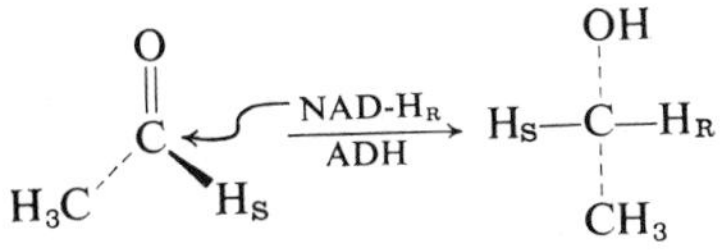

Figure 14

the stereospecificity of the ADH–NAD$^+$ system. Since the reduction of acetaldehyde-1-d by NAD–H and ADH leads to the formation of (S)-$(-)$-ethanol-1-d, it is clear that the acetaldehyde accepts the hydrogen from the re side,* Figure 14. In the oxidation of ethanol by this enzyme, it is the pro-R hydrogen of the methylene group which is transferred.

An independent determination of the configuration of ethanol-1-d was reported by Weber, Seibl, and Arigoni (28) utilizing the synthetic scheme shown in Figure 15. Asymmetric deuteroboration-oxidation of cis-2-butene with diisopinocampheyldeuteroborane (see footnote on p. 66) gave $(-)$-$erythro$-2-butanol-3-d (36), $[\alpha]_D^{25}$ $-8.1°$ (0.89 D/molecule). Assuming the rotatory contribution of the deuterated carbon to be negligibly small allows comparison of 36 with R-$(-)$-2-butanol (26) and, hence, the determination that 36 possesses the R-configuration at C-2 with an optical purity of 56%. Since the $(+)$-α-pinene utilized in the hydroboration reaction was 87.5% optically pure, asymmetric induction actually occurred to the extent of 64% (compare the analogous hydroboration of cis-1-butene-1-d, p. 66).

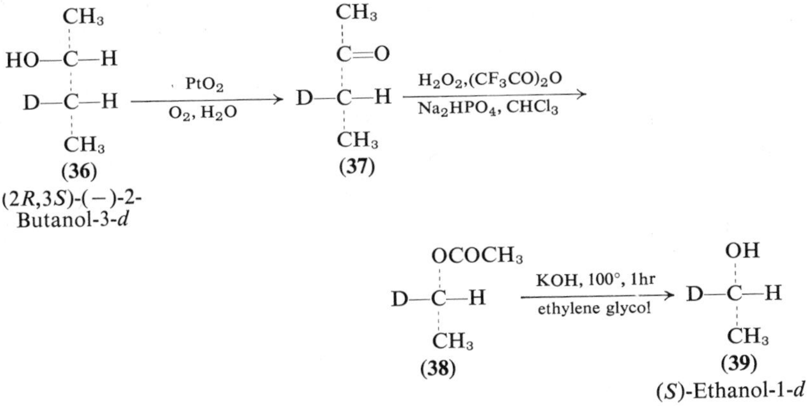

Figure 15

---

* The re side of a trigonal planar system Xabc is that side which, from consideration of the three groups abc according to the sequence rule (13), is clockwise. The si side is designated analogously: re and si, from $rectus$ and $sinister$ (14).

Authentic samples of racemic *threo-* and *erythro*-2-butanol-3-*d* were prepared in order to ascertain that *cis*-addition occurred during the hydroboration reaction. The IR spectra of the *threo-* and *erythro-N*-phenylurethane derivatives were sufficiently different so that the 2-butanol-3-*d* obtained in the hydroboration reaction could be assigned unequivocally the *erythro*-configuration **36**. Since the absolute configuration at C-2 is *R*, then that at C-3 must be *S*, and from the *cis*-addition mechanism the optical purity of the deuterated carbon is also expected to be 56%.

Oxidation of **36** led to the ketone (**37**) which was not isolated but was subjected directly to the Baeyer-Villiger rearrangement using trifluoroperacetic acid. The crude reaction product, chiefly the acetate (**38**), was hydrolyzed to yield ethanol-1-*d* (**39**).

Mass spectral investigation of the *N*-phenylurethanes of **36** and **39** showed that only 2.5% deuterium had been lost during the reaction sequence, indicating that significant racemization of the ketone (**37**) may be ruled out. Since the Baeyer-Villiger oxidation has been shown to proceed with retention of configuration (57), the ethanol-1-*d* obtained in this synthesis may be assigned the *S*-configuration.

Availability of only a small quantity of ethanol-1-*d* and the possibility of traces of optically active impurities precluded a polarimetric investigation of the alcohol. Instead the method of enzymatic investigation utilized earlier by Loewus, Westheimer, and Vennesland (32) was applied. A sample of (*S*)-ethanol-1-*d* (**39**) was oxidized with NAD$^+$–ADH and the acetaldehyde formed was isolated as the 2,4-dinitrophenylhydrazone derivative. As a further control, the hydrogen which had been transferred to the coenzyme during oxidation was transferred to pyruvate by means of lactic dehydrogenase and the *S*-lactic acid formed isolated as the phenacyl ester, Figure 16. Table II gives the results of the mass spectrometric

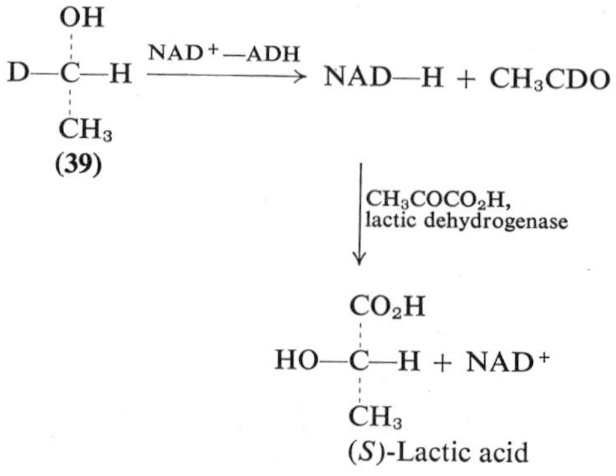

$$\text{OH}$$
$$\text{D—}\overset{|}{\underset{|}{\text{C}}}\text{—H} \xrightarrow{\text{NAD}^+\text{—ADH}} \text{NAD—H} + \text{CH}_3\text{CDO}$$
$$\text{CH}_3$$
(**39**)

CH$_3$COCO$_2$H,
lactic dehydrogenase

$$\text{CO}_2\text{H}$$
$$\text{HO—}\overset{|}{\underset{|}{\text{C}}}\text{—H} + \text{NAD}^+$$
$$\text{CH}_3$$
(*S*)-Lactic acid

Figure 16

TABLE II
Results of Mass Spectrometric Deuterium Determinations (28)

Starting compound	Configuration	Acetaldehyde: Per cent deuterium, Found	Acetaldehyde: Per cent deuterium, Calc.	Lactate: Per cent deuterium, Found	Lactate: Per cent deuterium, Calc.
Ethanol-1-*d* (**39**), 0.865 D/molecule 56% optically pure	*S*	67	67.5	16	19
Ethanol-1-*d*, 0.93 D/molecule	Racemic	48	46.5	45	46.5

deuterium determinations of the derivatives of ethanol-1-*d* (**39**) acetaldehyde, and lactic acid. The theoretical values were calculated on the assumption that **39** possesses the same optical purity as the (−)-2-butanol-3-*d* (**36**), from which it was synthesized, i.e., **39** is 56% optically pure so that the sample contains 78% of one enantiomer and 22% of the other.

The results of the enzymatic investigation of the (*S*)-ethanol-1-*d* obtained by Weber, Seibl, and Arigoni (28) show that in the oxidation by means of yeast ADH, it is chiefly the hydrogen atom of the alcohol which is transferred to the coenzyme. Since it has been demonstrated (5) that this transfer occurs in the case of levorotatory ethanol-1-*d*, the *S*-configuration may be assigned unequivocally to this enantiomer, in agreement with the independent establishment of the configuration of ethanol-1-*d* by Lemieux and Howard (p. 70).

The configuration of ethanol-1-*d* was also determined (58) by relating it to pentane-2-*d* of known absolute configuration, Figure 17. (−)-Ethanol-1-*d* was prepared by the reduction of acetaldehyde with

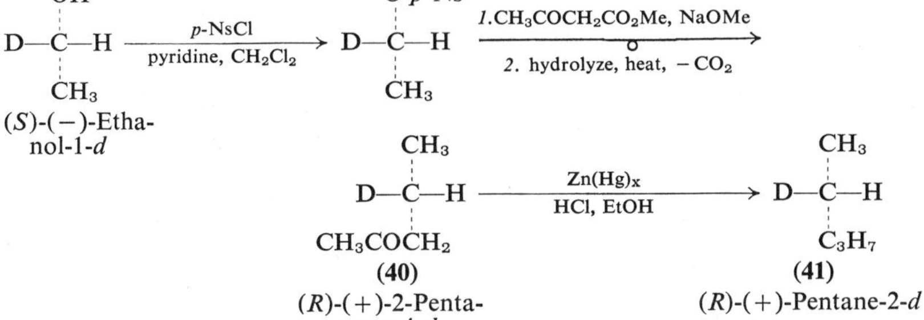

(*S*)-(−)-Ethanol-1-*d*

(**40**) (*R*)-(+)-2-Pentanone-4-*d*

(**41**) (*R*)-(+)-Pentane-2-*d*

Figure 17

the deuterated isobornyloxymagnesium bromide reagent. From mechanistic considerations (p. 62) the configuration of the alcohol is predicted to be **S**. The ethanol-1-*d*, produced in poor yield, contained 0.37 D/molecule (see footnote on p. 63 for the side reactions occurring during this reduction), and had $[\alpha]_D$ −0.123 ± 0.025° (corrected for 1 D/molecule). The rotation corresponds to an optical purity of 44 ± 9% based on the rotation of enzymatically prepared material (p. 80).

The partially deuterated (−)-ethanol-1-*d* was converted to the *p*-nitrobenzenesulfonate followed by displacement with methylaceto acetate in the presence of methoxide ion, Figure 17. Hydrolysis of the ester and subsequent decarboxylation afforded 2-pentanone-4-*d* (**40**), $[\alpha]_D$ +0.25 ± 0.03° (corrected for 1 D/molecule). Clemmensen reduction of the ketone (**40**) yielded (+)-pentane-2-*d* (**41**), $[\alpha]_D$ + 0.19 ±0.06° (corrected for 1 D/molecule). Based on the 44% optical purity of the starting ethanol-1-*d* and assuming no racemization in the synthetic route (however, see p. 65), optically pure pentane-2-*d* is calculated to have a minimum specific rotation of 0.43 ± 0.16°.

The absolute configuration of pentane-2-*d* was determined (58) by its preparation from (*R*)-(−)-2-pentanol as shown in Figure 18. The resultant (*S*)-pentane-2-*d* had $[\alpha]_D$ −0.067 ± 0.004°. Based on the optical purity (59,60) of the starting pentanol and the results of Helmkamp and Rickborn (p. 59) that the comparable LiAlD₄ reduction in the 2-butyl system proceeds with inversion of configuration and about 20% racemization, an $[\alpha]_D$ of 0.50° may be calculated for optically pure pentane-2-*d*, a value consistent with that arrived at above.

The conversion of (−)-ethanol-1-*d* to (*R*)-(+)-pentane-2-*d* by the series of reactions of known stereochemistry shown in Figure 17 allows assignment of the *S*-configuration to the (−)-alcohol and provides independent evidence for the structure of optically active ethanol-1-*d*.

The configuration of (*S*)-(−)-pentane-2-*d* was corroborated by the synthesis of this hydrocarbon via the asymmetric hydroboration reaction (61). The synthetic route started from methyl pentanoate whose α-hydrogens were exchanged for deuterium by treatment with CH₃OD in the presence of ⁻OCH₃. Reduction of the α,α-dideuterated ester with LiAlH₄ afforded 1-pentanol-2-*d₂*, which was converted to the acetate and pyrolyzed

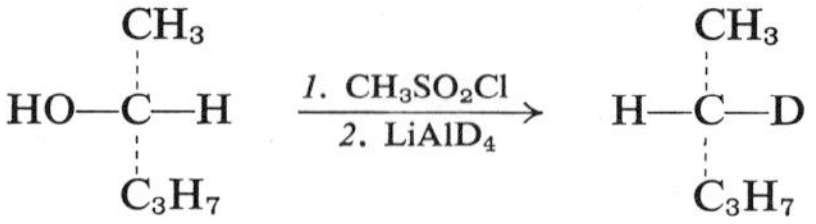

CH₃   CH₃

HO—C—H $\xrightarrow[\text{2. LiAlD}_4]{\text{1. CH}_3\text{SO}_2\text{Cl}}$ H—C—D

C₃H₇   C₃H₇

(*R*)-(−)-2-Pentanol   (*S*)-(−)-Pentane-2-*d*

Figure 18

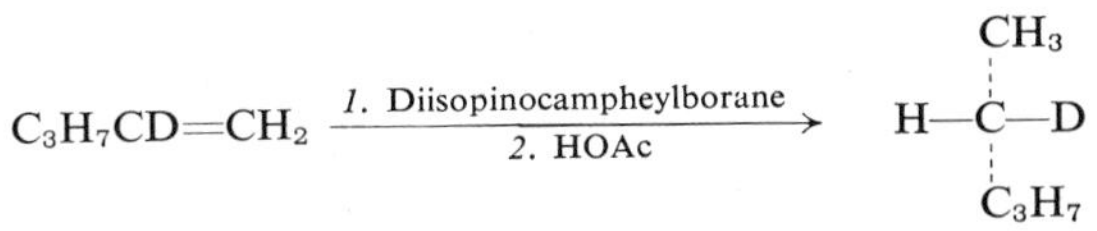

$$C_3H_7CD=CH_2 \xrightarrow[\text{2. HOAc}]{\textit{1. Diisopinocampheylborane}} \quad H-\overset{\displaystyle CH_3}{\underset{\displaystyle C_3H_7}{C}}-D$$

1-Pentene-2-d   (S)-(−)-Pentane-2-d

Figure 19

to give 1-pentene-2-d. This olefin was subjected to the asymmetric hydroboration reaction, p. 66, and the intermediate organoborane protonalyzed to yield pentane-2-d, $[\alpha]_D$ −0.25° (after correction for the 75.8% optical purity of the α-pinene utilized), Figure 19.

The hydroboration results were confirmed by the reduction of 1-pentene with diisopinocampheyldeuteroborane which gave pentane-2-d having $[\alpha]_D$ +0.25° (after correction to optically pure α-pinene) (61). From the observed rotations and estimates of probable errors, Streitwieser and co-workers (61) deduce a specific rotation of −0.52 ± 0.07° for optically pure (S)-pentane-2-d.

Based on his semiempirical rules of atomic and conformational asymmetry and on the molecular rotation and configuration of (+)-butane-2-d (p. 57), Brewster (62,63), in 1959, had calculated that the molecular rotation of 1-butanol-1-d should be 0.33° more positive than that of ethanol-1-d of the same configuration. This difference was attributable to the extra conformational dissymmetry contributed by the additional two methylene groups of the butanol. The magnitude of the observed molecular rotation of (−)-ethanol-1-d, $[\phi]_D$ −0.13 ± 0.02° (p. 68), would require that the configurationally related alcohols have opposite signs of rotation.

Streitwieser and co-workers (61) confirmed Brewster's conclusion by relating 1-butanol-1-d to (R)-(−)-1-bromopentane-2-d by a sequence of known stereochemistry, Figure 20. The interrelation was extended to (S)-(−)-ethanol-1-d which had been related to (R)-(+)-pentane-2-d, Figure 17, by relating 1-bromopentane-2-d to this hydrocarbon. (+)-1-Butanol-1-d, 34% optically pure (p. 80), was converted to the 2-nitrobenzenesulfonate which was then subjected to displacement with sodiomalonic ester. The resulting butylmalonic ester (42) exhibited no rotation but hydrolysis yielded optically active hexanoic-3-d acid (43), $[\alpha]_D$ −0.23 ± 0.04°. The modified Hunsdiecker reaction (64) afforded the corresponding 1-bromopentane-2-d (44), $[\alpha]_D$ −1.98 ± 0.20°, whose configuration was shown to be R by use of the asymmetric hydroboration reaction.

1-Pentene-2-d (p. 74) was hydroborated with diisopinocampheylborane and the intermediate deuterated pentylborane protonolyzed with

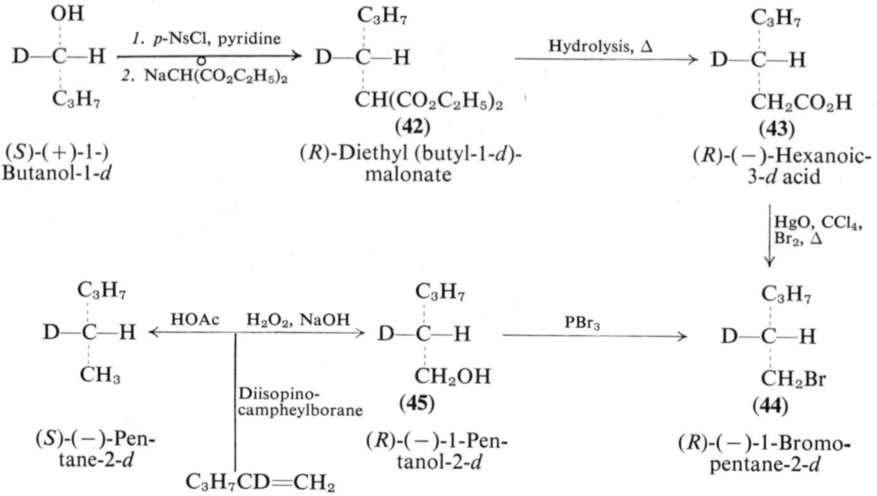

Figure 20

acetic acid in one run to give (S)-$(-)$-pentane-2-d (p. 74), and in another oxidized with alkaline H_2O_2 to give $(-)$-1-pentanol-2-d (**45**), $[\alpha]_D$ $-0.60 \pm$ $0.07°$. Since the stereochemistry of the reaction is determined during formation of the organoborane (51,52), $(-)$-pentane-2-d and $(-)$-1-pentanol-2-d obtained from the same organoborane possess the same absolute configuration, which corresponds to R for $(-)$-1-pentanol-2-d.

The deuteroalcohol (**45**) was converted to 1-bromopentane-2-d (**44**), $[\alpha]_D$ $-0.98°$ (after correction for the 75.8% optical purity of the α-pinene utilized in the previous hydroboration step). Note that this bromide obtained from the hydroboration reaction has only 50% of the rotation of the bromide obtained from 1-butanol-1-d (based on optically pure material). Hence, the pentane-2-d, $[\alpha]_D$ $-0.25°$ (p. 75), obtained from the same hydroboration reaction is probably no greater than about 50% optically pure. On this basis, Streitwieser et al. (61) estimate the specific rotation of optically pure pentane-2-d to be $0.52 \pm 0.07°$.

Since $(-)$-pentane-2-d obtained via the hydroboration reaction has the S-configuration, Figure 19, the $(-)$-1-bromopentane-2-d (**44**), obtained from this reaction possesses the same configuration at the deuterated carbon, to which the sequence rule assigns the R-designation. Hence, $(+)$-1-butanol-1-d, related to the (R)-$(-)$-bromide (**44**) by the series of reactions of Figure 20, is assigned the S-configuration.

Thus, the chemical conversions, Figures 17 and 20, confirm that $(-)$-ethanol-1-d and $(+)$-1-butanol-1-d both possess the S-configuration.

VII. PREPARATION OF 1-DEUTEROALCOHOLS BY YEAST REDUCTIONS

Mosher and his co-workers at Stanford University have developed convenient procedures for the reduction of 1-deuteroaldehydes by actively fermenting yeast. As will be discussed below, the yeast reduction possesses the advantage of yielding essentially optically pure 1-deuteroalcohols.

Reduction of trimethylacetaldehyde-1-*d* with actively fermenting Baker's yeast produced neopentyl-1-*d* alcohol having no observable rotation at the Na D line (65). However, the acid phthalate derivative showed a specific rotation of $-1.14 \pm 0.04°$. The configuration of this neopentyl-1-*d* alcohol has been demonstrated to be *S* by several methods, each of which will be discussed in turn.

The reduction of hindered ketones with optically active Grignard reagents yields optically active secondary alcohols and the mechanism of this reaction is considered to be understood (66). Reduction of trimethylacetaldehyde-1-*d* by the Grignard reagent from (+)-1-chloro-2-methylbutane (67), Figure 21, affords neopentyl-1-*d* alcohol (**46**), whose acid phthalate derivative is levorotatory and hence configurationally identical to **46** produced enzymatically (46,65). However, the neopentyl-1-*d* alcohol (**46**), from the Grignard reduction (Fig. 21, top line) possessed only 12% of the optical activity of the alcohol obtained from the yeast

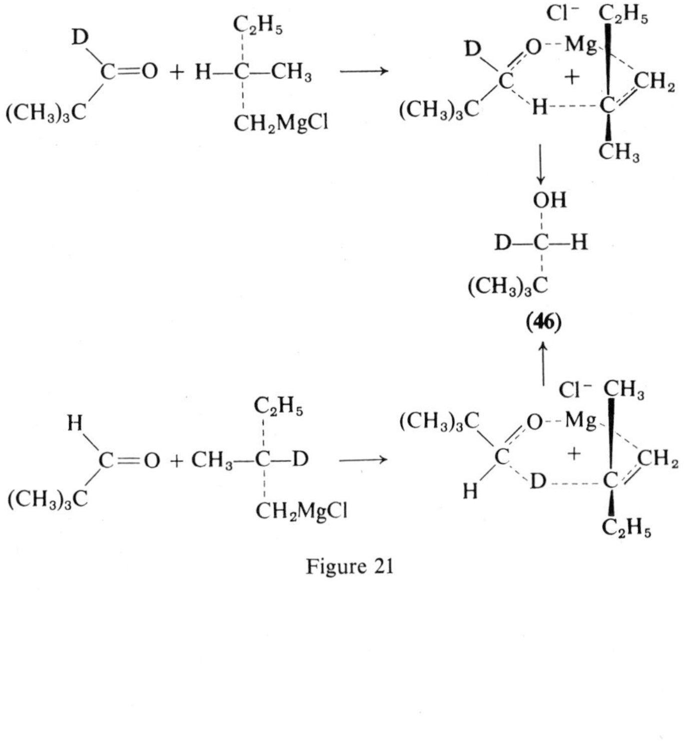

Figure 21

reduction (65), as determined from comparison of the respective acid phthalates.

The same enantiomer of **46** was also prepared (67) by reduction of nondeuterated trimethylacetaldehyde with the Grignard reagent from (−)-1-chloro-2-methylbutane-2-*d* (Fig. 21, bottom line). An interesting isotope effect was observed in this latter reaction. The neopentyl-1-*d* alcohol produced possessed 36% of the rotation (comparison of acid phthalate derivatives) of enzymatically produced alcohol. Hence, the reduction in which deuterium is transferred resulted in three times greater stereoselectivity than the analogous hydrogen transfer reaction.

Consideration of the two possible transition states (66) for the asymmetric reduction leads to the prediction that the one shown in Figure 21 is favored, i.e., the *t*-butyl group of the aldehyde is "*trans*" to the ethyl group at the chiral center of the Grignard reagent.* In an analogous manner the transition state shown in the bottom line of Figure 21 is predicted to be favored for the deuterium transfer reaction. The postulated transition states lead to the prediction of the *S*-configuration for the resulting neopentyl-1-*d* alcohol (67). Since the chemical reduction and the enzymatic reaction both yield neopentyl-1-*d* alcohol having a levorotatory acid phthalate, the enzymatically produced alcohol must also possess the *S*-configuration.

An independent proof of the configuration and of the optical purity of neopentyl-1-*d* alcohol was given by Sanderson and Mosher (69). By their chemical interconversion they also obtained evidence on the mechanism of the neopentyl rearrangement.

Neopentyl-1-*d* alcohol (**47**), from yeast fermentation of the deuterated aldehyde (65), was dehydrated as shown in Figure 22 to give 2-methyl-1-butene-3-*d* (**48**), $[\alpha]_D^{17}$ +1.01 ± 0.01° and the isomeric olefin 2-methyl-2-butene in a ratio of 70:30. Recovered starting material gave an acid phthalate of unchanged optical rotation. The configuration of **48** was shown to be *S* by diimide reduction to give 2-methylbutane-3-*d* (**49**), $[\alpha]_D^{17}$ −0.85 ± 0.03°, and subsequent synthesis of the enantiomer of **49** from optically pure (*S*)-(+)-3-methyl-2-butanol, Figure 22, bottom line. The (*R*)-(+)-2-methylbutane-3-*d* obtained in this manner had a specific rotation approximately 15% lower than the above prepared material. This value

---

* Evidence exists that this model for the transition state is considerably oversimplified. Other factors than the difference in effective size of the two carbonyl substituents must be important since the reduction of *t*-butyl phenyl ketone (68) with the Grignard reagent from (+)-1-chloro-2-methylbutane gives alcohol of the same *optical purity* as that obtained from the reduction of benzaldehyde-1-*d* (46) with the same reagent.

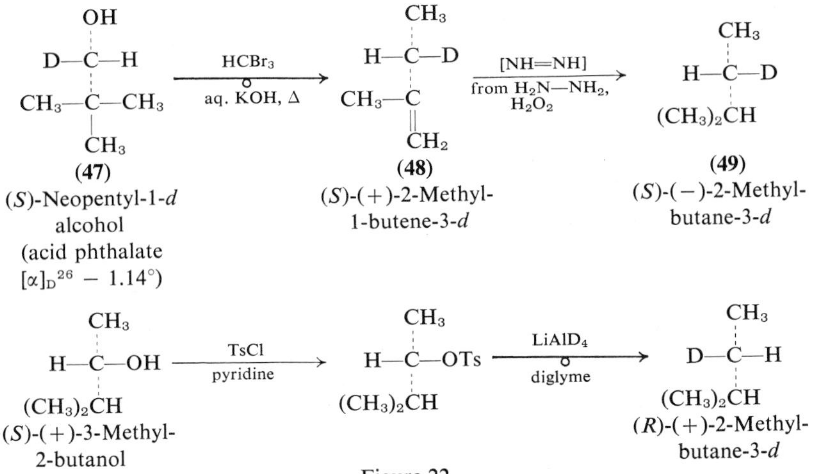

$$
\begin{array}{ccc}
\text{OH} & & \text{CH}_3 \\
\text{D—C—H} & \xrightarrow[\text{aq. KOH, }\Delta]{\text{HCBr}_3} & \text{H—C—D} \\
\text{CH}_3\text{—C—CH}_3 & & \text{CH}_3\text{—C} \\
\text{CH}_3 & & \text{CH}_2
\end{array}
$$

(47)
(S)-Neopentyl-1-d
alcohol
(acid phthalate
$[\alpha]_D^{26} - 1.14°$)

(48)
(S)-(+)-2-Methyl-
1-butene-3-d

$\xrightarrow[\substack{\text{from H}_2\text{N—NH}_2, \\ \text{H}_2\text{O}_2}]{[\text{NH}=\text{NH}]}$

CH$_3$
H—C—D
(CH$_3$)$_2$CH

(49)
(S)-(−)-2-Methyl-
butane-3-d

CH$_3$
H—C—OH
(CH$_3$)$_2$CH
(S)-(+)-3-Methyl-
2-butanol

$\xrightarrow[\text{pyridine}]{\text{TsCl}}$

CH$_3$
H—C—OTs
(CH$_3$)$_2$CH

$\xrightarrow[\text{diglyme}]{\text{LiAlD}_4}$

CH$_3$
D—C—H
(CH$_3$)$_2$CH
(R)-(+)-2-Methyl-
butane-3-d

Figure 22

agrees well with the amount of racemization observed by Helmkamp and Rickborn (p. 59) for the deuteride reduction of secondary tosylates.

Comparison of the rotations of the two samples of 2-methylbutane-3-d prepared by two different routes indicates that the neopentyl-1-d alcohol produced enzymatically is enantiomerically pure and possesses the S-configuration. Sanderson and Mosher further conclude (69) that the neopentyl rearrangement, **47** → **48**, is highly stereoselective and proceeds with inversion of configuration at the deuterium-substituted carbon. A free neopentyl cation is not involved.

Confirmation of the optical purity of enzymatically synthesized neopentyl-1-d alcohol is provided by analysis of the NMR spectra of its esters with R, S, and racemic O-methylmandelyl chloride (70–72), p. 105, as well as by Horeau's partial resolution method (117), p. 104, both of which indicate the alcohol to be 99–100% optically pure.

Althouse, Ueda, and Mosher (65) had concluded earlier that the use of actively fermenting yeast provides a reliable means for the establishment of absolute configuration. This conclusion has been criticized by Lemieux and Howard (56) on the basis of results of Lemieux and Giguere (73) who showed that actively fermenting yeast gave β-hydroxy acids of *opposite* configuration when 3-ketohexanoic and 3-ketoöctanoic acids were reduced. Hence, the possibility exists that at least two mechanisms for the reduction of a carbonyl group may be operative in actively fermenting yeast.

In order to furnish proof of the stereospecificity of carbonyl group reduction in this medium, Mosher and co-workers (74) investigated the yeast reduction of a series of ketones which included all combinations of the substituents methyl, ethyl, n-propyl, n-butyl, and phenyl. Eight of the nine secondary alcohols produced possessed the S-configuration (the configuration of the ninth compound, $(-)$-3-heptanol, is in doubt but probably could be determined by ORD comparison with the other alcohols in the series).

These results, together with the observations that the reduction of acetaldehyde-1-d by *purified* yeast ADH-NADH (55,56) and trimethylacetaldehyde-1-d by *actively fermenting* yeast (65) give alcohols of the S-configuration agree with the assumption that the same enzyme system is responsible for the *in vitro* and the *in vivo* reductions. However, the *caveat* afforded by the work of Lemieux and Giguere (73) indicates the necessity of an independent verification of conclusions based solely on yeast reductions.

In a key paper Mosher and co-workers (46) describe the reduction of butyraldehyde-1-d and benzaldehyde-1-d with actively fermenting yeast to yield the corresponding deuterated alcohols. The results, together with data from other work, are shown in Table III. The 1-butanol-1-d, $[\alpha]_D^{27.5}$ $+0.471 \pm 0.005°$, and the benzyl-α-d alcohol, $[\alpha]_D^{25}$ $+1.58 \pm 0.01°$, are presumed to be optically pure and to have the S-configuration. Both

TABLE III

Comparisons of Enzymatically and Chemically Produced Alcohols,[a]

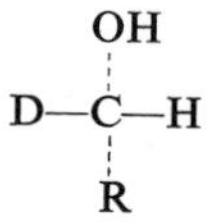

$$\begin{array}{c} OH \\ | \\ D-C-H \\ | \\ R \end{array}$$

R	Enzymatically produced, $[\alpha]_D$	Chemically produced, $[\alpha]_D$	Enzymatic/chemical
CH_3	$-0.28 \pm 0.03°$ [b]	$-0.123 \pm 0.025°$ [c]	2.3
n-C_3H_7	$+0.471 \pm 0.005°$ [d,e]	$+0.185 \pm 0.018°$ [f,g]	2.5
C_6H_5	$+1.58 \pm 0.01°$ [d]	$+0.715 \pm 0.002°$ [f,g,h]	2.2

[a] Unless otherwise noted, all specific rotations are at 25° for the compound containing 1 D/molecule.
[b] Ref. 55.
[c] Ref. 58.
[d] Ref. 46.
[e] Measured at 27.5°.
[f] Ref. 40.
[g] Enantiomer actually prepared.
[h] Measured at 30°.

alcohols from the yeast reduction have the opposite configuration to those synthesized by Streitwieser (40) from the corresponding 1-deuteroaldehydes and isobornyloxymagnesium bromide with the specific rotations of the enzymatically produced alcohols being about 2.3 times larger than those synthesized via the isobornyloxy reduction.

Asymmetric Grignard reduction of benzaldehyde-1-d with the reagent prepared from (+)-1-chloro-2-methylbutane (p. 77) yields benzyl-α-d alcohol, $[\alpha]_D^{24}$ +0.29 ± 0.02° (46). The postulated transition state for this reaction (Fig. 21) is predicted to yield the S-enantiomer, in agreement with the results from the enzymatic reduction, Table III. Comparison of the specific rotation of this benzyl-α-d alcohol with that produced enzymatically indicates that the Grignard reduction proceeds with about 18% asymmetric induction.

Varma and Caspi (110) have reduced isobutyraldehyde-1-d with actively fermenting yeast and obtained (S)-(+)-2-methyl-1- propanol-1-d, α_D^{25} +0.49° (neat, ℓ 1, 0.8 D/molecule).

VIII. 1-DEUTEROAMINES

1-Aminobutane-1-d, $[\alpha]_D^{25}$ −0.009 ± 0.003°, has been prepared from (R)-(−)-1-butanol-1-d, $[\alpha]_D^{25}$ −0.043 ± 0.001°, by conversion of the alcohol to the brosylate followed by treatment of the latter with sodium azide and subsequent reduction with LiAlH$_4$ (33). Since the corresponding series of reactions on optically active 2-octanol gave 2-aminoöctane with over 97% inversion of configuration (33), (−)-1-aminobutane-1-d may be tentatively assigned the S-configuration. In view of the very small observed rotation of the aminobutane and the fact that many examples are known in which amine and alcohol of the same configuration possess the same sign of rotation (compare also benzylamine-α-d and benzyl-α-d alcohol below), a verification of the rotation of (S)-1-aminobutane-1-d would be desirable.

Nitrous acid deamination of (−)-1-aminobutane-1-d in acetic acid gave, in addition to a small amount of alkyl nitrites, a mixture of esters with the ratio of 1-butyl-1-d acetate to 2-butyl-1-d acetate being 2:1 (75). The (R)-(+)-1-butyl-1-d acetate formed in this reaction was demonstrated to be 69 ± 7% inverted and 31 ± 7% racemized.

Streitwieser and Wolfe (76) reported the synthesis of benzylamine-α-d from benzyl-α-d alcohol using the reaction scheme shown in Figure 23. The alcohol, $[\alpha]_D^{25}$ −0.215° (13.5% optically pure, Table III) was converted to the tosylate which was then subjected to S$_N$2 displacement with sodium azide. Reduction of the deuterated azide with LiAlH$_4$ gave benzyl-amine-α-d, $[\alpha]_D$ +0.24°. Since the reaction sequence involved one inversion

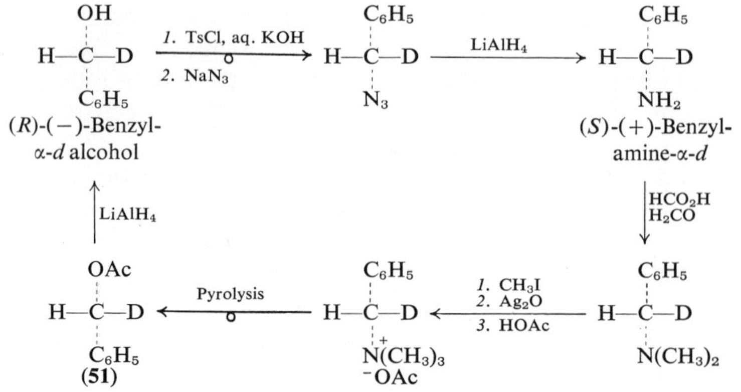

OH C_6H_5 C_6H_5

H—C—D   *1.* TsCl, aq. KOH   H—C—D   LiAlH$_4$ H—C—D

C_6H_5 *2.* NaN$_3$ N$_3$ NH$_2$

(R)-(−)-Benzyl-   (S)-(+)-Benzyl-
α-d alcohol   amine-α-d

LiAlH$_4$ HCO$_2$H / H$_2$CO

OAc C_6H_5 C_6H_5

H—C—D   Pyrolysis   H—C—D   *1.* CH$_3$I   H—C—D

C_6H_5 $\overset{+}{N}(CH_3)_3$ *2.* Ag$_2$O   N(CH$_3$)$_2$

(51) $^-$OAc   *3.* HOAc

Figure 23

of configuration, the dextrorotatory amine has the *S*-configuration. Hence, benzyl-α-d alcohol and benzylamine-α-d of the same sign of rotation have the same configuration.

In order to compare the optical purity of the amine relative to the starting alcohol, the cyclic method of Snyder and Brewster (77) was utilized, Figure 23. Recovered benzyl-α-d alcohol had a specific rotation of −0.180°, indicating that 18% racemization had occurred. Since Snyder and Brewster have shown (77) that a similar conversion of amine to acetate, **50 → 51**, occurs with essentially complete inversion of configuration, the observed racemization probably took place in the preparation of the benzylamine-α-d. It seems highly likely that the Schotten-Baumann preparation of the tosylate accounted for most of the racemization since reaction of OH$^-$ with benzyl-α-d tosylate as it forms would generate alcohol of inverted configuration, eventually resulting in tosylate of lower optical purity than the starting alcohol. Hence, based on the value of $[\alpha]_D$ 1.58° for optically pure benzyl-α-d alcohol (Table III), and the 18% racemization occurring in the preparation of the amine, Figure 23, optically pure benzylamine-α-d is calculated to have $[\alpha]_D$ 2.15°.

However, using an NMR technique, Gerlach (p. 105) has determined the specific rotation of optically pure benzylamine-α-d to be 1.77°. It is interesting that this value is almost exactly 18% less than the one calculated from the results of Streitwieser and Wolfe and seems to indicate the unlikely fact that no racemization occurred in the preparation of the deuterated amine from benzyl-α-d alcohol as shown in Figure 23.

Guthrie, Meister, and Cram (78) reported the synthesis of optically active neopentylamine-α-d via a novel method which promises to be useful for the preparation of other primary deuterated amines. In their method,

Figure 24, trimethylacetaldehyde-1-*d* and (*S*)-(−)-α-phenylethylamine, $[\alpha]_D^{25}$ −40.6° (optically pure), were converted to the Schiff's base (52), α_D^{25} −42.48° (ℓ 1, neat). Isomerization of 52 afforded as the chief product α-methylbenzylidineneopentylamine-α-*d* (53), α_D^{25} +5.24° (ℓ 1, neat). The rotation is unusually large for a compound which owes its optical activity to hydrogen–deuterium dissymmetry and may be due to a relatively intense positive Cotton effect of the optically active imine chromophore near 245 nm* (79) which tails off slowly in the visible region. It will be interesting to see if the ORD curve of 53 agrees with this prediction.

Hydrolysis of the imine (53) yielded neopentylamine-α-*d*, α_D^{25} +0.20° (1 1, contaminated with some Et_2O). The maximum rotation is estimated by the authors (78) to be about +0.30°. Based on Brewster's rules (62,63) and the assumption that methyl and *t*-butyl groups have similar polarizabilities relative to the amine group, the *R*-configuration is assigned to the (+)-amine (54).

The proton transfer in the base-catalyzed reaction 52 → 53 was demonstrated to be essentially completely stereospecific by $LiAlH_4$ reduction of 53 to the dextrorotatory mixture of diastereomers and resolution with (+)-camphorsulfonic acid. NMR analysis (p. 105) of the resulting (+)-(*R*,*R*)- and (−)-(*S*,*R*)-diastereomers indicated the starting imine (53) to have been about 98% optically pure.

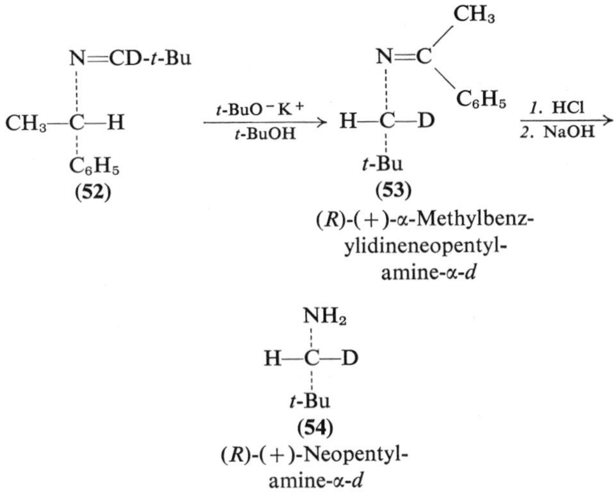

(*R*)-(+)-α-Methylbenz-
ylidineneopentyl-
amine-α-*d*

(54)
(*R*)-(+)-Neopentyl-
amine-α-*d*

Figure 24

* Nanometer, nm = 10^{-9} m; replaces mμ.

As part of a stereochemical investigation of enzymatic decarboxylation, Belleau and Burba (80) synthesized (R)-tyramine-1-d (56), (and its enantiomer) by a series of reactions of known stereochemistry. Reduction of p-methoxyphenylacetaldehyde with isobornyloxy-2-d-magnesium bromide gave a 40% yield of 2-(p-methoxyphenyl)-ethanol-1-d, (55), $[\alpha]_D^{24}$ $-1.44°$ (neat). From the mechanism of this reduction (p. 62) the S-configuration may be assigned to the alcohol. By analogy with similar reductions the optical purity of the alcohol is probably about 40–50%.

The tosylate of 55 was treated with excess sodium azide, Figure 25, under conditions where complete inversion of configuration has been demonstrated (33). Reduction of the crude azide with LiAlH$_4$, followed by hydrolysis of the p-methyl ether with HBr gave tyramine-1-d whose configuration is assigned as R based on the method of synthesis.

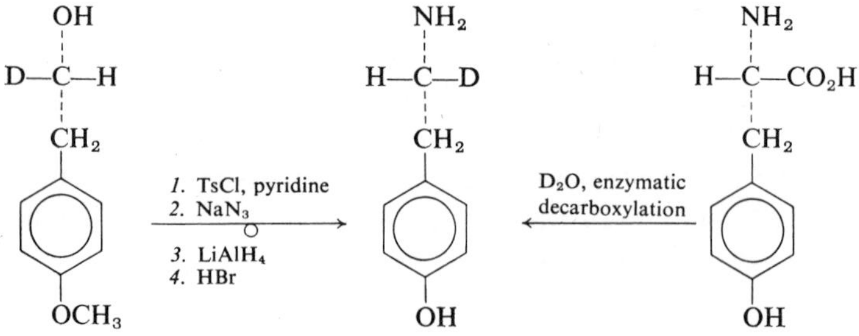

Figure 25

The enantiomer, (S)-tyramine-1-d, was prepared by treating the $(-)$-alcohol (55) with PBr$_3$ in the presence of collidine; a reaction involving inversion of configuration (29,31). The resulting bromide was then subjected to the same reaction sequence as the tosylate of 55, Figure 25, to afford (S)-tyramine-1-d.

The absolute configuration of the tyramine-1-d formed in the enzymatic decarboxylation of (S)-tyrosine in D$_2$O was determined by comparison of the relative rates of oxidation by monoamine oxidase of the synthetic tyramines with the enzymatically prepared one.

Enzymatic decarboxylation of (S)-tyrosine (57), in D$_2$O gave tyramine-1-d whose rate of oxidation was identical to (R)-tyramine-1-d. Hence, the enzymatic decarboxylation of tyrosine (and presumably of other α-amino acids as well) proceeds with retention of configuration.

IX. MISCELLANEOUS

Optically active propionic-2-d acid has been an important reference in the study of enzymatic reactions and several syntheses have been reported within the past few years.

Starting from enzymatically prepared (S)-$(-)$-ethanol-1-d (p. 58), (R)-$(-)$-propionic-2-d acid (**58**) was obtained as shown in Figure 26 (81). The sodium salt of **58** exhibited a plain negative ORD curve in the region from 300 to 220 nm.

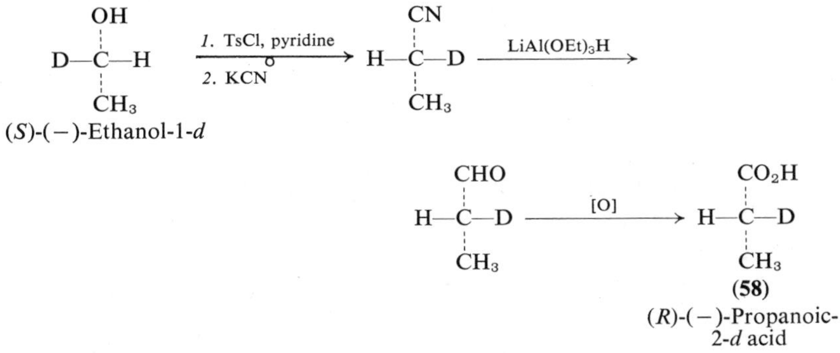

OH
|
D—C—H $\xrightarrow[\text{2. KCN}]{\text{1. TsCl, pyridine}}$ H—C—D $\xrightarrow{\text{LiAl(OEt)}_3\text{H}}$
|
CH₃

(S)-$(-)$-Ethanol-1-d

CN
|
H—C—D
|
CH₃

CHO
|
H—C—D $\xrightarrow{\text{[O]}}$ H—C—D
|
CH₃

CO₂H
|
H—C—D
|
CH₃
(**58**)
(R)-$(-)$-Propanoic-2-d acid

Figure 26

The (S)-$(+)$-enantiomer of propanoic-2-d acid has been prepared by hypobromite oxidation of optically pure $(2R,3S)$-$(-)$-2-butanol-3-d (p. 58), Figure 27 (82). ORD measurements in aqueous solution gave the following results: $[\alpha]_{500}$ $+1°$, $[\alpha]_{400}$ $+2°$, $[\alpha]_{250}$ $+27.5°$, $[\alpha]_{230}$ $+80°$.

Both enantiomers of propanoic-2-d acid have been obtained by the reaction sequence shown in Figure 28 for the R-acid (83). Use of the sequence starting from (R)-$(-)$-alanine afforded the (S)-propionic-2-d acid.

Formation of (S)-$(-)$-α-bromopropanoic acid (**59**), Figure 28, $[\alpha]_D^{25}$ $-23.3°$, involved approximately 50% racemization based on $[\alpha]_D$ $-45.2°$ reported for optically pure material (84). The (R)-propanoic-2-d acid isolated

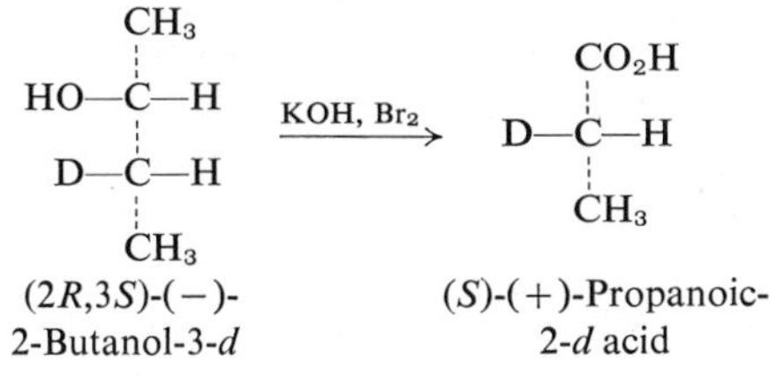

CH₃
|
HO—C—H
|
D—C—H $\xrightarrow{\text{KOH, Br}_2}$
|
CH₃

$(2R,3S)$-$(-)$-
2-Butanol-3-d

CO₂H
|
D—C—H
|
CH₃

(S)-$(+)$-Propanoic-
2-d acid

Figure 27

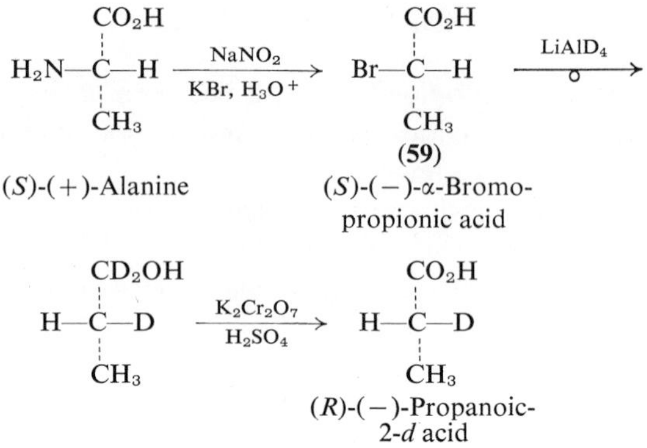

$$\underset{\substack{\text{CH}_3 \\ \text{(S)-($+$)-Alanine}}}{\text{H}_2\text{N}-\overset{\text{CO}_2\text{H}}{\underset{|}{\text{C}}}-\text{H}} \xrightarrow[\text{KBr, H}_3\text{O}^+]{\text{NaNO}_2} \underset{\substack{\text{CH}_3 \\ \textbf{(59)} \\ \text{(S)-($-$)-α-Bromo-} \\ \text{propionic acid}}}{\text{Br}-\overset{\text{CO}_2\text{H}}{\underset{|}{\text{C}}}-\text{H}} \xrightarrow[0]{\text{LiAlD}_4}$$

$$\underset{\substack{\text{CH}_3}}{\text{H}-\overset{\text{CD}_2\text{OH}}{\underset{|}{\text{C}}}-\text{D}} \xrightarrow[\text{H}_2\text{SO}_4]{\text{K}_2\text{Cr}_2\text{O}_7} \underset{\substack{\text{CH}_3 \\ \text{(R)-($-$)-Propanoic-} \\ \text{2-d acid}}}{\text{H}-\overset{\text{CO}_2\text{H}}{\underset{|}{\text{C}}}-\text{D}}$$

Figure 28

from this sequence had $[\alpha]_{675}$ $-0.39°$ and $[\alpha]_{400}$ $-1.54°$. If one assumes that no further racemization occurred in the deuteride reduction and in the oxidation step, then optically pure (R)-propanoic-2-d acid may be calculated to have a minimum rotation of $[\alpha]_{400}$ $-3°$.

In his epoch-making work *La Chimie dans l'Espace*, van't Hoff (85) pointed out that, in addition to compounds possessing asymmetric carbon atoms, symmetrically disubstituted allenes of the type $XYC=C=CXY$ containing an odd number of carbon atoms should also exhibit optical activity. It was not until 1935, however, that this prediction was verified (86). Long before this, in 1909, Perkin, Pope, and Wallach (87) had resolved 4-methylcyclohexylidineacetic acid as the first example of an optically active molecule which contains no asymmetric carbon atom in the formal sense. The absolute configuration of this compound was determined in 1966 utilizing optically active deuterium compounds (42).

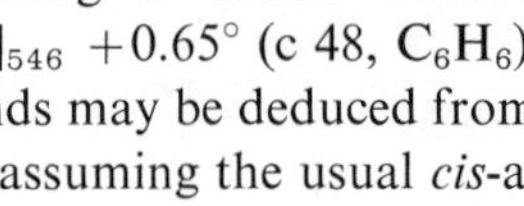

Catalytic hydrogenation of ($+$)-α-deutero-4-methylcyclohexylidine-acetic acid (**60**), Figure 29, (resolved using α-phenylethylamine) leads to two saturated acids having a chiral α-carbon (**61**), $[\alpha]_{546}$ $+0.44°$ (c 26, C_6H_6) and (**62**), $[\alpha]_{546}$ $+0.65°$ (c 48, C_6H_6). The *relative* configurations of the two compounds may be deduced from the two possible modes of addition of hydrogen, assuming the usual *cis*-addition mechanism (15). If the addition of hydrogen occurs from the *re* side (p. 71) of the α-carbon, corresponding to the addition of an axial hydrogen atom to the ring, then the *trans*-diastereomer is obtained. Analogous *cis*-addition of

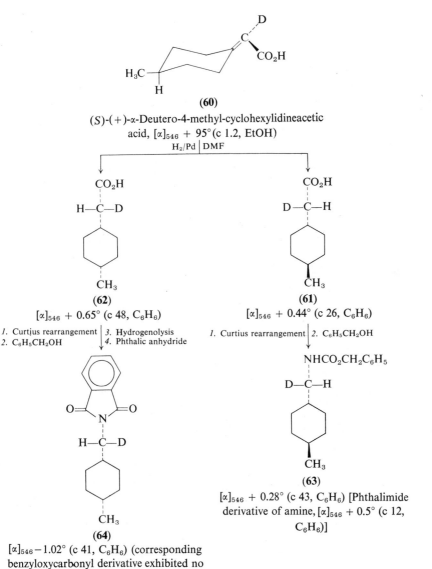

(60)

(S)-(+)-α-Deutero-4-methyl-cyclohexylidineacetic
acid, $[\alpha]_{546} + 95°$ (c 1.2, EtOH)

H_2/Pd | DMF

(62)
$[\alpha]_{546} + 0.65°$ (c 48, C_6H_6)

1. Curtius rearrangement | 3. Hydrogenolysis
2. $C_6H_5CH_2OH$ | 4. Phthalic anhydride

(61)
$[\alpha]_{546} + 0.44°$ (c 26, C_6H_6)

1. Curtius rearrangement | 2. $C_6H_5CH_2OH$

(63)
$[\alpha]_{546} + 0.28°$ (c 43, C_6H_6) [Phthalimide
derivative of amine, $[\alpha]_{546} + 0.5°$ (c 12,
C_6H_6)]

(64)
$[\alpha]_{546} - 1.02°$ (c 41, C_6H_6) (corresponding
benzyloxycarbonyl derivative exhibited no
rotation at 546 nm)

Figure 29

hydrogen to the *si* side of **60** leads to the *cis*-isomer. Hence, determination of the absolute configuration of *one* of the hydrogenation products allows the specification of configuration of the other, as well as that of the precursor (**60**).

Gerlach (42) was able to assign the *trans*- and *cis*-geometry to **61** and **62**, respectively, by stereospecific synthesis of the corresponding non-deuterated analogs via the Arndt–Eistert homologation reaction of *trans*- and *cis*-4-methylcyclohexanecarboxylic acid of known absolute configuration (88) and comparison of physical properties with the respective deuterated 4-methylcyclohexaneacetic acids **61** and **62**. The introduction of a single deuterium atom does not significantly affect properties such as melting or boiling points.

Determination of the absolute configuration at the α-carbon of *trans*-**61** and *cis*-**62** was effected by relating these acids to the corresponding

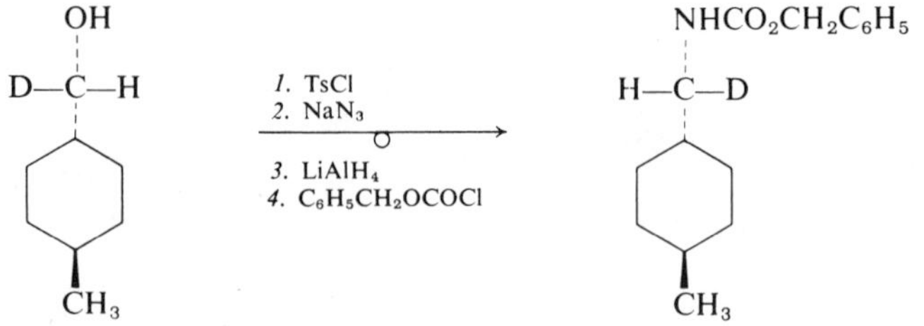

(**65**)

(S)-(+)-*trans*-4-Methylcyclohexyl-carbinol-
α-d α_{546} + 0.18° (ℓ 1, neat, 0.65 D/molecule)
[hydrogen phthalate [α]$_{546}$ − 0.1°
(c 4.5, EtOH)]

(**67**)

[α]$_{546}$ − 0.24° (c 33, C$_6$H$_6$) [phthalimide
derivative [α]$_{546}$ − 0.15° (c 33, C$_6$H$_6$)]

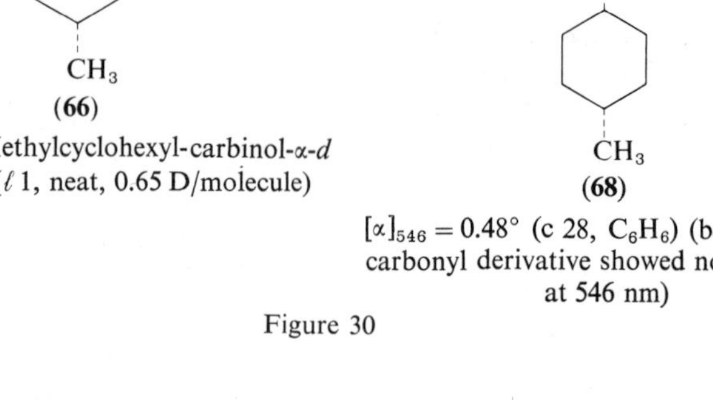

(**66**)

(S)-(+)-*cis*-4-Methylcyclohexyl-carbinol-α-d
α_{546} + 0.11° (ℓ 1, neat, 0.65 D/molecule)

(**68**)

[α]$_{546}$ = 0.48° (c 28, C$_6$H$_6$) (benzyloxycarbonyl derivative showed no rotation
at 546 nm)

Figure 30

amines. The configurations of the amines, in turn, could be related to the corresponding alcohols, available by use of Streitwieser's asymmetric reduction method (p. 61). Accordingly, **61** and **62** were converted to derivatives of α-deutero-4-methycyclohexylmethylamine, **63** and **64**, respectively, via the Curtius rearrangement, Figure 29.

The alcohols corresponding to these compounds were then prepared by isobornyloxy-2-*d*-magnesium bromide reduction of the appropriate aldehydes. From the mechanism of the reduction (p. 62) the *S*-configuration is expected for the resulting primary deuterated alcohols. The resultant *S*-alcohols **65** and **66** were related to the amine derivatives **63** and **64** by the series of reactions shown in Figure 30. Only one inversion of configuration is involved; the azide displacement of the tosylate. Hence, the absolute configurations of **67** and **68** are both *R*. Therefore, in the original series of reactions, Figure 29, *trans*-**63** and *cis*-**64** possess the *S*- and *R*-chirality, respectively, so that the acids from which the amines were derived may be assigned the same configurations: (*S*)-**61** and (*R*)-**62**. From the now-established *R*-configuration of the *cis*-acid (**62**), the *S*-chirality (13) may be assigned to that enantiomer of α-deutero-4-methylcyclohexyl-idineacetic acid (**60**), which upon hydrogenation gives a *cis*-(+)-acid.

Caspi and Varma (136) have reported the preparation of (*S*)-(−)-4-methylpentanol-3-*d* (**69**), $[\alpha]_D^{25}$ −0.168° (c 31.6, CHCl$_3$). The configuration of the alcohol, whose optical purity was in the 30% range, was determined by Horeau's method (p. 104). Treatment of **69** with the Jones reagent oxidized it to (*S*)-(−)-4-methylpentanoic-3-*d* acid (**70**), $[\alpha]_D^{25}$ −0.453° (c 30, CHCl$_3$), Figure 31.

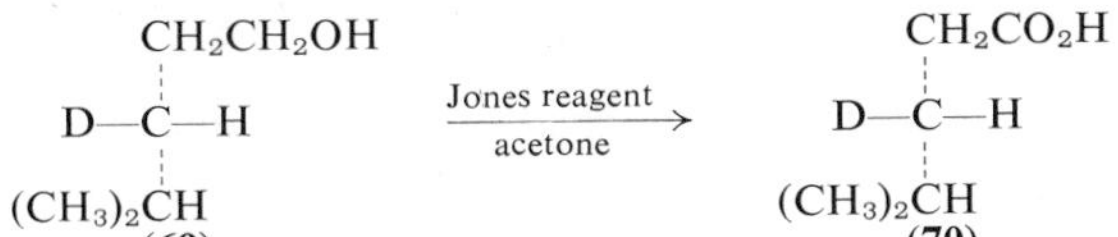

$$CH_2CH_2OH \qquad\qquad CH_2CO_2H$$
$$D—C—H \quad\xrightarrow[\text{acetone}]{\text{Jones reagent}}\quad D—C—H$$
$$(CH_3)_2CH \qquad\qquad (CH_3)_2CH$$
$$\textbf{(69)} \qquad\qquad\qquad \textbf{(70)}$$

Figure 31

X. OPTICALLY ACTIVE DEUTERIUM COMPOUNDS NOT HAVING DEUTERIUM ATTACHED TO THE CHIRAL CENTER

In a landmark investigation of the citric acid cycle, Martius and Schorre (89,90) prepared optically active α,α-dideuterocitric acid (**69**), $[\alpha]_{546}^{20}$ −1.03 ± 0.01° (c 12.6, water), by resolution of oxalcitramalic acid lactone followed by treatment of the (−)-lactone with D$_2$O to replace the

$$\underset{(\mathbf{71})}{\overset{\displaystyle CO_2H}{\underset{\displaystyle CO_2H}{\overset{\displaystyle |}{\underset{\displaystyle |}{\overset{\displaystyle CH_2}{\underset{\displaystyle CD_2}{\overset{\displaystyle |}{\underset{\displaystyle |}{HO_2C\!-\!C\!-\!OH}}}}}}}} \xrightarrow{\text{Aconitase}} \overset{\displaystyle CO_2H}{\underset{\displaystyle CO_2H}{\overset{\displaystyle |}{\underset{\displaystyle |}{\overset{\displaystyle C=O}{\underset{\displaystyle CD_2}{\overset{\displaystyle |}{\underset{\displaystyle |}{CH_2}}}}}}} \quad \overset{but}{\underset{no}{}} \quad \overset{\displaystyle CO_2H}{\underset{\displaystyle CO_2H}{\overset{\displaystyle |}{\underset{\displaystyle |}{\overset{\displaystyle CH_2}{\underset{\displaystyle C=O}{\overset{\displaystyle |}{\underset{\displaystyle |}{CH_2}}}}}}}$$

Figure 32

exchangeable hydrogens, and then oxidative ring opening to the product. The ammonium molybdate complex of **69** exhibited a specific rotation at 546 nm of -34 ± 2. This is a striking example of the introduction of an optically active absorption band (in this case the metal ion) near the region of measurement to increase the magnitude of rotation. It is interesting to note that, even today, this method seems to be little appreciated and seldom used. A recent paper by Djerassi and co-workers (137) utilizes this effect to determine CD spectra in the sugar series.

When the $(-)$-dideuterocitric acid was subjected to enzymatic oxidation, Figure 32, it was found that the α-ketoglutaric acid formed had retained all the deuterium. In contrast, use of $(+)$-dideuterocitric acid gave α-ketoglutaric acid containing no deuterium (89).

The experiments clearly demonstrate that the symmetric molecule citric acid is degraded asymmetrically and confirmed the postulate put forth earlier by Ogston (91) that, although the two $-CH_2CO_2H$ groups of citric acid are chemically equivalent, an enzyme can differentiate between them because of their diastereomeric environments in the presence of the chiral substance. This concept is illustrated schematically in Figure 33.

The simplest example of the class of optically active deuterium compounds not having deuterium attached to the chiral center is 2-propanol-1-

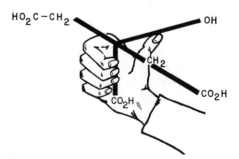

Fig. 33. Schematic representation of the $-CH_2CO_2H$ groups of citric acid rendered nonequivalent by combination with an enzyme whose chirality is represented by a right hand. Reproduced with permission from Roberts and Caserio (133).

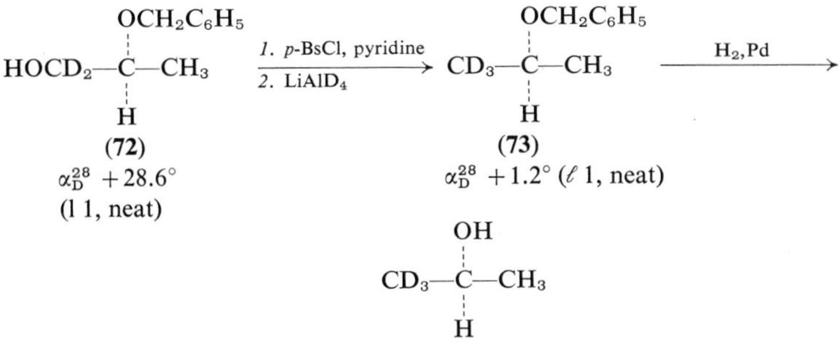

$$\underset{\substack{\text{(72)}\\ \alpha_D^{28} +28.6^\circ\\ \text{(l 1, neat)}}}{\underset{\substack{|\\ \text{H}}}{\overset{\substack{\text{OCH}_2\text{C}_6\text{H}_5\\ |}}{\text{HOCD}_2\text{—C—CH}_3}}} \xrightarrow[\substack{2.\ \text{LiAlD}_4}]{\substack{1.\ p\text{-BsCl, pyridine}}} \underset{\substack{\text{(73)}\\ \alpha_D^{28} +1.2^\circ\ (\ell\ 1,\ \text{neat})}}{\underset{\substack{|\\ \text{H}}}{\overset{\substack{\text{OCH}_2\text{C}_6\text{H}_5\\ |}}{\text{CD}_3\text{—C—CH}_3}}} \xrightarrow{\substack{\text{H}_2,\text{Pd}}}$$

$$\underset{\substack{\text{(74)}\\ (S)\text{-}(+)\text{-2-Propanol-1-}d_3\\ \alpha_D^{25} +0.27^\circ\ (\ell\ 1,\ \text{neat})\\ 2.9\ \text{D/molecule}}}{\underset{\substack{|\\ \text{H}}}{\overset{\substack{\text{OH}\\ |}}{\text{CD}_3\text{—C—CH}_3}}}$$

Figure 34

d_3 (**74**), synthesized by Mislow, O'Brien, and Schaefer (92). The synthesis started from (S)-methyl lactate which was treated with benzyl bromide in the presence of silver oxide to give (−)-methyl O-benzyllactate, followed by LiAlD$_4$ reduction to afford the alcohol (**72**). Conversion of **72** to the brosylate, Figure 34, followed by LiAlD$_4$ reduction gave the (+)-benzyl ether (**73**). Hydrogenolysis of the benzyl group afforded 2-propanol-1-d_3 (**74**), $\alpha_D^{25} +0.27^\circ$ (ℓ 1, neat), 2.9 D/molecule, whose absolute configuration is deduced to be S from the method of synthesis. Rotational data at several wavelengths are given in Table IV.

TABLE IV

Optical Rotations at Several Wavelengths for (S)-(+)-2-Propanol-1-d_3 (**74**), and its Hydrogen Phthalate Derivative[a]

λ, nm	(74) α^{25}, deg. (ℓ 1, neat)	Hydrogen phthalate of 74 $[\alpha]^{27}$, deg. (c 4.0, CHCl$_3$)
589	+0.27	+0.2
578	0.28	0.5
546	0.31	0.6
435	0.58	0.8
365	0.98	1.2

[a] Ref. 92.

Optically active 2-propanol-1-d_3 was utilized by Streitwieser and Stang (93) in a mechanistic study of the BF$_3$-catalyzed alkylation of aromatics with alcohols:

$$\text{ROH} + \text{ArH} \xrightarrow{\text{BF}_3} \text{R—Ar} + \text{H}_2\text{O}$$

Previous work (94) utilizing optically active 2-butanol, benzene, and BF$_3$ had given *sec*-butylbenzene with 99% racemization. However, the results were ambiguous because of the known rapid rearrangement of secondary systems under the reaction conditions. Hence, Streitwieser and Stang chose for their alkylating agent 2-propanol-1-d_3 (74), a system with only one secondary position.

In order to ascertain that the expected alkylation product, 2-phenyl-propane-1-d_3, possessed measurable optical activity, the hydrocarbon was prepared starting from the methyl ester of optically pure (R)-$(-)$-3-phenyl-butanoic acid, Figure 35. Since no bonds to the chiral center are broken in the reaction sequence, the resultant $(+)$-2-phenylpropane-1-d_3 (75) is assigned the R-configuration and is presumed to be optically pure.

Alkylation of benzene with (S)-$(+)$-2-propanol-1-d_3 and BF$_3$ at 5° gave 75 having α_D^{25} $+0.009 \pm 0.005°$ (ℓ 1, neat). A similar reaction in 60:40 benzene:nitromethane gave 75 having a rotation of $0.033 \pm 0.005°$ (ℓ 1, neat). The large amount of racemization ($> 93\%$) and small amount of net inversion of configuration indicate that the isopropyl cation inter-mediate in the alkylation is essentially free with the reaction resembling an S$_N$1 solvolysis (93).

The first example of a chiral sulfur atom which is optically active because of isotopic dissymmetry has been recently reported. (R)-$(-)$-Dimethyl-1-d_3 sulfoxide (76), was synthesized by Grignard reaction of methyl-d_3 magnesium iodide with (R)-$(-)$-menthyl methanesulfinate of 28% diastereomeric purity, Figure 36 (95).

The optical purity of 76 was determined by NMR examination of the sulfoxide in the chiral solvent (R)-$(-)$-1-phenyl-2,2,2-trifluoroethanol.*

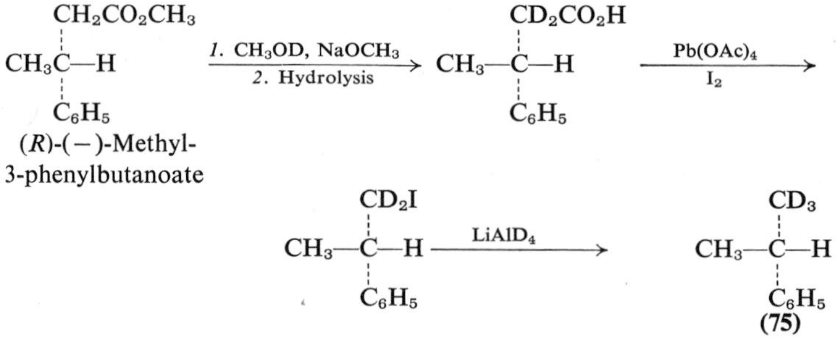

Figure 35

* The absolute configuration of 76 is not determined directly from the NMR analysis but follows from the known stereochemistry of the reaction (96,97).

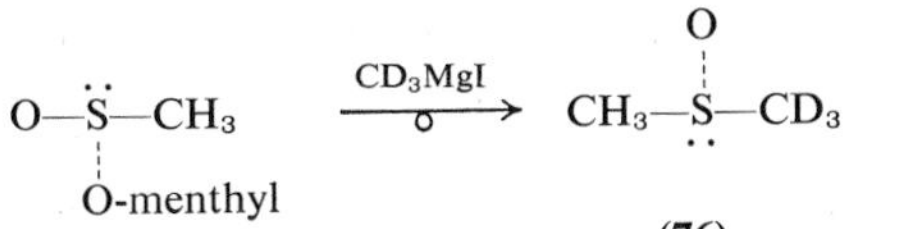

O—S̈—CH$_3$ $\xrightarrow[\text{O}]{\text{CD}_3\text{MgI}}$ CH$_3$—S̈—CD$_3$

Ö-menthyl

(76)

(R)-(−)-Menthyl   (R)-(−)-Dimethyl-1-d_3
methanesulfinate   sulfoxide

Figure 36

In chiral solvents, enantiomeric nuclei exist in diastereomeric environ-ments and therefore exhibit chemical shift differences (see Section XII for further examples of related uses of NMR). The NMR spectrum of **76** shows two unequally intense methyl resonances whose ratios allow the direct determination of optical purity as 28%. Based on the observed rotation of **76** optically pure (R)-dimethyl-1-d_3 sulfoxide would have $[\alpha]_D^{25}$ −3 8° (neat).

The dissymmetric dideuterobenzene hexachloride isomer (**77**), α_D^{26} +0.50 ± 0.002° (ℓ 1, c 66, acetone) has been reported by Riemschneider (99).

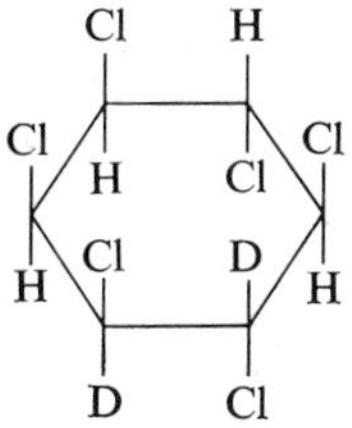

(77)

XI. INVESTIGATIONS OF REACTION MECHANISMS UTILIZING OPTICALLY ACTIVE DEUTERIUM COMPOUNDS

This section describes research whose principal aim was not the study of optically active deuterium compounds *per se*, but rather the use of this type of compound to obtain stereochemical information on the mechanisms of organic reactions.

Studies of the acetolysis reaction at primary carbon have been carried out by Streitwieser and co-workers (100–102). Acetolysis of (R)-(+)-1-butyl-1-d p-nitrobenzenesulfonate gave (S)-(−)-1-butyl-1-d acetate with 85% inversion of configuration (100). Results for the acetolysis of (R)-benzyl-α-d

tosylate were similar; resultant (S)-$(+)$-benzyl-α-d acetate exhibited approximately 80% net inversion (101). The respective acetates were shown to be optically stable to the reaction conditions but unreacted starting sulfonate was not examined. Results obtained in the 2-octyl system (101,102) indicate that the 15–20% apparent racemization occurring in the primary system is due mainly to racemization of starting sulfonate with the sulfonic acid formed in the reaction. The acetolysis reaction itself appears to involve essentially complete inversion of configuration (101).

Streitwieser and Wolfe (43) have studied the ethanolysis reaction of benzyl-α-d tosylate under a variety of conditions. The tosylate, prepared from the R-alcohol having $[\alpha]_D$ $-0.715°$ (40) by treatment with tosyl chloride in pyridine at $-40°$ was not extensively purified because of its instability. Solvolysis of (R)-benzyl-α-d tosylate in absolute ethanol at 25° gave a product which contained chlorine. The authors conclude that ethanolysis of some contaminating tosyl chloride produces chloride ion which reacts with benzyl-α-d tosylate at a rate comparable to ethanolysis of the tosylate. The benzyl-α-d chloride had $[\alpha]_D$ $+0.6°$ and, since it is presumably formed by inversion of configuration, the $(+)$-chloride is assigned the S-configuration.

Removal of the benzyl-α-d chloride allowed the isolation of the main ethanolysis product of (R)-benzyl-α-d tosylate, benzyl-α-d ethyl ether, $[\alpha]_D$ $+0.098°$, Figure 37. Comparison of the rotation with the deuterated ether prepared from benzyl-α-d alcohol via the Williamson synthesis indicated that the ethanolysis of the tosylate apparently goes with complete inversion of configuration. Reaction of the tosylate with alcoholic sodium ethoxide also involves complete inversion. Ethanolysis and hydrolysis of the tosylate in 80% aqueous ethanol was shown to proceed with inversion also but to involve a few per cent racemization.

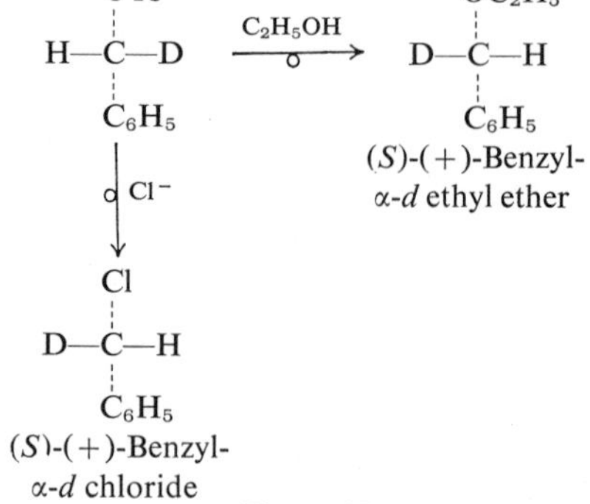

$$
\begin{array}{ccc}
\text{OTs} & & \text{OC}_2\text{H}_5 \\
| & \xrightarrow[\text{o}]{\text{C}_2\text{H}_5\text{OH}} & | \\
\text{H—C—D} & & \text{D—C—H} \\
| & & | \\
\text{C}_6\text{H}_5 & & \text{C}_6\text{H}_5
\end{array}
$$

(S)-$(+)$-Benzyl-α-d ethyl ether

$$
\begin{array}{c}
\Big\downarrow \text{Cl}^- \\
\text{Cl} \\
| \\
\text{D—C—H} \\
| \\
\text{C}_6\text{H}_5
\end{array}
$$

(S)-$(+)$-Benzyl-α-d chloride

Figure 37

As part of Cram's comprehensive studies of the mechanism of electrophilic substitution at saturated carbon (103), use was made of optically active ethylbenzene-α-d. Cram and Rickborn (104) found that the base-catalyzed cleavage of ($-$)-2-methyl-3-phenyl-2-butanol (**78**) in t-BuOD gave ethylbenzene-α-d with predominantly retention of configuration, Figure 38, while in ethylene glycol-O,O'-d_2 as solvent, cleavage occurs predominantly with inversion. The results are explained in terms of carbanions as discrete intermediates. In proton-donating solvents of high dielectric constant, cleavage occurs to give carbanions solvated at the front by the leaving group and at the rear by solvent. Proton (deuteron) capture occurs from the rear face of the anion and inverted product is obtained. The retention mechanism occurs in solvents of low dielectric constant in which the carbanion exists as an intimate ion pair surrounded by the proton-donating solvent. Because of the poor ion-solvating power of the medium, dissociation to free ions is much slower than is collapse to give product of retained configuration.

The stereochemistry of the deamination of (R)-($+$)-neopentylamine-α-d (p. 83) has been studied by Guthrie (105). Treatment of the amine of 93% optical purity (0.985 D/molecule) with n-butyl nitrite in acetic acid gave a mixture of (R)-($-$)-2-methyl-1-butene-3-d (**79**), 85% optically pure (p. 78), the corresponding (R)-($-$)-acetate (**80**), 2-methyl-2-butene-3-d of 58 % isotopic purity, and a trace of neopentyl-1-d acetate, Figure 39.

Pyrolysis of the acetate (**80**) gave the olefin (**79**) of the same sign and magnitude of rotation as that formed initially in the reaction. Since the configuration and minimum optical purity of the olefin were known from

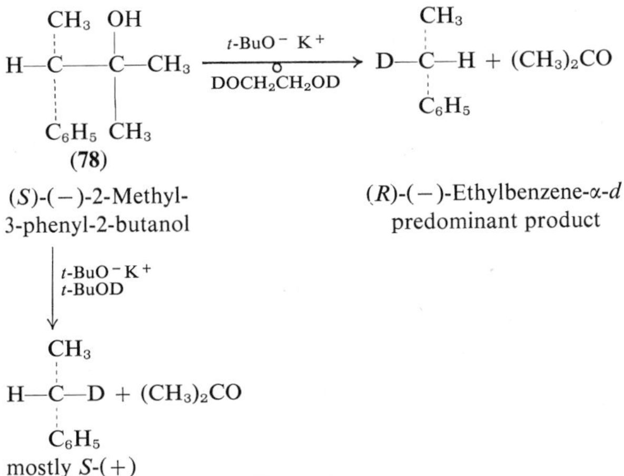

CH$_3$ OH

H—C——C—CH$_3$ $\xrightarrow[\text{DOCH}_2\text{CH}_2\text{OD}]{t\text{-BuO}^-\text{ K}^+}$ D—C—H + (CH$_3$)$_2$CO

C$_6$H$_5$ CH$_3$ CH$_3$ C$_6$H$_5$

(**78**)

(S)-($-$)-2-Methyl-   (R)-($-$)-Ethylbenzene-α-d
3-phenyl-2-butanol   predominant product

$\downarrow$ $\begin{array}{l} t\text{-BuO}^-\text{K}^+ \\ t\text{-BuOD} \end{array}$

CH$_3$

H—C—D + (CH$_3$)$_2$CO

C$_6$H$_5$

mostly S-($+$)

Figure 38

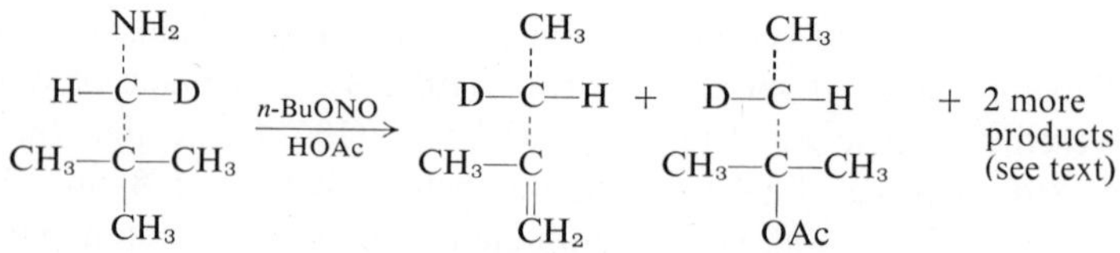

Figure 39

the work of Sanderson and Mosher (p. 78), it could be determined from the deamination results that methyl migration occurs with at least 85% inversion of configuration at the migration terminus.

It is unfortunate that more neopentyl-1-d acetate is not formed in the deamination since this material could yield information on the competing substitution reaction. If, by chance, this reaction were also found to proceed with a high degree of stereospecificity, the optically active neopentyl-1-d acetate could then be related to the enzymatically synthesized neopentyl-1-d alcohol (p. 77).

Solvolysis of (S)-neopentyl-1-d tosylate (no observable rotation 589–300 nm; prepared from optically pure alcohol) in ethanol yielded a variety

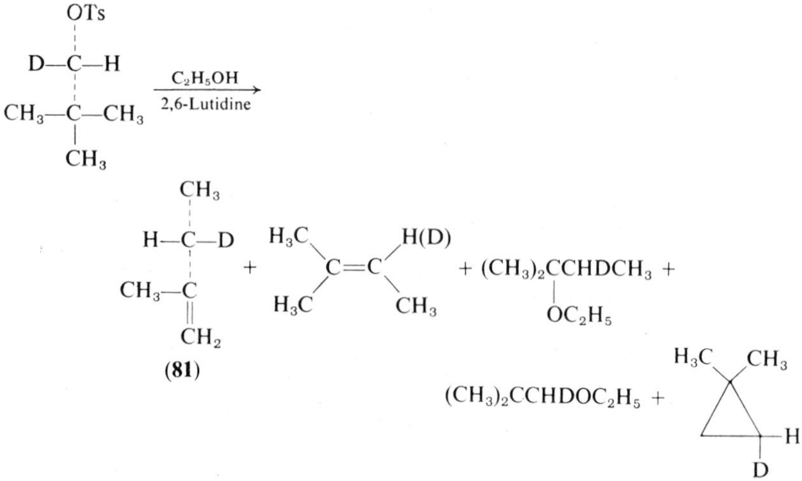

Figure 40

of products, Figure 40 (106). From the observed optical activity of the olefin (81) compared to the maximum reported $[\alpha]_D$ of $+1.01°$ (p. 78) the ethanolysis reaction was found to proceed with $91 \pm 7\%$ inversion at the migration terminus in agreement with the results for the dehydration of neopentyl-1-d alcohol (p. 78) and for the deamination of the amine (p. 96).

In an investigation of the stereochemistry of the S_E1 and S_E2 reactions using as substrate optically active *trans*-1-methyl-2-phenylcyclopropanol in deuterated media, Figure 41, the product of ring opening, optically

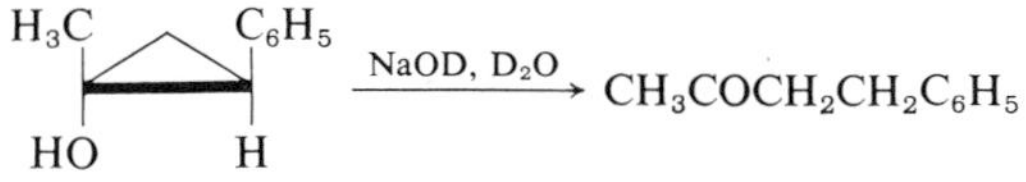

Figure 41

active 4-phenyl-2-butanone-4-d, was required (107). The compound had been reported by Streitwieser and co-workers (p. 63) as having the S-$(-)$-configuration. An alternative method of synthesis was used by DePuy and his co-workers, Figure 42, which confirmed the previous results.

The ethyl ester of (R)-$(-)$-mandelic acid was treated with $SOCl_2$ in ether-pyridine solution to yield (S)-$(+)$-ethyl α-chlorophenylacetate, $[\alpha]_D^{26}$ $+87.3°$ (c 2, EtOH). The acetate is 58% optically pure based on the value of $151°$ determined by Pasto, Cumbo, and Fraser (108). Reduction of the acetate with $LiAlH_4$ gave the chloroalcohol (82) which should also be 58% optically pure. Reduction of 82 with $LiAlD_4$ yielded (R)-2-phenylethanol-2-d, $[\alpha]_D^{25.6}$ $+1.74°$. A direct displacement reaction is indicated for this step since it has been shown (109) that neither styrene oxide nor phenylacetaldehyde is an intermediate in the reduction. Mass spectrometry and NMR results confirmed that the alcohol was completely monodeuterated at the β-carbon. The desired ketone (84) was obtained in a straightforward manner, Figure 42, with the sign in agreement with that expected from Streitwieser's work (p. 63).

As an independent check on the expected stereochemistry, the enantiomer* of the tosylate (83) was reduced with $LiAlH_4$ to give $(-)$-ethylbenzene-α-d of the expected R-configuration (p. 55).

The results of the study of the ring opening of optically active *trans*-1-methyl-2-phenylcyclopropanol indicate that in acid solution the S_E2 reaction proceeds with retention of configuration while in basic solution the cyclopropanol undergoes an S_E1 reaction with inversion of configuration (107).

* I thank Professor DePuy for communicating this correction of his original paper, ref. 107.

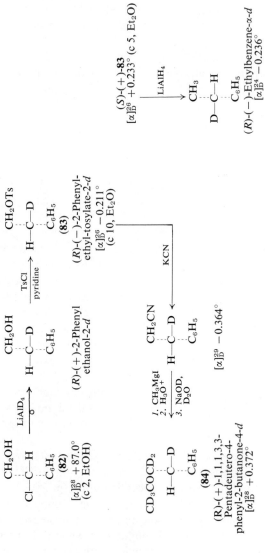

Figure 42

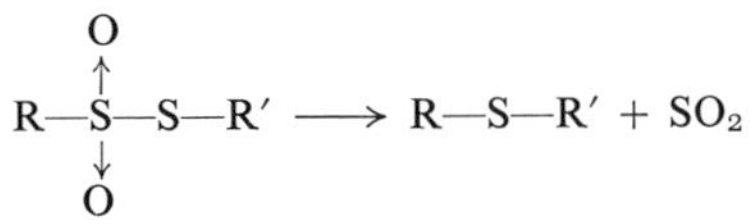

$$\text{R}\overset{\overset{\text{O}}{\uparrow}}{\underset{\underset{\text{O}}{\downarrow}}{\text{S}}}\text{—S—R}' \longrightarrow \text{R—S—R}' + \text{SO}_2$$

Figure 43

Optically active phenyl α-toluenethiolsulfonate-α-d, $C_6H_5SSO_2$-$CHDC_6H_5$, was synthesized by Kice, Engebrecht, and Pawlowski (111) in a stereochemical investigation of the thermal decomposition of thiolsulfonates, Figure 43.

The synthesis started from (R)-($-$)-benzyl-α-d alcohol, $[\alpha]_D^{25}$ $-0.694°$, 44% optically pure (p. 80), Figure 44. Treatment of the alcohol with phosgene gave the chlorocarbonate which was not isolated but was decomposed thermally in dioxane to yield (R)-($-$)-benzyl-α-d chloride (**85**). In addition to Streitwieser's serendipitous preparation of ($+$)-benzyl-α-d chloride and his assignment of its configuration as S (p. 94), there is indirect but confirmatory evidence that the *levorotatory* benzyl-α-d chloride (**85**) is R. Wiberg and Shryne (112) have demonstrated that the corresponding decomposition of α-phenylethyl chlorocarbonate proceeds with a high degree of *retention* so that assignment of the R-configuration to ($-$)-benzyl-α-d chloride is consistent with the above results.

The displacement reaction with sulfite, **85** → **86**, is expected to proceed with inversion of configuration and the next three steps, leading to the desired thiolsulfonate (**89**), should not alter the configuration at the chiral center. As a check on the stereochemistry of the steps **86** → **89**, a portion of the α-toluenesulfinate-α-d (**88**) was converted to benzyl-α-d ethyl sulfone (**91**), $[\alpha]_D^{25}$ $-0.19°$ (c 20, dioxane, 0.69 $\pm$ 0.02 D/molecule). A sample of the same sulfone was also synthesized from the (R)-($-$)-chloride (**85**) by reaction with EtS$^-$ (inversion of configuration) followed by permanganate oxidation of the resulting sulfide to the sulfone (**90**), $[\alpha]_D^{26}$ $-0.26°$ (c 16, dioxane, 0.95 $\pm$ 0.05 D/molecule). The 27% lower optical activity of the sulfone prepared from the α-toluenesulfinate-α-d (**88**) was found to be due to loss of deuterium. The authors (111) suggest that deuterium exchange occurs in the α-toluenesulfinate-α-d anion before the reaction solution is neutralized, **87** → **88**. Note that **91** possesses 73% of the rotation of **90** and 72% of the deuterium content. The observation that the amount of racemization is essentially equal to the amount of deuterium exchange indicates that the exchange reaction proceeds with retention of configuration, in agreement with the results of Cram (103) in other sulfur systems.

Hence, the reaction sequence, Figure 44, of presumed known stereochemistry leads to the assignment of the S-configuration to ($+$)-phenyl

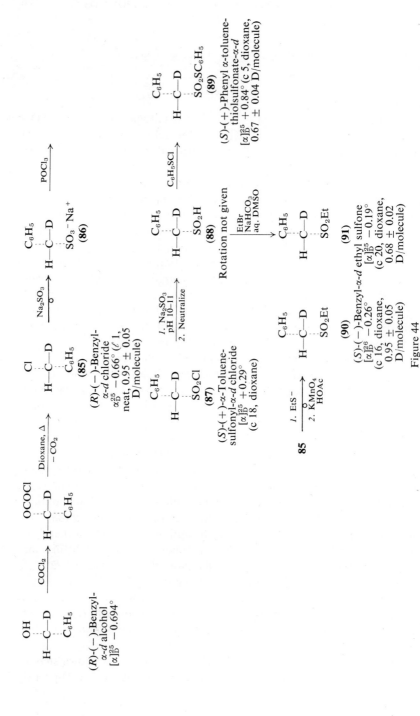

Figure 44

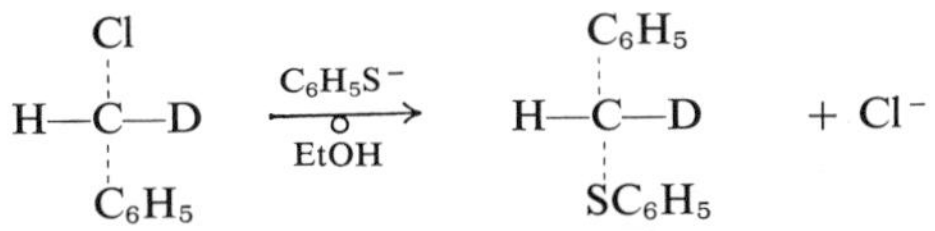

$$\underset{\underset{\text{(85)}}{\overset{\displaystyle Cl}{\underset{\displaystyle C_6H_5}{H-C-D}}}}{} \xrightarrow[\text{EtOH}]{C_6H_5S^-} \underset{\underset{\text{(92)}}{\overset{\displaystyle C_6H_5}{\underset{\displaystyle SC_6H_5}{H-C-D}}}}{} + Cl^-$$

(R)-$(-)$-Benzyl- (S)-$(+)$-Benzyl-α-d
α-d chloride   phenyl sulfide

Figure 45

α-toluenethiolsulfonate-α-d (**89**). Based on the 44% optical purity of the starting benzyl-α-d alcohol, the 27% racemization occurring in the steps **87** → **88**, and correcting for 1 D/molecule, allows the calculation of a minimum specific rotation of $+3.9°$ for optically pure (R)-phenyl α-toluene-thiolsulfonate-α-d.

Benzyl-α-d phenyl sulfide (**92**), $[\alpha]_D^{25}$ $+0.384°$ (c 15, dioxane, 0.88 ± 0.06 D/molecule), the expected product of the thermal decomposition of **89**, Figure 43, was prepared from (R)-$(-)$-benzyl-α-d chloride (**85**) by reaction with sodium thiophenolate in ethanol solution, Figure 45. Since this reaction is assumed to involve inversion of configuration, the S-configuration may be assigned to the $(+)$-sulfide.

Decomposition of **89** in several solvents according to Figure 43 gave sulfide **92** which was extensively racemized. Control experiments established that **92** is optically stable under the conditions of the thiolsulfonate decomposition and that **89** did not lose deuterium or undergo racemization prior to its decomposition. Hence, the results of Kice and co-workers (111) indicate that in contrast to other $S_N i$ reactions, the reaction studied by them is characterized by an almost complete lack of stereoselectivity.

In studies of the thermal racemization of benzyl p-tolyl sulfoxide (**93**), and the thermal rearrangement of benzyl p-toluenesulfenate (**94**), Mislow

$$\overset{\displaystyle O}{\overset{\displaystyle \uparrow}{C_6H_5CH_2SC_6H_4\text{-}p\text{-}CH_3}}$$
(**93**)

$C_6H_5CH_2OSC_6H_4$-p-CH_3
(**94**)

and co-workers (113) prepared the corresponding optically active benzyl -α-d analogs. The synthesis of the sulfoxide was patterned after Kice's procedure above (Fig. 44) for the preparation of optically active benzyl-α-d phenyl sulfide. Oxidation of (R)-$(-)$-benzyl-α-d p-tolyl sulfide (**95**) (obtained from enzymatically synthesized (S)-$(+)$-benzyl-α-d alcohol) with sodium metaperiodate, Figure 46, afforded (R)-benzyl-α-d p-tolyl sulfoxide (**96**), $[\alpha]_D^{27}$ $+1.97 ± 0.08°$ (c 1.32, EtOH, 0.92 ± 0.01 D/molecule) which proved to be an equimolar mixture of enantiomers *at sulfur* as demonstrated by the observation that exchange α-deuterium for hydrogen yielded racemic benzyl p-tolyl sulfoxide.

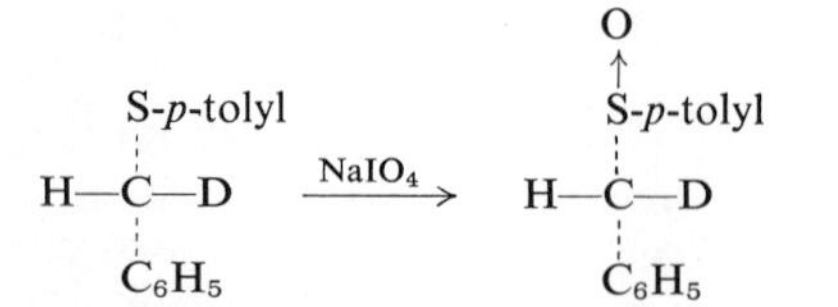

$$
\begin{array}{ccc}
\text{S-}p\text{-tolyl} & & \overset{O}{\underset{\uparrow}{\text{S-}p\text{-tolyl}}} \\
| & & | \\
\text{H---C---D} & \xrightarrow{\text{NaIO}_4} & \text{H---C---D} \\
| & & | \\
\text{C}_6\text{H}_5 & & \text{C}_6\text{H}_5 \\
(\textbf{95}) & & (\textbf{96})
\end{array}
$$

(R)-(−)-Benzyl-   (R-)(+)-Benzyl-α-d
α-d p-tolyl   p-tolyl sulfoxide
sulfide

Figure 46

The *enantiomer* of **96** was synthesized starting from (R)-(−)-benzyl-α-d alcohol of 56% optical purity. Conversion of the alcohol to the chloride via the chlorocarbonate decomposition (p. 99) gave (R)-(−)-benzyl-α-d chloride, $[\alpha]_D^{22}$ −0.70 ± 0.04° (neat), which was reacted with potassium p-toluenethiolate in ethanol solution to afford (S)-(+)-benzyl-α-d p-tolyl sulfide, $[\alpha]_D^{27}$ +0.92 ± 0.09° (c 1.2, EtOH). Oxidation with sodium metaperiodate gave the sulfoxide having $[\alpha]_D^{27}$ −1.22 ± 0.08° (c 1.3, EtOH).

(R)-(−)-Benzyl-α-d p-toluenesulfenate was prepared from the above (R)-(−)-benzyl-α-d alcohol as shown in Figure 47.

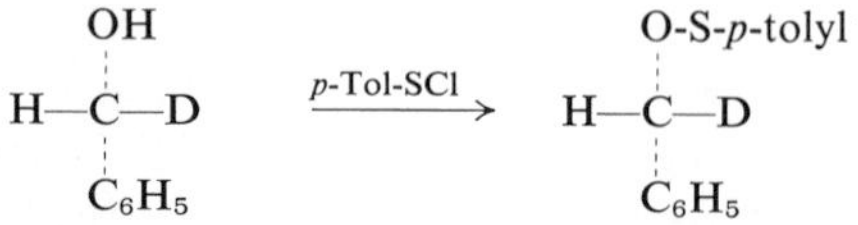

$$
\begin{array}{ccc}
\text{OH} & & \text{O-S-}p\text{-tolyl} \\
| & & | \\
\text{H---C---D} & \xrightarrow{p\text{-Tol-SCl}} & \text{H---C---D} \\
| & & | \\
\text{C}_6\text{H}_5 & & \text{C}_6\text{H}_5
\end{array}
$$

(R)-(−)-Benzyl-α-d   (R)-(−)-Benzyl-α-d
alcohol   p-toluenesulfenate
$[\alpha]_D^{27}$ −0.89 ± $[\alpha]_D^{26}$ −0.3 ± 0.1°
0.01° (neat)   (c 8.2, EtOH)

Figure 47

King and Smith have recently reported (114) that alkanesulfonyl halides may be smoothly converted to the corresponding alkyl halides by treatment with halide ion as exemplified in Figure 48.

$$
\text{C}_6\text{H}_5\text{CH}_2\text{SO}_2\text{Br} \xrightarrow[\text{CH}_2\text{Cl}_2]{\text{Br}^-} \text{C}_6\text{H}_5\text{CH}_2\text{Br} + \text{SO}_2
$$

Figure 48

Several lines of evidence, among them kinetic results (including ^{13}C kinetic isotope effect studies), indicated the reaction to be of the S_N2 type. Stereochemical evidence for the S_N2 mechanism was obtained by the use of optically active phenylmethanesulfonyl-1-d bromide (**97**), whose synthesis and subsequent conversion to the corresponding bromide is shown in

Figure 49. Starting from enzymatically obtained (S)-$(+)$-benzyl-α-d alcohol (p. 80), $[\alpha]_{400}$ $+2.1°$ (c 9, CH_2Cl_2), $(+)$ phenylmethanesulfonyl-1-d bromide (97), was obtained having $[\alpha]_{400}$ $+6.2°$, 0.935 D/molecule (115). From the method of synthesis the absolute configuration of dextrorotatory 97 is assigned as S. Although starting benzyl-α-d alcohol was optically pure, resultant (S)-$(+)$-97 is probably only 60–75%, optically pure, since the PBr_3 reaction gives about 20% racemization (p. 58) and the ensuing sulfite displacement of the benzyl-α-d bromide also involves some degree of racemization, presumably by attack of liberated bromide ion on the starting material (115).

Treatment of 97 with tetraethylammonium bromide gave (R)-$(—)$-benzyl-α-d bromide, the stereochemical product expected if the attack by bromide ion proceeds with inversion of configuration.

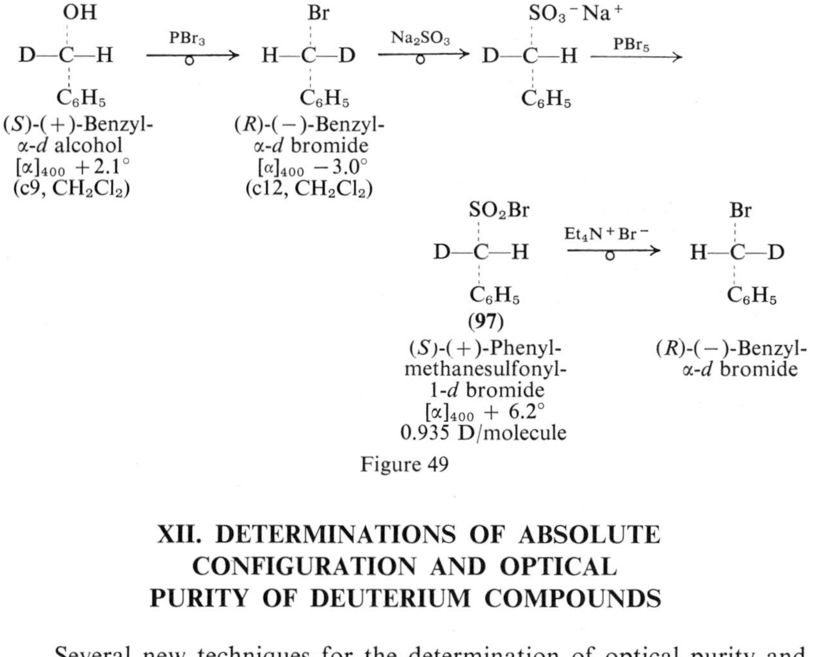

OH	Br	$SO_3{}^- Na^+$	
D—C—H	H—C—D	D—C—H	
C_6H_5	C_6H_5	C_6H_5	
(S)-$(+)$-Benzyl- α-d alcohol $[\alpha]_{400}$ $+2.1°$ (c9, CH_2Cl_2)	(R)-$(-)$-Benzyl- α-d bromide $[\alpha]_{400}$ $-3.0°$ (c12, CH_2Cl_2)		

$\xrightarrow[\hphantom{o}]{PBr_3}$ $\xrightarrow[\hphantom{o}]{Na_2SO_3}$ $\xrightarrow{PBr_5}$

SO_2Br

D—C—H $\xrightarrow[\hphantom{o}]{Et_4N^+Br^-}$ H—C—D

C_6H_5 C_6H_5

(97)

(S)-$(+)$-Phenyl-methanesulfonyl-1-d bromide $[\alpha]_{400}$ $+ 6.2°$ 0.935 D/molecule

(R)-$(-)$-Benzyl-α-d bromide

Figure 49

XII. DETERMINATIONS OF ABSOLUTE CONFIGURATION AND OPTICAL PURITY OF DEUTERIUM COMPOUNDS

Several new techniques for the determination of optical purity and configuration have been developed within the past few years and applied to the case of optically active deuterium compounds. A comprehensive review of techniques for the determination of optical purity has recently appeared (72).

Horeau (116) has developed a method for determining the configuration of secondary alcohols based on their preferential rate of esterification with α-phenylbutyric anhydride in pyridine solution. A kinetic resolution occurs since the transition states for esterification of a partially resolved alcohol with racemic α-phenylbutyric anhydride are diastereomeric; hence the activation energies are unequal and one diastereomer will be formed faster.

The esterification is allowed to proceed partway and the unreacted α-phenylbutyric acid is isolated. An empirical rule has been devised which states that if the unreacted acid is levorotatory, the alcohol has the configuration **98**. The magnitude of rotation of the recovered acid is related to the optical purity of the alcohol in question.

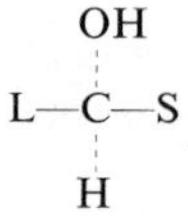

$$
\begin{array}{c}
\text{OH} \\
| \\
\text{L—C—S} \quad \longrightarrow \quad (-)\text{-α-Phenylbutyric acid} \\
| \\
\text{H} \\
\textbf{(98)}
\end{array}
$$

L = larger group
S = smaller group

The partial resolution method was extended to 1-deuteroalcohols when it was shown that the steric size of D is less than that of H (117). The empirical rule for 1-deuteroalcohols states that if the isolated α-phenylbutyric acid is levorotatory, the deuterated alcohol possesses the R-configuration. Application of the method to some optically active primary alcohols gave the results shown in Table V.

The partial asymmetric alcoholysis (116) was applied (118) to (S)-(+)-2-propanol-1-d_3 (p. 90). The isolated α-phenylbutyric acid was dextrorotatory, indicating the operation of a steric isotope effect, i.e., a

TABLE V
Determination of Configuration and Optical Purity by
Horeau's Method (117)

Alcohol	D/molecule	$[\alpha]_D$	Configuration	Optical purity, %
Neopentyl-1-d alcohol	1.0	No observable rotation; acid phthalate −1.12°	S	100
Benzyl-α-d alcohol	0.95	+1.43° (24°)	S	100
Benzyl-α-d- alcohol		−0.234° (30°)	R	14 ± 4
1-Butanol-1-d		−0.185° (27.5°)	R	35 ± 4

secondary deuterium isotope effect due to the difference between CH_3 and CD_3. Isolation of the *dextrorotatory* acid indicates that the alcohol must have the structure **99**, i.e., enantiomeric to that depicted on p. 104, cor-

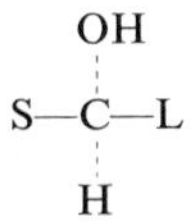

$$OH$$
$$S—C—L$$
$$H$$

(99)

responding to isolation of $(-)$-α-phenylbutyric acid. The known *S*-configuration of $(+)$-2-propanol-1-d_3 will correspond to this geometry only if the larger group is taken as CH_3 and the smaller one as CD_3. This assignment agrees with an earlier conclusion that CD_3 is smaller than CH_3 in effective steric size (119).

The NMR method of Raban and Mislow (70) for the determination of optical purity involves the conversion of the partially resolved mixture A and A′ into a mixture of diastereomers AB and A′B by reaction with an enantiomerically pure reagent B. The ratio of diastereomers AB : A′B, which must equal the ratio A : A′, is determined from comparison of appropriate NMR peaks due to both stereoisomers (compare the similar method of Gerlach below).

Using enantiomerically pure *O*-methylmandelyl chloride is the diastereomer-forming reagent. Raban and Mislow (71) have determined the optical purity of (S)-$(+)$-2-propanol-1-d_3 (p. 90) and (S)-neopentyl-1-d alcohol (p. 77) to each be 99 $\pm$ 1% in agreement with the results of Horeau discussed above.

Gerlach has independently developed an NMR method for the determination of optical purity and applied it to the case of benzylamine-α-d, (p. 82) (120).

Starting with $(+)$-benzylamine-α-d, $[\alpha]_D^{25}$ $+1.10°$, $[\alpha]_{365}$ $+4.83°$ (121) (corrected for 3.8% nondeuterated material and assuming a density of 1.00), prepared by the method of Streitwieser and Wolfe (76), $(4R)$-$(+)$-3-(benzyl-α-d)4-phenyloxazolidine-2-thione (**100**), was synthesized as shown in Figure 50. The compound was shown to be optically pure at C-4 by the isotope dilution technique.

NMR examination of **100**, the $(4S)$-$(-)$-isomer, and the $(4R)$-*nondeu-terated* analog of **100** showed the benzylic protons to be nonequivalent and to give an AB quartet with the two doublets centered near 3.8 and 5.5 ppm (60 MHz) and cleanly separated from other signals. Considerations of deshielding by the thiocarbonyl group and the most probable conformation of the benzyl group in **100**, Figure 50, allowed the tentative assignment of

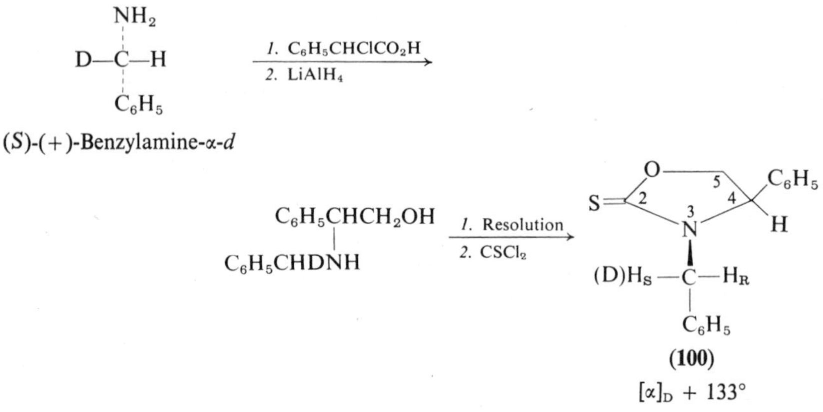

$$\begin{array}{c} NH_2 \\ | \\ D-C-H \\ | \\ C_6H_5 \end{array} \quad \xrightarrow[\text{2. LiAlH}_4]{\text{1. C}_6\text{H}_5\text{CHClCO}_2\text{H}}$$

(S)-(+)-Benzylamine-α-*d*

$$\begin{array}{c} C_6H_5CHCH_2OH \\ | \\ C_6H_5CHDNH \end{array} \quad \xrightarrow[\text{2. CSCl}_2]{\text{1. Resolution}}$$

(100)

$[\alpha]_D + 133°$

Figure 50

the low field resonance to the H_s benzylic proton which was confirmed in the deuterated analog since the absolute configuration of (+)-benzylamine-α-*d* is known to be *S* (p. 82).

The relative NMR intensities of the benzylic protons of the nondeuterated compound were 1:1 while for **100** the ratio was 81:19 (corrected for nondeuterated material) which corresponds to an optical purity of 62.5 ± 2% for the starting benzylamine-α-*d* of $[\alpha]_D + 1.10°$. By this method, then, optically pure benzylamine-α-*d* is calculated to have a specific rotation of 1.77°.

Some derivatives of (+)-benzylamine-α-*d* are shown in Table VI (121).

A new technique for establishing the absolute configuration of a compound whose chirality is due to hydrogen–deuterium dissymmetry has been used to determine the configuration of glycolic-*d* acid formed in the enzymatic reduction of glyoxylic-*d* acid, Figure 51 (122).

The method makes use of the markedly different neutron-scattering amplitudes of H and D and the anomalous neutron-scattering amplitude

TABLE VI

Derivatives of (S)-(+)-Benzylamine-α-*d*, 62.5 ± 2%
Optically Pure (121)

Derivative	$[\alpha]_{546}$
Benzyloxycarbonyl, $C_6H_5CHDNCCO_2CH_2C_6H_5$	+0.22° (c 30, C_6H_6)
Phthalimide, $C_6H_5CHDNCOCOC_6H_4$	−0.26° (c 15, C_6H_6)
	−0.10° (c 21, dioxane)
Thiobenzoyl, $C_6H_5CHDNHCSC_6H_5$	+0.1° (c 10, EtOH)

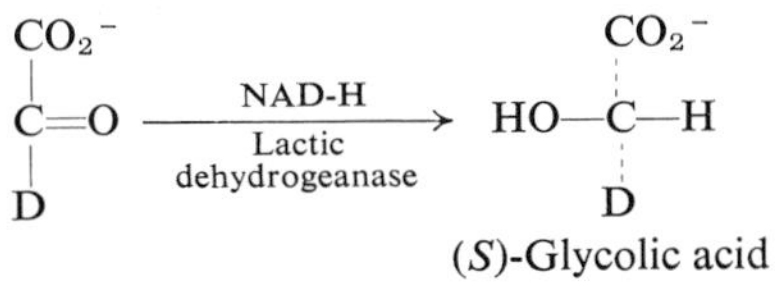

$$\begin{array}{ccc}
CO_2^- & & CO_2^- \\
| & \xrightarrow[\text{dehydrogeanase}]{\text{NAD-H}} & | \\
C{=}O & \xrightarrow{\text{Lactic}} & HO{-}C{-}H \\
| & & | \\
D & & D
\end{array}$$

(S)-Glycolic acid

Figure 51

of ^{6}Li in the lithium glycolate crystal. By this means, the glycolic-d acid was shown to be S, in agreement with the earlier prediction based on the stereospecificity of the enzyme for (S)-lactic acid (123).

XIII. OPTICAL ROTATORY DISPERSION STUDIES OF DEUTERIUM COMPOUNDS

The shape of an ORD curve, particularly in the region of an absorption band (the Cotton effect) may be related to the absolute configuration of a molecule. Enantiomeric compounds will exhibit ORD curves which are mirror images of each other.

Streitwieser, Verbit, and Andreades in 1965 (124) reported ORD measurements of a series of deuterium compounds, Table VII. The

TABLE VII

Optical Rotatory Dispersion Data for Some Deuterium Compounds[a]

	Observed rotation, α, in degrees[b] λ, nm			
Compound	589	500	400	300
(S)-(+)-1-Propanol-1-d[c]	0.01	0.015	0.026	0.059
(S)-(−)-1-Propyl-1-d xanthate[c]	−0.073	−0.114	—	—
(R)-(−)-1-Butanol-1-d	−0.016	−0.023	−0.042	−0.094
(R)-(+)-1-Butyl-1-d acetate[d]	0.071	0.104	0.180	—
(S)-(+)-Benzyl-α-d-alcohol	0.025	0.038	0.074	0.196[e]
(S)-(+)-Ethyl-benzene-α-d	0.046	0.070	0.137	0.435
(S)-(+)-Benzylamine-α-d	0.007	0.010	0.021	0.06[e]
(R)-(−)-n-Butylbenzene-α-d	−0.10	−0.18	−0.366	—

[a] Ref. 124.

[b] Cary Model 60 Spectropolarimeter, 0.1 dm cell, 26°, neat, except where noted.

[c] Prepared by reduction of propionaldehyde with isobornyloxy-2-d-magnesium bromide; configuration assigned by analogy with the corresponding reduction of butyraldehyde (p. 62), A. Streitwieser, Jr. and V. Sarich, unpublished results quoted in ref. 124.

[d] A Rudolph photoelectric polarimeter was used, 2 dm. cell.

[e] At 310 nm.

compounds were synthesized chemically and were not optically pure. The small magnitude of the rotations and the relatively large molar absorptivities of the compounds precluded measurements in the region of electronic absorption. Nevertheless, the work confirmed that deuterium behaves as an ordinary substituent as far as dispersion of the optical rotation is concerned.

An attempt to assess the rotational contribution of a deuterium atom at a chiral CHD center was made by Djerassi and Tursch (125) who prepared chiral cyclopentanone-3-d (101), whose optical purity was probably in the range of 50–80%. ORD measurements in isooctane solution gave no measurable rotation at 322 nm under conditions where $[\alpha]_D$ 42° would have been detected. The authors conclude that since (+)-3-methyl-cyclopentanone (102) exhibits a specific rotation of over 4000° at 322 nm, the rotational contribution of a deuterium is at most 1/100 that of a methyl group.

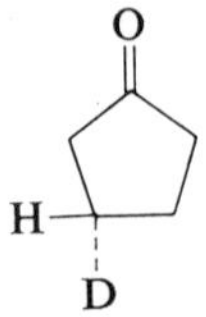

(101) 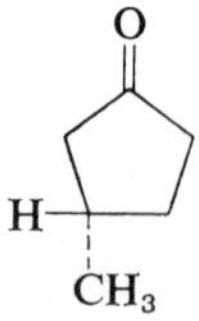   (102)

Since the optical purity of some deuterium compounds are now known, it is interesting to compare the difference in rotational contribution of D and CH$_3$ in some open chain systems, Figure 52. For the examples in the benzyl system cited in Figure 52, the deuterated compounds have a rotation approximately 5% as great as the corresponding methyl analogs.

The deuterated patulin (103), was reported in a short communication (126) to have $[\phi]_{320}$ +8°, $[\phi]_{300}$ +16°* whereas naturally occurring (nondeuterated) patalin is optically inactive.

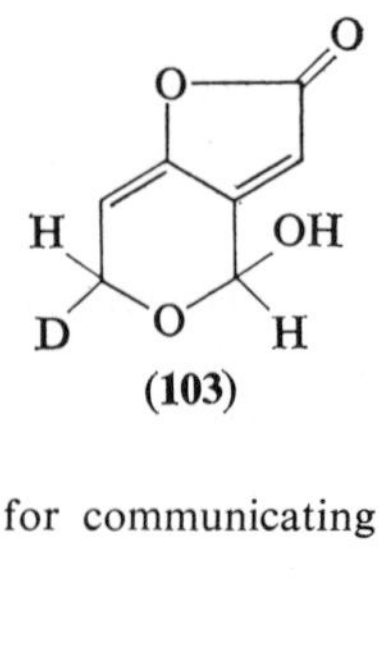

(103)

* I thank Professor Scott for communicating this correction of his original paper, ref. 126.

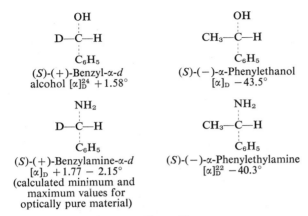

(S)-$(+)$-Benzyl-α-d
alcohol $[\alpha]_D^{24}$ $+1.58°$

(S)-$(-)$-α-Phenylethanol
$[\alpha]_D$ $-43.5°$

(S)-$(+)$-Benzylamine-α-d
$[\alpha]_D$ $+1.77 - 2.15°$
(calculated minimum and
maximum values for
optically pure material)

(S)-$(-)$-α-Phenylethylamine
$[\alpha]_D^{22}$ $-40.3°$

Figure 52

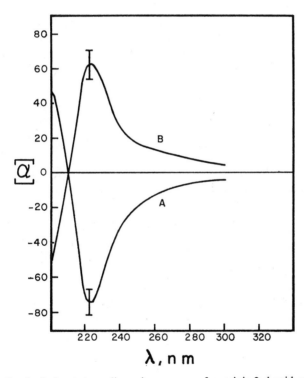

Fig. 53. Optical rotatory dispersion curves of succinic-2-d acid obtained by reduction of the product of maleate hydration in D_2O (curve A) and (S)-succinic-2-d acid (curve B). Rotations are corrected to correspond to succinic acid containing 1 D/molecule. Reproduced with permission from ref. 128.

The first published observation of a Cotton effect in a molecule which is optically active because of hydrogen–deuterium dissymmetry was reported for (R)-(+)-1-butyl-1-d acetate (127) which exhibited a positive Cotton effect with the first extremum at 226 nm.

Shortly thereafter, a Cotton effect in the carboxyl absorption region of enzymatically prepared succinic-2-d acid was measured, Figure 53 (128). The plain portion of the ORD curve had previously been reported in the literature (129–132).

XIV. FUTURE DEVELOPMENTS

The unique properties of chiral deuterium compounds make it clear that increasing use will be made of them in investigations of stereochemistry and reaction mechanisms.

With the rapidly accelerating pace of research in the field of inorganic chemistry, one soon expects reports to be forthcoming of inorganic compounds such as transition metal complexes which are optically active because of H–D dissymmetry.

The first report has appeared of a sulfur atom which is optically active because of isotopic substitution not at the chiral center (p. 92). It seems clear that we shall see further examples of this kind as well as examples of chiral H–X–D centers where X is an element such as B, N, P, Si, Ge, Al, Sn, etc.

With improved instrumentation becoming available, more use will be made of the techniques of optical rotatory dispersion and circular dichroism to establish correlations for the rapid determination of absolute configurations of deuterium compounds. Of particular interest will be ORD and CD investigations in the *infrared region* of the spectrum where the mass difference between hydrogen and deuterium might be expected to give rise to discrete Cotton effects.

Last, but probably most important, one looks forward to advances in theory which will allow the *a priori* calculation of rotatory dispersion curves based on assumed molecular geometries for a given compound.

TABLE VIII

Chiral Deuterium Compounds Discussed in the Present Review

Compound	Section numbers in this chapter	Highest reported rotation	Reference
Hydrocarbons			
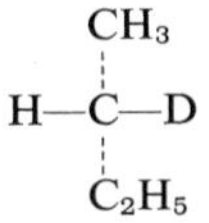 CH$_3$ │ H—C—D │ C$_2$H$_5$	IV, V	$[\alpha]_D^{25}$ −0.60°,[a] neat (1.1° calculated)	21,25,30
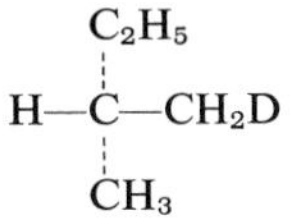 C$_2$H$_5$ │ H—C—CH$_2$D │ CH$_3$	II	No detectable rotation	11
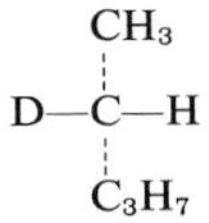 CH$_3$ │ D—C—H │ C$_3$H$_7$	VI	$[\alpha]_D$ +0.19 ± 0.06°,[a] (0.52 ± 0.07° calculated for optically pure material)	58,61
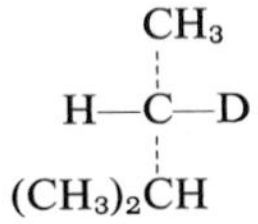 CH$_3$ │ H—C—D │ (CH$_3$)$_2$CH	VII	$[\alpha]_D^{17}$ −0.85 ± 0.03° [a]	65
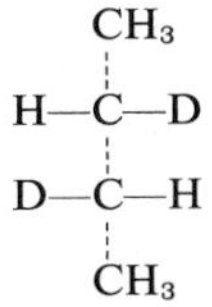 CH$_3$ │ H—C—D │ D—C—H │ CH$_3$	V	$[\alpha]_D^{25}$ −1.01°	30
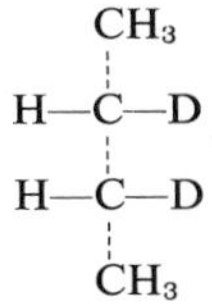 CH$_3$ │ H—C—D │ H—C—D │ CH$_3$	V	See text	30
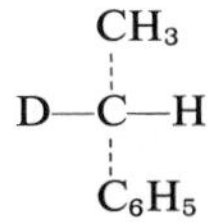 CH$_3$ │ D—C—H │ C$_6$H$_5$	III, IV, VI, XI, XIII	$[\alpha]_D$ −0.63° (calculated value)	16,21,104, 124
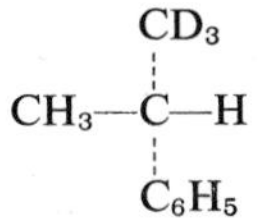 CD$_3$ │ CH$_3$—C—H │ C$_6$H$_5$	X	α_D^{22} +0.48 ± 0.02° (*l* 1, neat)	93

[a] Enantiomer also prepared.

Compound	Section numbers in this chapter	Highest reported rotation	Reference
C$_6$H$_5$$\overset{*}{C}$H(CH$_3$)CH$_2$D	II	No detectable rotation	10
D—C—H with C$_3$H$_7$ and C$_6$H$_5$	VI, XIII	$[\alpha]_D$ $-0.78 \pm 0.04°$ [a] (2.2° calculated for optically pure material)	40,46,124
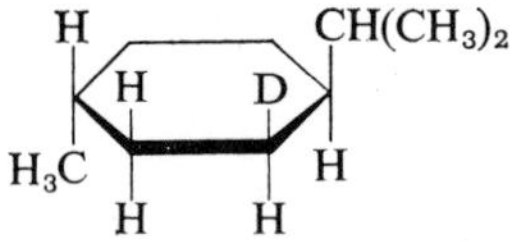	III	$[\alpha]_D$ $-0.09 \pm 0.02°$ (neat)	12
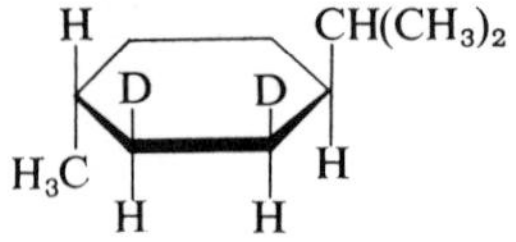	III	$[\alpha]_D^{25}$ $-0.09 \pm 0.01°$ (neat)	9

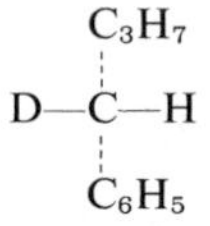

Alcohols

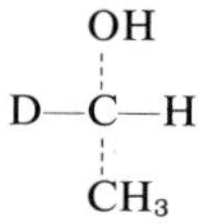

Compound	Section numbers in this chapter	Highest reported rotation	Reference
OH D—C—H CH$_3$	VI, VII, IX	$[\alpha]_D^{28}$ $-0.28 \pm 0.03°$ [a]	28,55,56,58
OH D—C—H C$_2$H$_5$	XIII	α_D^{26} $+0.01°$ (l 0.1, neat)	124
CD$_2$OH H—C—D CH$_3$	IX	Rotation not reported	83
OH CD$_3$—C—CH$_3$ H	X, XII	α_D^{25} $+0.27°$ (l 1, neat)	71,92,118
OH D—C—H C$_3$H$_7$	VI, XII, XIII	$[\alpha]_D^{27.5}$ $+0.47 \pm 0.005°$ [a]	31,33,46,47, 61,124,127

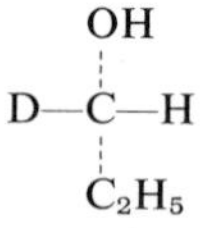

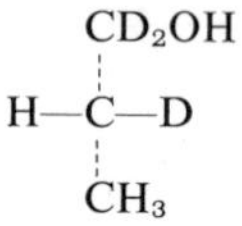

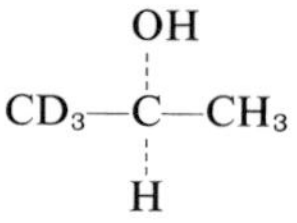

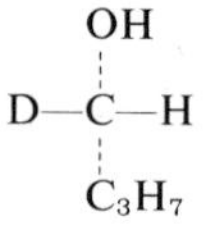

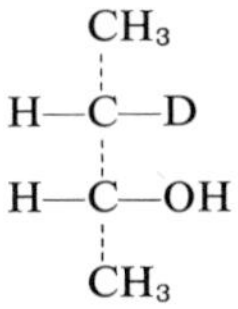

CH₃
|
H—C—D
|
H—C—OH
|
CH₃

$$CH_3$$
$$H-\overset{|}{\underset{|}{C}}-D$$
$$H-\overset{|}{\underset{|}{C}}-OH$$
$$CH_3$$

V, VI $[\alpha]_D^{25} -13.59°$ 25,28,30,82

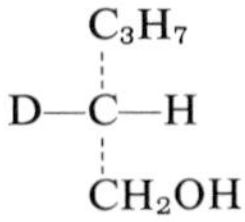

$$C_3H_7$$
$$D-\overset{|}{\underset{|}{C}}-H$$
$$CH_2OH$$

VI $[\alpha]_D -0.60 \pm 0.007°$ 61

$$OH$$
$$D-\overset{|}{\underset{|}{C}}-H$$
$$(CH_3)_2CH$$

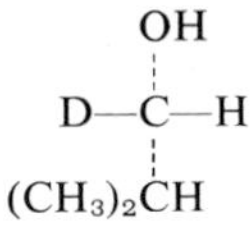

VII $\alpha_D^{25} +0.49°$ (neat, l 1, 0.8 D/molecule)   110

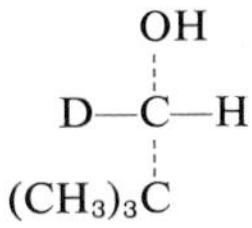

$$OH$$
$$D-\overset{|}{\underset{|}{C}}-H$$
$$(CH_3)_3C$$

VII, XII   Not observable (acid phthalate derivative $[\alpha]_D -1.14 \pm 0.04°$)   46,65,69, 71,117

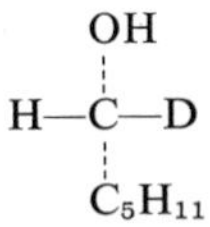

$$OH$$
$$H-\overset{|}{\underset{|}{C}}-D$$
$$C_5H_{11}$$

VI   Rotation not reported   54

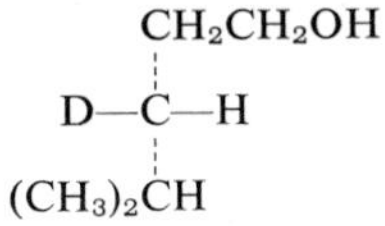

$$CH_2CH_2OH$$
$$D-\overset{|}{\underset{|}{C}}-H$$
$$(CH_3)_2CH$$

IX $[\alpha]_D^{25} -0.168°$ (c 31.6, CHCl₃)   136

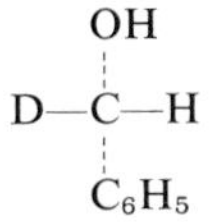

$$OH$$
$$D-\overset{|}{\underset{|}{C}}-H$$
$$C_6H_5$$

VI, VIII, XI, XII, XIII $[\alpha]_D^{25} +1.58 \pm 0.01°$ [a]   36,40,46, 111,113,114, 115,117,124

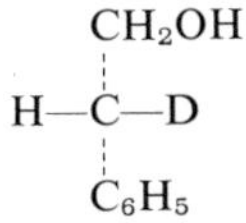

$$CH_2OH$$
$$H-\overset{|}{\underset{|}{C}}-D$$
$$C_6H_5$$

XI $[\alpha]_D^{25.6} +1.74°$ 107

$$OH$$
$$D-\overset{|}{\underset{|}{C}}-H$$
$$CH_2$$

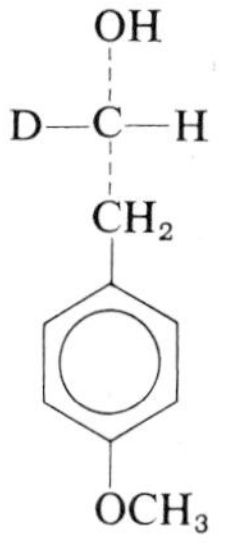

VIII $[\alpha]_D^{24} -1.44°$ (ca. 40% optically pure)   80

OCH₃

Compound	Section numbers in this chapter	Highest reported rotation	Reference
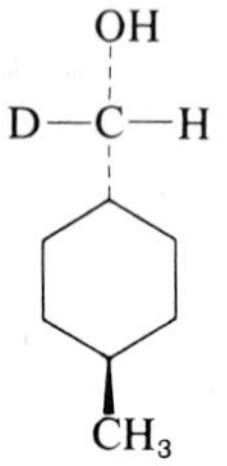 OH D—C—H (cyclohexyl ring) CH$_3$	IX	α_{546} +0.18° (l 1, neat, 0.65 D/molecule)	42
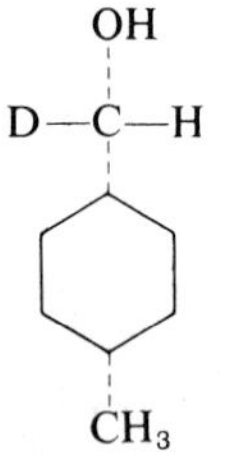 OH D—C—H (cyclohexyl ring) CH$_3$	IX	α_{546} +0.11° (l 1, neat, 0.65 D/molecule)	42
Alkyl and aryl halides			
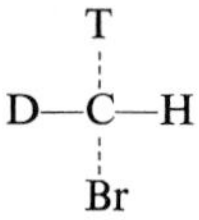 T D—C—H Br	IV	$[\alpha]_D$ −0.0003° (theoretical value)	23
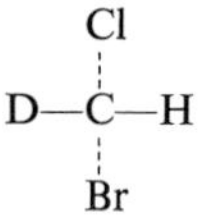 Cl D—C—H Br	IV	$[\alpha]_D$ +0.0002° (theoretical value)	23
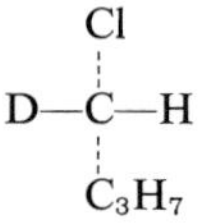 Cl D—C—H C$_3$H$_7$	VI	$[\alpha]_D$ + 0.39 ± 0.03° (ca. 8% optically pure)	40,44
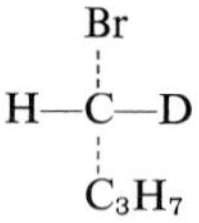 Br H—C—D C$_3$H$_7$	VI	α_D −0.16 ± 0.01° (l 2)	31
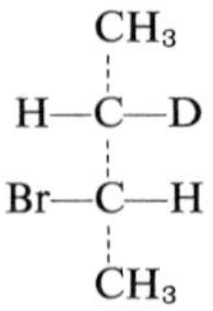 CH$_3$ H—C—D Br—C—H CH$_3$	V	$[\alpha]_D^{25}$ +29.34°	25,30

C$_3$H$_7$ D—C—H CH$_2$Br	VI	$[\alpha]_D$ −1.98 ± 0.20°	61
Cl H—C—D C$_6$H$_5$	XI	$[\alpha]_D^{22}$ − 0.70 ± 0.04° [a] (ca. 56% optically pure)	43,111,113
Br H—C—D C$_6$H$_5$	XI	$[\alpha]_D$ − 0.72° $[\alpha]_{400}$ − 3.0° (c 12, CH$_2$Cl$_2$)	114,115

Aldehydes and ketones

CHO H—C—D CH$_3$	IX	Rotation not reported	81
CH$_3$ C=O D—C—H CH$_3$	VI	Not measured	28
D OHC—C—O—CHOBu H CHO	VI	Not measured	56
CH$_3$ D—C—H CH$_3$COCH$_2$	VI	$[\alpha]_D$ +0.25 ± 0.03°	58
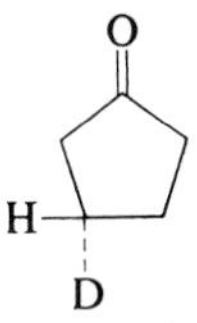 O (cyclopentanone ring, H—C—D)	XIII	No observable rotation	125
CD$_3$COCD$_2$ H—C—D C$_6$H$_5$	VI, XI	$[\alpha]_D^{28}$ + 0.372° [a]	40,107

Compound	Section numbers in this chapter	Highest reported rotation	Reference
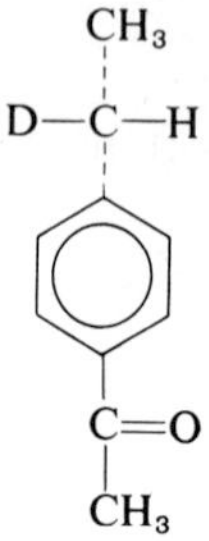 CH₃ D—C—H (benzene ring) C=O CH₃	III	$[\alpha]_D^{23}$ −0.27°	16

$$\begin{array}{c} CH_3 \\ | \\ D-C-H \\ \end{array}$$

Olefins

$\begin{array}{c} CH_3 \\ | \\ H-C-D \\ CH_3-C \\ \| \\ CH_2 \end{array}$	VII, XI	$[\alpha]_D^{17}$ +1.01 ± 0.01° [a]	69,105,106

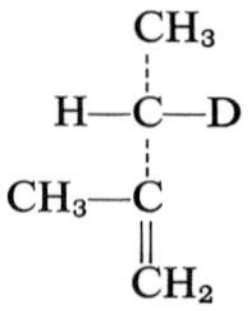

Carboxylic acids and esters

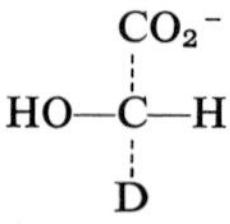

$\begin{array}{c} CO_2^- \\ | \\ HO-C-H \\ | \\ D \end{array}$	XII	Not measured	122
$\begin{array}{c} OCOCH_3 \\ | \\ D-C-H \\ | \\ CH_3 \end{array}$	VI	Not measured	28
$\begin{array}{c} CO_2H \\ | \\ H-C-D \\ | \\ CH_3 \end{array}$	IX	$[\alpha]_{400}$ +2° [a]	81–83
$\begin{array}{c} CO_2H \\ | \\ D-C-H \\ | \\ CH_2CO_2H \end{array}$	XIII	$[\alpha]_{223}^{27}$ +60 ± 10° [a] (D_2O)	128–132
$\begin{array}{c} OCOCH_3 \\ | \\ H-C-D \\ | \\ C_3H_7 \end{array}$	XIII	$[\alpha]_D^{27}$ + 0.687° [a]	100,124,127

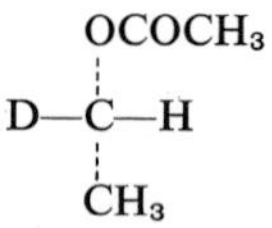

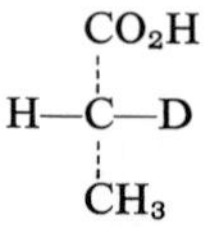

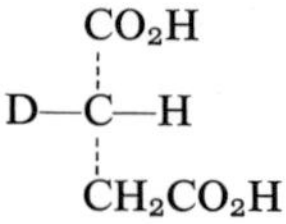

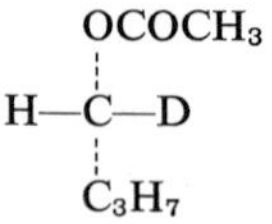

$(CH_3)_2CCHDOC_2H_5$	XI	Rotation not reported	106
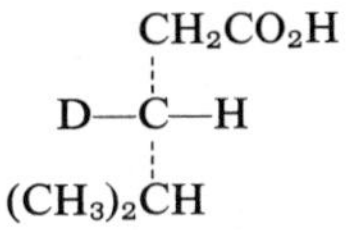 CH_2CO_2H D—C—H $(CH_3)_2CH$	IX	$[\alpha]_D^{25}$ $-0.453°$ (c 30, $CHCl_3$)	136
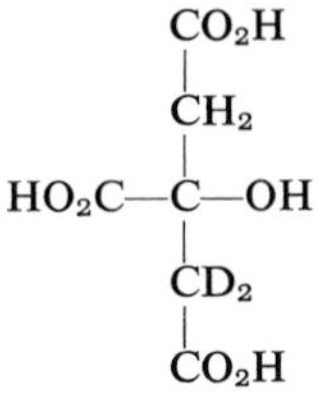 CO_2H CH_2 HO_2C—C—OH CD_2 CO_2H	X	$[\alpha]_{546}^{20}$ $-1.03 \pm 0.01°$ (c 12, water)	89,90
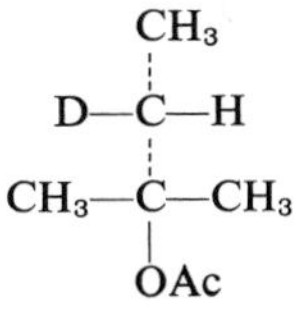 CH_3 D—C—H CH_3—C—CH_3 OAc	XI	Rotation not reported	105
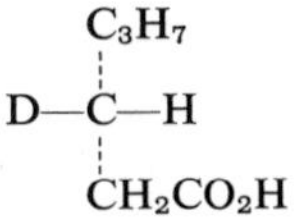 C_3H_7 D—C—H CH_2CO_2H	VI	$[\alpha]_D$ $-0.23 \pm 0.04°$	61
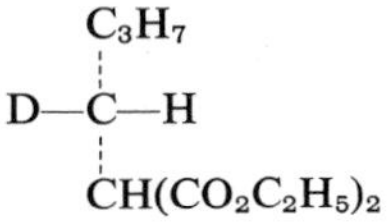 C_3H_7 D—C—H $CH(CO_2C_2H_5)_2$	VI	No observable rotation	61
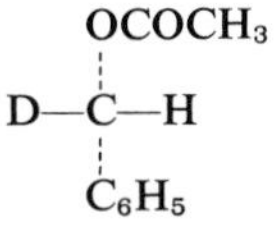 $OCOCH_3$ D—C—H C_6H_5	VIII	$[\alpha]_D$ $+0.172 \pm$ $0.004°$	40,76,101
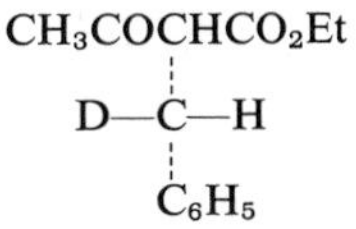 $CH_3COCHCO_2Et$ D—C—H C_6H_5	VI	$[\alpha]_D$ $-0.39 \pm 0.01°$	40
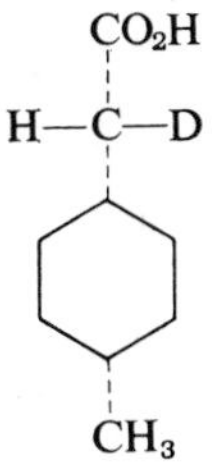 CO_2H H—C—D CH_3	IX	$[\alpha]_{546}$ $+0.65°$ (c 48, C_6H_6)	42

Compound	Section numbers in this chapter	Highest reported rotation	Reference
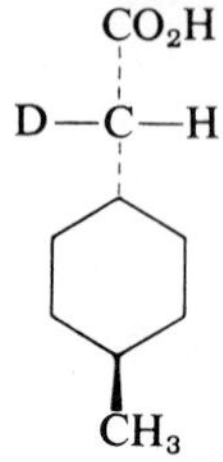 CO_2H $D-C-H$ (cyclohexyl ring) CH_3	IX	$[\alpha]_{546} +0.44°$ (c 26, C_6H_6)	42
Amines			
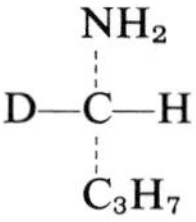 NH_2 $D-C-H$ C_3H_7	VIII	$[\alpha]_D^{25} -0.009 \pm 0.003°$	33
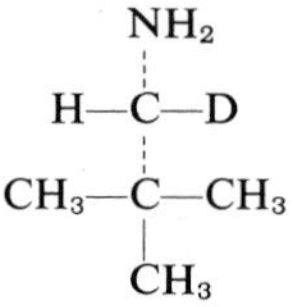 NH_2 $H-C-D$ CH_3-C-CH_3 CH_3	VIII	$[\alpha]_D^{25} +0.20°$ (l 1)	78,105,120
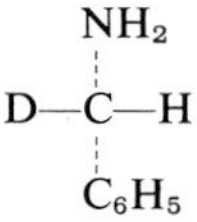 NH_2 $D-C-H$ C_6H_5	VIII, XII, XIII	$[\alpha]_D^{25} +1.10°,$[a] $[\alpha]_{365} +4.83°$	76,120,121, 124
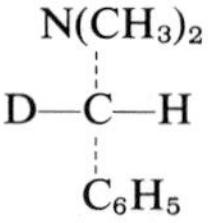 $N(CH_3)_2$ $D-C-H$ C_6H_5	VIII	Rotation not reported	76
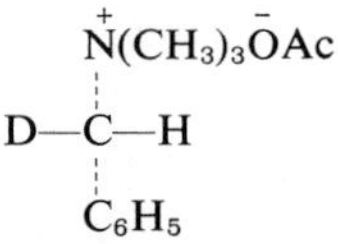 $\overset{+}{N}(CH_3)_3\bar{O}Ac$ $D-C-H$ C_6H_5	VIII	Rotation not reported	76
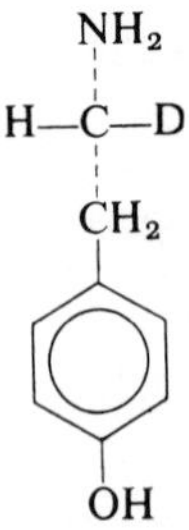 NH_2 $H-C-D$ CH_2 (phenyl ring) OH	VIII	Rotation not reported[a]	80

Sulfur compounds

CH_3—$\overset{\overset{\displaystyle O}{\|}}{\underset{\cdot\cdot}{S}}$—$CD_3$	X	$[\alpha]_D^{25}$ $-3.8°$ (Calculated for optically pure material)	95
$\overset{\overset{\displaystyle C_6H_5}{\|}}{\underset{\underset{\displaystyle SO_2Cl}{\|}}{H—C—D}}$	XI	$[\alpha]_D^{25}$ $+0.29°$ (c 18, dioxane)	111
$\overset{\overset{\displaystyle C_6H_5}{\|}}{\underset{\underset{\displaystyle SO_3^-Na^+}{\|}}{H—C—D}}$	XI	Rotation not reported	111,114,115
$\overset{\overset{\displaystyle C_6H_5}{\|}}{\underset{\underset{\displaystyle SO_2Et}{\|}}{H—C—D}}$	XI	$[\alpha]_D^{26}$ $-0.26°$ (c 16, dioxane)	111
$\overset{\overset{\displaystyle C_6H_5}{\|}}{\underset{\underset{\displaystyle SC_6H_5}{\|}}{H—C—D}}$	XI	$[\alpha]_D^{25}$ $+0.384°$ (c 15, dioxane, 0.88 ± 0.06 D/molecule)	111
$\overset{\overset{\displaystyle C_6H_5}{\|}}{\underset{\underset{\displaystyle SO_2SC_6H_5}{\|}}{H—C—D}}$	XI	$[\alpha]_D^{25}$ $+0.84°$ (c 5, dioxane, 0.67 ± 0.04 D/molecule) ($+3.9°$ calculated for optically pure material)	111
$\overset{\overset{\displaystyle SO_2Br}{\|}}{\underset{\underset{\displaystyle C_6H_5}{\|}}{D—C—H}}$	XI	$[\alpha]_{400}$ $+6.2°$ 0.935 D/molecule (ca. 60% optically pure)	114,115
$\overset{\overset{\displaystyle O—S—p\text{-tolyl}}{\|}}{\underset{\underset{\displaystyle C_6H_5}{\|}}{H—C—D}}$	XI	$[\alpha]_D^{26}$ $-0.3 \pm 0.1°$ (c 8.2, EtOH)	113
$\overset{\overset{\displaystyle S—p\text{-tolyl}}{\|}}{\underset{\underset{\displaystyle C_6H_5}{\|}}{D—C—H}}$	XI	$[\alpha]_D^{27}$ $+0.92 \pm 0.09°$ [a] (c 1.2, EtOH) (ca. 56% optically pure)	113

Compound	Section numbers in this chapter	Highest reported rotation	Reference
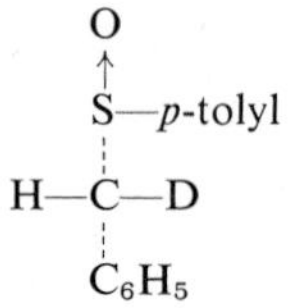 $\overset{\underset{\uparrow}{S}-p\text{-tolyl}}{\underset{\vert}{O}}$ H—C—D $\vert$ C_6H_5	XI	$[\alpha]_D^{27} +1.97 \pm 0.08°$ [a] (c 1.32, EtOH, 0.92 $\pm$ 0.01 D/molecule)	113
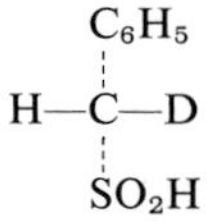 C_6H_5 $\vert$ H—C—D $\vert$ SO_2H	XI	Rotation not reported	111
Miscellaneous			
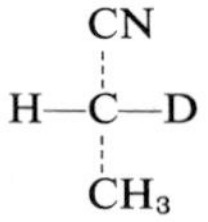 CN $\vert$ H—C—D $\vert$ CH_3	IX	Rotation not reported	81
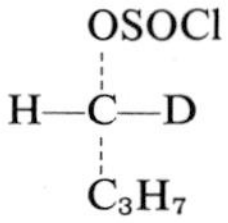 OSOCl $\vert$ H—C—D $\vert$ C_3H_7	VI	Rotation not reported	40,44
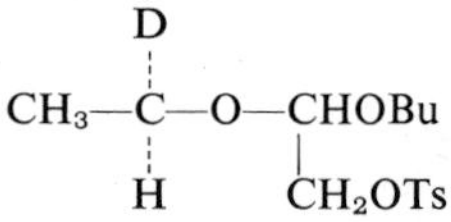 D $\vert$ CH_3—C—O—CHOBu $\vert$ $\quad\;$ $\vert$ H $\quad$ CH_2OTs	VI	Rotation not reported	56
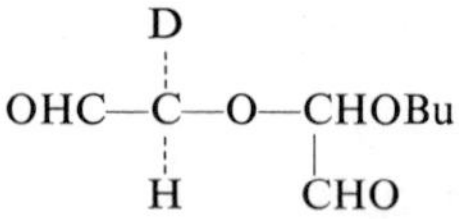 D $\vert$ OHC—C—O—CHOBu $\vert$ $\quad\;$ $\vert$ H $\quad$ CHO	VI	Rotation not reported	56
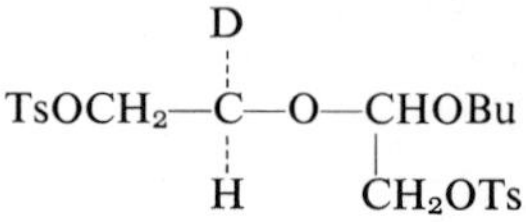 D $\vert$ $TsOCH_2$—C—O—CHOBu $\vert$ $\quad\;$ $\vert$ H $\quad$ CH_2OTs	VI	Rotation not reported	56
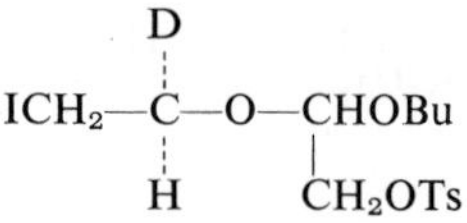 D $\vert$ ICH_2—C—O—CHOBu $\vert$ $\quad\;$ $\vert$ H $\quad$ CH_2OTs	VI	Rotation not reported	56
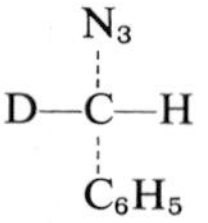 N_3 $\vert$ D—C—H $\vert$ C_6H_5	VIII	Rotation not reported	76

Structure			
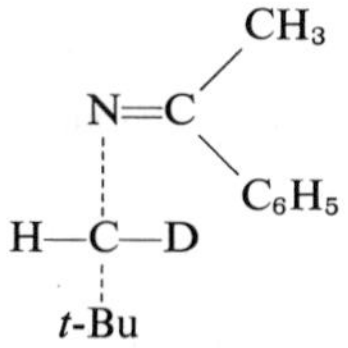 CH_3 $N=C$ $\quad C_6H_5$ H—C—D t-Bu	VIII	$[\alpha]_D^{25} +5.24°$ (l 1, neat)	78
(S)-$C_6H_5CHDNHCSC_6H_5$	XII	$[\alpha]_{546} +0.1°$ (c 10, EtOH)	121
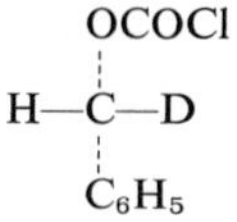 OCOCl H—C—D C_6H_5	XI	Rotation not measured	111
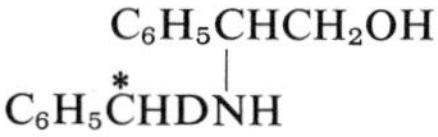 $C_6H_5CHCH_2OH$ $C_6H_5\overset{*}{C}HDNH$	XII	Rotation not reported	120
(S)-$C_6H_5CHDNCOCOC_6H_4$	XII	$[\alpha]_{546} -0.26°$ (c 15 C_6H_6), $[\alpha]_{546} -0.10°$ (c 21, dioxane)	121
(S)-$C_6H_5CHDNCCO_2CH_2C_6H_5$	XII	$[\alpha]_{546} +0.22°$ (c 30, C_6H_6)	121
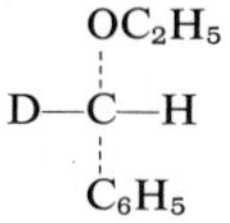 OC_2H_5 D—C—H C_6H_5	XI	$[\alpha]_D +0.098°$	43
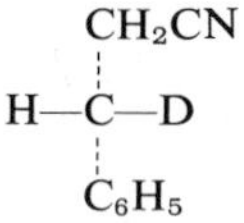 CH_2CN H—C—D C_6H_5	XI	$[\alpha]_D^{29} -0.364°$	107
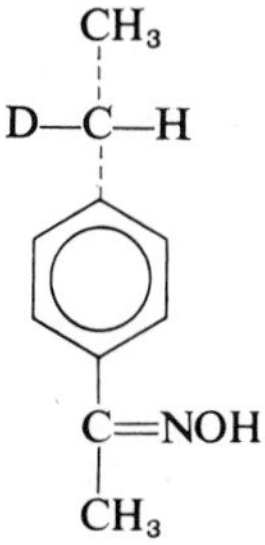 CH_3 D—C—H (ring) C=NOH CH_3	III	$[\alpha]_D^{28} -0.17°$	16
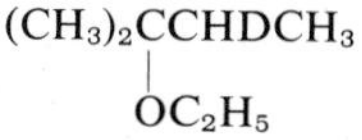 $(CH_3)_2CCHDCH_3$ OC_2H_5	XI	Rotation not reported	106

Compound	Section numbers in this chapter	Highest reported rotation	Reference

$$\begin{array}{c} N_3 \\ | \\ D-C-H \\ | \\ CH_2 \\ \end{array}$$

[benzene ring with OCH$_3$]

| 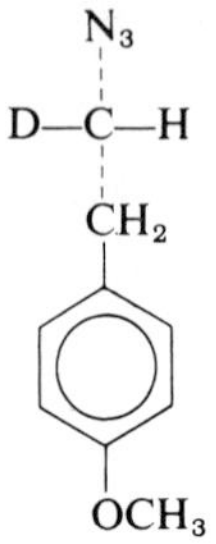 | VIII | Rotation not reported | 80 |

$$\begin{array}{c} O\text{-}p\text{-Ns} \\ | \\ D-C-H \\ | \\ CH_3 \end{array}$$

| 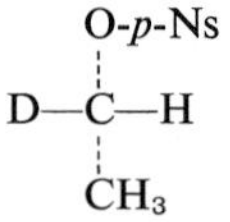 | VI | Rotation not reported | 58 |

$$\begin{array}{c} OTs \\ | \\ D-C-H \\ | \\ CH_3-C-CH_3 \\ | \\ CH_3 \end{array}$$

| 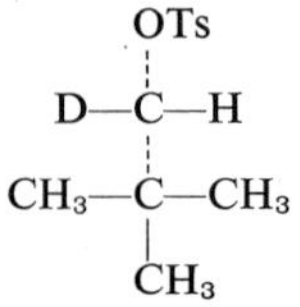 | XI | No observable rotation 589–300 nm | 106 |

$$\begin{array}{c} CH_2OTs \\ | \\ H-C-D \\ | \\ C_6H_5 \end{array}$$

| 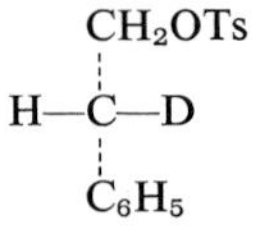 | XI | $[\alpha]_D^{26} -0.211°$ [a] (c 10, Et$_2$O) | 107 |

$$\begin{array}{c} OTs \\ | \\ H-C-D \\ | \\ C_6H_5 \end{array}$$

| 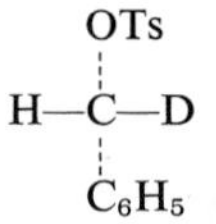 | XI | Rotation not measured [a] | 43 |

[cyclohexane ring: Cl, H, Cl, Cl, H, Cl, D, Cl, H, H, D, Cl]

| 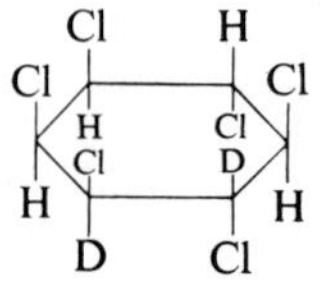 | X | $\alpha_D^{26} +0.050 \pm 0.002°$ (l 1, c 66, acetone) | 99 |

$$\begin{array}{c} OTs \\ | \\ D-C-H \\ | \\ CH_2 \end{array}$$

[benzene ring with OCH$_3$]

| 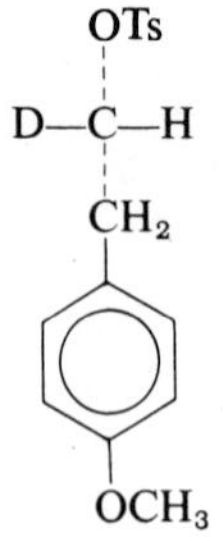 | VIII | Rotation not reported | 80 |

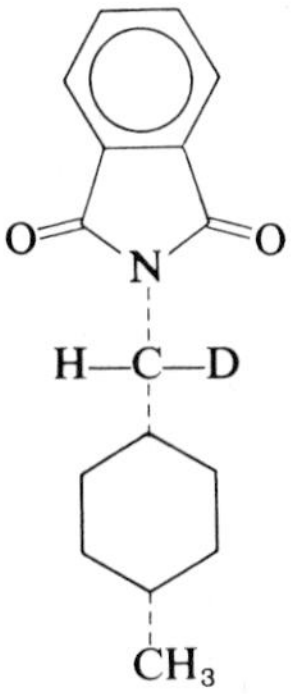	IX	$[\alpha]_{546}$ $-1.02°$ (c 41, C_6H_6)	42
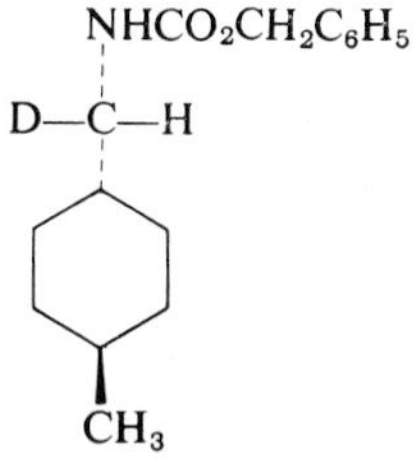	IX	$[\alpha]_{546}$ $+0.28°$ [a] (c 43, C_6H_6)	42
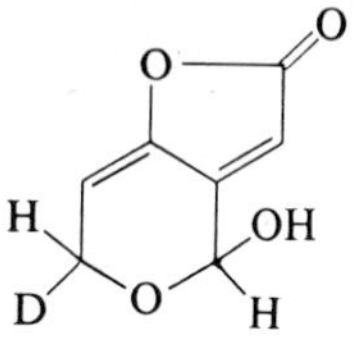	XIII	$[\phi]_{300}$ $+16°$ (correction of literature value)	126

Acknowledgment

I am indebted to the following colleagues for sending material in advance of publication or for helpful discussions: Drs. E. Caspi, J. W. Cornforth, D. J. Cram, C. H. DePuy, E. L. Eliel, H. Gerlach, P. J. Heffron, J. L. Kice, J. F. King, H. S. Mosher, A. I. Scott, A. Streitwieser, Jr., and S. Wolfe.

References

1. S. Englard and K. R. Hanson, in *Methods in Enzymology*, Vol. 13, S. P. Colowick and N. O. Kaplan, Eds., Academic Press, New York, 1969.
2. D. Arigoni and E. L. Eliel, in *Topics in Stereochemistry*, Vol. 4, N. L. Allinger and E. L. Eliel, Eds., Interscience, New York, 1969.
3. I. A. Rose, *Ann. Revs. Biochem.*, 35, 23 (1966).
4. J. W. Cornforth and G. Ryback, *Ann. Repts. Progr. Chem. 1965*, (*Chem. Soc. London*), 62, 428 (1966).

5. H. R. Levy, P. Talalay, and B. Vennesland, in *Progress in Stereochemistry*, Vol. 3, P. B. D. de la Mare and W. Klyne, Eds., Butterworths, Washington, D.C., 1962, ch. 8.
6. E. M. Kosower, *Molecular Biochemistry*, McGraw-Hill, New York, 1962.
7. H. C. Urey, F. Brickwedde, and G. M. Murphy, *Phys. Rev.*, *39*, 164 (1932); E. W. Washburn and H. C. Urey, *Proc. Natl. Acad. Sci. U.S.*, *18*, 496 (1932).
8. C. Buchanan, *Chem. Ind. (London)*, *1938*, 748.
9. E. R. Alexander and A. G. Pinkus, *J. Am. Chem. Soc.*, *71*, 1786 (1949).
10. R. L. Burwell, F. Hummel, and E. S. Wallis, *J. Org. Chem.*, *1*, 332 (1936).
11. H. C. Brown and C. Groot, *J. Am. Chem. Soc.*, *64*, 2563 (1942).
12. E. R. Alexander, *J. Am. Chem. Soc.*, *72*, 3796 (1950).
13. R. S. Cahn, C. K. Ingold, and V. Prelog, *Angew. Chem.*, *78*, 413 (1966); *Angew. Chem. Intern. Ed. Engl.*, *5*, 385 (1966).
14. K. R. Hanson, *J. Am. Chem. Soc.*, *88*, 2731 (1966).
15. R. L. Burwell, *Chem. Rev.*, *57*, 895 (1957).
16. E. L. Eliel, *J. Am. Chem. Soc.*, *71*, 3970 (1949).
17. H. J. Dauben and L. L. McCoy, *J. Am. Chem. Soc.*, *81*, 5404 (1959).
18. E. L. Eliel, *Stereochemistry of Carbon Compounds*, McGraw-Hill, New York, 1962.
19. S. F. Boys, *Proc. Roy. Soc. (London)*, Ser. A, *144*, 655 (1934).
20. P. Freon, *J. Phys. Radium*, [8], *1*, 374 (1940).
21. W. Fickett, *J. Am. Chem. Soc.*, *74*, 4204 (1952).
22. J. G. Kirkwood, *J. Chem. Phys.*, *5*, 479 (1937).
23. N. V. Cohan and H. F. Hameka, *J. Am. Chem. Soc.*, *88*, 2136 (1966).
24. H. F. Hameka, *J. Chem. Phys.*, *41*, 3612 (1964).
25. G. K. Helmkamp, C. D. Joel, and H. Sharman, *J. Org. Chem.*, *21*, 844 (1956).
26. J. Kenyon, H. Phillips, and V. P. Pittman, *J. Chem. Soc.*, *1935*, 1077; a specific rotation of 13.1° has also been reported: K. Mislow, R. E. O'Brien, and H. Schaeffer, *J. Am. Chem. Soc.*, *84*, 1940 (1962).
27. D. K. Murphy, R. L. Alumbaugh, and B. Rickborn, *J. Am. Chem. Soc.*, *91*, 2649 (1969).
28. H. Weber, J. Seibl, and D. Arigoni, *Helv. Chim. Acta*, *49*, 741 (1966).
29. P. S. Skell, R. G. Allen, and G. K. Helmkamp, *J. Am. Chem. Soc.*, *82*, 410 (1960); D. G. Goodwin and H. R. Hudson, *J. Chem. Soc. (B)*, *1968*, 1333.
30. G. K. Helmkamp and B. F. Rickborn, *J. Org. Chem.*, *22*, 479 (1957).
31. A. Streitwieser, Jr., *J. Am. Chem. Soc.*, *75*, 5014 (1953).
32. F. A. Loewus, F. H. Westheimer, and B. Vennesland, *J. Am. Chem. Soc.*, *75*, 5018 (1953).
33. A. Streitwieser, Jr. and W. D. Schaeffer, *J. Am. Chem. Soc.*, *78*, 5597 (1956).
34. L. J. Le Roux and S. Sugden, *J. Chem. Soc.*, *1939*, 1279.
35. For a summary of the work of Hughes, Ingold, and their school, see for example C. K. Ingold, *Structure and Mechanism in Organic Chemistry*, Cornell University Press, Ithaca, N.Y., 2nd Ed., 1969.
36. A. Streitwieser, Jr., and J. R. Wolfe, Jr., *J. Am. Chem. Soc.*, *79*, 903 (1957).
37. G. Vavon and A. Antonini, *Compt. Rend. Acad. Sci. Paris*, *232*, 1120 (1951).
38. D. S. Noyce and D. B. Denney, *J. Am. Chem. Soc.*, *72*, 5743 (1950).
39. W. Kuhn and H. K. Gore, *Z. Physik. Chem.*, *B12*, 389 (1931).
40. A. Streitwieser, Jr., J. R. Wolfe, Jr., and W. D. Schaeffer, *Tetrahedron*, *6*, 338 (1959).

41. M. S. Kharasch and O. Reinmuth, *Grignard Reactions of Nonmetallic Substances*, Prentice-Hall, New York, 1954, p. 160.
42. H. Gerlach, *Helv. Chim. Acta*, *49*, 1291 (1966).
43. A. Streitwieser, Jr., and J. R. Wolfe, Jr., *J. Am. Chem. Soc.*, *81*, 4912 (1959).
44. A. Streitwieser, Jr., and W. D. Schaeffer, *J. Am. Chem. Soc.*, *79*, 379 (1957).
45. S. J. Cristol, J. W. Ragsdale, and J. S. Meeks, *J. Am. Chem. Soc.*, *73*, 810 (1951).
46. V. E. Althouse, D. M. Feigl, W. A. Sanderson, and H. S. Mosher, *J. Am. Chem. Soc.*, *88*, 3595 (1966).
47. A. Streitwieser, Jr., L. Verbit, and R. Bittman, *J. Org. Chem.*, *32*, 1530 (1967).
48. G. Zweifel, N. R. Ayyangar, T. Munekata, and H. C. Brown, *J. Am. Chem. Soc.*, *86*, 1076 (1964).
49. A. A. Bothner-By, and C. A. Naar-Colin, *J. Am. Chem. Soc.*, *83*, 231 (1961).
50. D. R. Brown, S. F. A. Kettle, J. McKenna, and J. M. McKenna, *Chem. Commun.*, *1967*, 667.
51. H. C. Brown, N. R. Ayyangar, and G. Zweifel, *J. Am. Chem. Soc.*, *86*, 397, 1071 (1964).
52. H. C. Brown and G. Zweifel, *J. Am. Chem. Soc.*, *83*, 2544 (1961).
53. K. Mislow, R. E. O'Brien, and H. Schaefer, *J. Am. Chem. Soc.*, *82*, 5512 (1960).
54. H. Weber, P. Loew, and D. Arigoni, *Chimia*, *19*, 595 (1965).
55. H. R. Levy, F. A. Loewus, and B. Vennesland, *J. Am. Chem. Soc.*, *79*, 2949 (1957).
56. R. U. Lemieux and J. Howard, *Can. J. Chem.*, *41*, 308 (1963).
57. K. Mislow and J. Brenner, *J. Am. Chem. Soc.*, *75*, 2318 (1953).
58. A. Streitwieser, Jr., and M. R. Granger, *J. Org. Chem.*, *32*, 1528 (1967).
59. R. H. Pickard and J. Kenyon, *J. Chem. Soc.*, *99*, 45 (1911).
60. J. A. Mills and W. Klyne, in *Progress in Stereochemistry*, Vol. 1, W. Klyne, Ed., Butterworths, London, 1954, ch. 5.
61. A. Streitwieser, Jr., I. Schwager, L. Verbit, and H. Rabitz, *J. Org. Chem.*, *32*, 1532 (1967).
62. J. H. Brewster, *Tetrahedron Letters*, *1959*, No. 20, 23.
63. For a critique of Brewster's theory, see J. H. Brewster, in *Topics in Stereochemistry*, Vol. 2, N. L. Allinger and E. L. Eliel, Eds., Interscience, New York, 1967, p. 1.
64. S. J. Cristol and W. C. Firth, Jr., *J. Org. Chem.*, *26*, 280 (1961).
65. V. E. Althouse, K. Ueda, and H. S. Mosher, *J. Am. Chem. Soc.*, *82*, 5938 (1960).
66. D. R. Boyd and M. A. McKervey, *Quart. Revs.*, *22*, 95 (1968).
67. V. E. Althouse, E. Kaufmann, P. Loeffler, K. Ueda, and H. S. Mosher, *J. Am. Chem. Soc.*, *83*, 3138 (1961).
68. W. A. Sanderson and H. S. Mosher, *J. Am. Chem. Soc.*, *83*, 5033 (1961).
69. W. A. Sanderson and H. S. Mosher, *J. Am. Chem. Soc.*, *88*, 4185 (1966).
70. M. Raban and K. Mislow, *Tetrahedron Letters*, *1965*, 4249.
71. M. Raban and K. Mislow, *Tetrahedron Letters*, *1966*, 3961.
72. M. Raban and K. Mislow, in *Topics in Stereochemistry*, Vol. 2, N. L. Allinger and E. L. Eliel, Eds., Interscience, New York, 1967, p. 199.
73. R. U. Lemieux and J. Giguere, *Can. J. Chem.*, *29*, 678 (1951).
74. R. MacLeod, H. Prosser, L. Finkentscher, J. Lanyi, and H. S. Mosher, *Biochemistry*, *3*, 838 (1964).
75. A. Streitwieser, Jr. and W. D. Schaeffer, *J. Am. Chem. Soc.*, *79*, 2888 (1957).
76. A. Streitwieser, Jr. and J. R. Wolfe, Jr., *J. Org. Chem.*, *28*, 3263 (1963).

77. H. Snyder and J. Brewster, *J. Am. Chem. Soc.*, *71*, 291 (1949).
78. R. D. Guthrie, W. Meister, and D. J. Cram, *J. Am. Chem. Soc.*, *89*, 5288 (1967).
79. H. E. Smith, S. L. Cook, and M. E. Warren, Jr., *J. Org. Chem.*, *29*, 2265 (1964).
80. B. Belleau and J. Burba, *J. Am. Chem. Soc.*, *82*, 5751 (1960).
81. B. Zagalak, P. A. Frey, G. L. Karabatsos, and R. A. Abeles, *J. Biol. Chem.*, *241*, 3029 (1966).
82. J. Rétey, A. Umani-Ronchi, and D. Arigoni, *Experentia*, *22*, 72 (1966).
83. D. J. Prescott and J. L. Rabinowitz, *J. Biol. Chem.*, *243*, 1551 (1968).
84. E. Fischer, *Ber.*, *40*, 489 (1907); E. Fischer and K. Raska, *Ber.*, *39*, 3981 (1907).
85. J. H. van't Hoff, *La Chimie dans l'Espace*, P. M. Bazendijk, Rotterdam, 1875, p. 29.
86. P. Maitland and W. H. Mills, *Nature*, *135*, 994 (1935); *J. Chem. Soc.*, *1936*, 987.
87. W. H. Perkin, Jr., W. J. Pope, and O. Wallach, *Ann. Chem.*, *371*, 180 (1909); *J. Chem. Soc.*, *1909*, 95, 1789.
88. R. G. Cooke and A. K. MacBeth, *J. Chem. Soc.*, *1939*, 1245.
89. C. Martius and G. Schorre, *Ann. Chem.*, *570*, 140, 143 (1950).
90. C. Martius and G. Schorre, *Z. Naturforsch.*, *5B*, 170 (1950).
91. A. G. Ogston, *Nature*, *162*, 963 (1948); *167*, 693 (1951); *181*, 1420 (1958).
92. K. Mislow, R. E. O'Brien, and H. Schaefer, *J. Am. Chem. Soc.*, *84*, 1940 (1962).
93. A. Streitwieser, Jr. and P. J. Stang, *J. Am. Chem. Soc.*, *87*, 4953 (1965).
94. R. L. Burwell and S. Archer, *J. Am. Chem. Soc.*, *64*, 1032 (1942); C. C. Price and M. Lund, *ibid.*, *62*, 3105 (1940).
95. W. H. Pirkle and S. D. Beare, *J. Am. Chem. Soc.*, *90*, 6250 (1968).
96. K. K. Anderson, *Tetrahedron Letters*, *1962*, 93.
97. M. Axelrod, P. Bickart, J. Jacobus, M. M. Green, and K. Mislow, *J. Am. Chem. Soc.*, *90*, 4835 (1968).
98. W. L. Waters, W. S. Linn, and M. C. Caserio, *J. Am. Chem. Soc.*, *90*, 6741 (1968).
99. R. Riemschneider, *Chem. Ber.*, *89*, 2713 (1956).
100. A. Streitwieser, Jr. and W. D. Schaeffer, *J. Am. Chem. Soc.*, *79*, 6233 (1957).
101. A. Streitwieser, Jr., T. D. Walsh, and J. R. Wolfe, Jr., *J. Am. Chem. Soc.*, *87*, 3682 (1965).
102. A. Streitwieser, Jr. and T. D. Walsh, *J. Am. Chem. Soc.*, *87*, 3686 (1965).
103. D. J. Cram, *Fundamentals of Carbanion Chemistry*, Academic Press, New York, 1965.
104. D. J. Cram and B. F. Rickborn, *J. Am. Chem. Soc.*, *83*, 2178 (1961).
105. R. D. Guthrie, *J. Am. Chem. Soc.*, *89*, 6718 (1967).
106. G. Solladié, M. Muskatirovic, and H. S. Mosher, *Chem. Commun.*, *1968*, 809.
107. C. H. DePuy, F. W. Breitbeil, and K. R. DeBruin, *J. Am. Chem. Soc.*, *88*, 3347 (1966).
108. D. J. Pasto, C. C. Cumbo, and J. Fraser, *J. Am. Chem. Soc.*, *88*, 2194 (1966).
109. E. L. Eliel and M. H. Rerick, *J. Am. Chem. Soc.*, *82*, 1362 (1960).
110. K. R. Varma and E. Caspi, *Tetrahedron*, *24*, 6365 (1968).
111. J. L. Kice, R. H. Engebrecht, and N. E. Pawlowski, *J. Am. Chem. Soc.*, *87*, 4131 (1965).
112. K. B. Wiberg and T. M. Shryne, *J. Am. Chem. Soc.*, *77*, 2774 (1955).
113. E. G. Miller, D. R. Rayner, H. T. Thomas, and K. Mislow, *J. Am. Chem. Soc.*, *90*, 4861 (1968).
114. J. F. King and D. J. H. Smith, *J. Am. Chem. Soc.*, *89*, 4803 (1967).

115. J. F. King, private communication.
116. A. Horeau and H. B. Kagan, *Tetrahedron*, *20*, 2431 (1964); A. Horeau, *J. Am. Chem. Soc.*, *86*, 3171 (1964); *Bull. Soc. Chim. France*, *1964*, 2673.
117. A. Horeau and A. Nouaillé, *Tetrahedron Letters*, *1966*, 3953.
118. A. Horeau, A. Nouaillé, and K. Mislow, *J. Am. Chem. Soc.*, *87*, 4957 (1965).
119. K. Mislow, R. Graeve, A. J. Gordon, and G. H. Wahl, Jr., *J. Am. Chem. Soc.*, 1199 (1963); *86*, 1733 (1964).
120. H. Gerlach, *Helv. Chim. Acta*, *49*, 2481 (1966).
121. H. Gerlach, private communication.
122. C. K. Johnson, E. J. Gabe, M. R. Taylor, and I. A. Rose, *J. Am. Chem. Soc.*, *87*, 1802 (1965).
123. I. A. Rose, *J. Am. Chem. Soc.*, *80*, 5835 (1958).
124. A. Streitwieser, Jr., L. Verbit, and S. Andreades, *J. Org. Chem.*, *30*, 2078 (1965)
125. C. Djerassi and B. Tursch, *J. Am. Chem. Soc.*, *83*, 4609 (1961).
126. A. I. Scott and M. Yalpani, *Chem. Commun.*, *1967*, 945.
127. L. Verbit, *J. Am. Chem. Soc.*, *89*, 167 (1967).
128. S. Englard, J. S. Britten, and I. Listowsky, *J. Biol. Chem.*, *242*, 2255 (1967).
129. J. W. Cornforth, G. Ryback, G. Popják, C. Donninger, and G. Schroepfer, Jr., *Biochem. Biophys. Res. Commun.*, *9*, 371 (1962).
130. J. W. Cornforth, R. H. Cornforth, C. Donninger, G. Popják, G. Ryback, and G. J. Schroepfer, Jr., *Proc. Roy. Soc.*, *Ser. B, London*, *163*, 436 (1966).
131. M. Sprecher, R. Berger, and D. B. Sprinson, *J. Biol. Chem.*, *239*, 4268 (1964).
132. M. Sprecher, R. L. Switzer, and D. B. Sprinson, *J. Biol. Chem.*, *241*, 864 (1966).
133. J. D. Roberts and M. C. Caserio, *Basic Principles of Organic Chemistry*, W. A. Benjamin, New York, 1964.
134. S. Wolfe and A. Rauk, *Can. J. Chem.*, *44*, 2591 (1966).
135. K. Mislow, M. M. Green, P. Laur, J. T. Melillo, T. Simmons, and A. L. Ternay, Jr., *J. Am. Chem. Soc.*, *87*, 1958 (1965).
136. E. Caspi and K. R. Varma, *J. Org. Chem.*, *33*, 2181 (1968).
137. W. Voelter, E. Bayer, R. Records, E. Bunnenberg, and C. Djerassi, *Chem. Ber.*, *102*, 1005 (1969); see also L. Velluz and M. Legrand, *Compt. Rend. Acad. Sci, Paris*, *263*, 1429 (1966).

Protonated Cyclopropanes

By C. C. Lee

The University of Saskatchewan, Saskatoon, Saskatchewan, Canada

CONTENTS

I. INTRODUCTION

In order to explain the observed isotope position rearrangements accompanying the solvolyses of 2,3-$^{14}C_2$-2-norbornyl brosylates, which included the appearance of some of the ^{14}C-label at the C-5,6 positions of the products, Roberts and co-workers (1,2), initially in 1951, proposed the formation of the "nortricyclonium" ion (**1a**) as one of the two nonclassical carbonium ion intermediates involved in the reaction. Structure **1a**, which may be alternatively represented as **1b**, was apparently the first suggested "face-protonated" cyclopropane, although it was not named as such when it was proposed. As an alternative to the nortricyclonium ion, Winstein and Trifan (3) suggested the equilibration between nonclassical

(1a)   (1b)

norbornonium ions **2a–c**, presumably via "edge-protonated" cyclopropanes
such as **3**. Since these early studies, the norbornyl system has been sub-

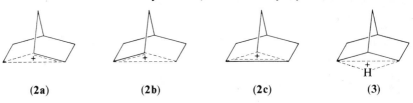

(2a)   (2b)   (2c)   (3)

jected to very intensive investigations, especially in relation to the question
of nonclassical versus rapidly equilibrating classical ions. For discussions
on this subject from various points of view, the reader is referred to recent
reviews by Gream (more or less neutral on nonclassical ions) (4), Sargent
(definitely for nonclassical ions) (5), and Brown (emphatically against nonclassical ions) (6). In the present review, discussions on the norbornyl
system will be made only in pertinent cases in relation to protonated
cyclopropanes.

The unsubstituted, face-protonated cyclopropane, **4**, was invoked by
Skell and Starer (7) in 1960 as a possible explanation for the formation of
some cyclopropane among the products from deoxideation* and deamination reactions involving the 1-propyl system. Skell and Starer, however,

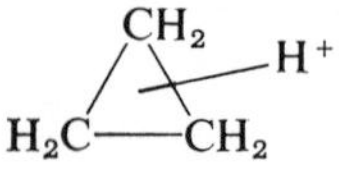

$$CH_2$$
$$\diagdown H^+$$
$$H_2C—CH_2$$

$$H_2C----H$$
$$\diagdown + \diagup$$
$$H_2C—CH_2$$

(4) 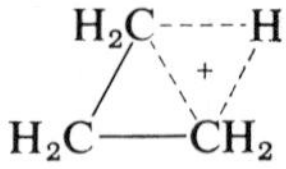   (5)

later modified the suggested intermediary role of protonated cyclopropane
to one of "1,3-interaction" (9) in order to conform to the reported finding
of Reutov and Shatkina (10) that the deamination of 1-^{14}C-1-propylamine
is accompanied by isotope position rearrangements of the label from C-1
solely to C-3. Compelling evidence for edge-protonated cyclopropane (**5**)
was reported in 1964 by Baird and Aboderin (11) in their studies on the

* This term was originally used to indicate the overall removal of the oxide ion
from an alkoxide ion (8), but some later workers have shortened it to deoxidation.
In a private communication to the author, Professor Skell has suggested that the
original spelling should be retained.

solvolysis of cyclopropane in D_2SO_4. In 1965, Lee and co-workers (12,13) and Karabatsos and co-workers (14) have independently obtained strong evidence indicating that protonated cyclopropane intermediates do play a role in the deamination of 1-propylamine. Since these demonstrations of the involvement of protonated cyclopropane species in such reactions as solvolysis and deamination, interest in them has been enhanced. In the present review, an attempt is made to summarize the experimental observations in support of protonated cyclopropanes and discuss some current views on protonated cyclopropanes as possible reaction intermediates or transition states.

II. THE STRUCTURE OF PROTONATED CYCLOPROPANES

In suggesting the equilibration of norbornonium ions **2a–c** via **3** as an alternative to the nortricyclonium ion **1a**, Winstein and Trifan (3) stated that considering the wave-mechanical description of the cyclopropane ring, structure **3** illustrates one of the attractive ways to protonate a cyclopropane. More recently, Winstein and co-workers (15) have quoted private communications from H. C. Longuet-Higgins and C. A. Coulsen suggesting that theoretical considerations favor edge protonation rather than face protonation of cyclopropanes. Experimentally, Berson and Grubb (16) have found that in the 2-carboxy-3-methyl-5-norbornyl cation, intramolecular transannular ("6,2") hydride shifts took place exclusively *endo→ endo*. For example, the D-labeled cation **6** gave the rearranged lactone **7** with the deuterium located solely at C-2. The mechanism for this transformation required *endo → endo* 6,2-shifts and it was shown that such

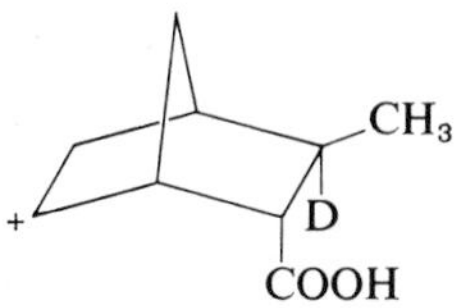

(6) 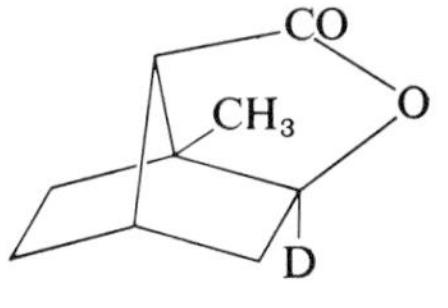   (7)

results were incompatible with a nortricyclonium ion-type of intermediate or transition state.

Hoffmann (17) has carried out calculations on the unsubstituted protonated cyclopropane, $C_3H_7^+$, based on the extended Huckel theory. By testing several likely routes for the extra proton while keeping the cyclopropane geometry intact, it was found that the preferred approach is edge protonation in the direction of the center of a C—C bond and in the plane of the ring. With the C—C distances fixed, this edge-protonated form was found to have lower energy than the methyl-bridged "ethylene alkonium ion" **8**. However, if the C—C distances were allowed to vary,

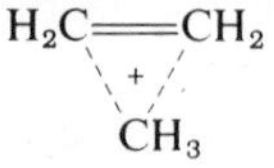

$$H_2C \!=\!\!=\! CH_2$$
$$\diagdown \overset{+}{} \diagup$$
$$CH_3$$

(8)

the edge-protonated structure appeared to be unstable with respect to ethylene and $CH_3{}^+$. Hoffmann (17) also stated that the results of his calculations supported to some extent the orbital picture of Walsh (18) on the distribution of electrons in cyclopropane itself. In recent reviews (19,20) on the various published theoretical studies on cyclopropane, including that of Walsh (18), it was indicated that there is general agreement that the C—C bonds of cyclopropane have a considerable p-component, and this would be in accord with a preference for edge protonation. Very recently, Petke and Whitten (20a) have carried out *ab initio* self-consistent-field calculations on the geometry of protonated cyclopropane and have concluded that $C_3H_7{}^+$ in the gas phase is stable relative to C_3H_6, that its equilibrium geometry corresponds to that of an edge-protonated species, and that this configuration is preferred over one of face protonation.

Recently, Joris, Schleyer, and Gleiter (21) have reported their observations, based on infrared absorption, indicating that cyclopropane rings can act as proton-acceptor groups in both intermolecular and intramolecular hydrogen bonding and that the preferred site for proton-donor interaction is the "edge" of the cyclopropane ring. For example, the infrared spectrum of the hydroxyl group of *endo-syn*-tricyclo[3.2.1.0^{2,4}]-octan-8-ol (**9a**) gave undisputable evidence for hydrogen bonding to the cyclopropane ring while the *anti*-isomer **9b** showed no hydrogen bonding.

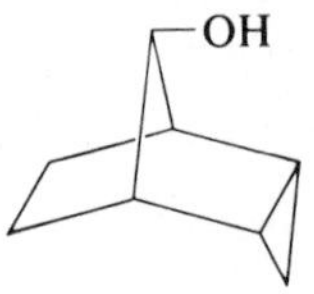

(9a) 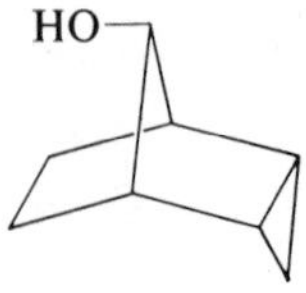   (9b)

Similarly, the axial alcohol **10** gave a spectrum which indicated hydrogen bonding to the cyclopropane ring, while the equatorial isomer **11**, which has the hydroxyl group lying over the cyclopropane ring in a bisected conformation, did not show any evidence of hydrogen bonding. Thus

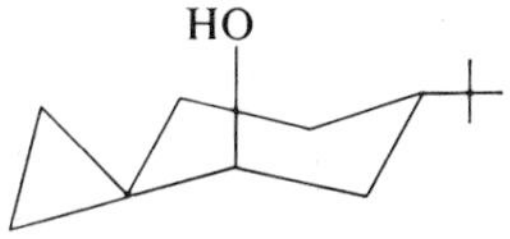

(10) 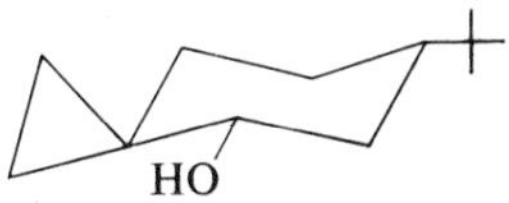   (11)

from both theoretical considerations and experimental evidence, edge protonation is apparently more favorable than face protonation in protonated cyclopropanes.

Bartlett has suggested the definition that an ion is nonclassical if its ground state has delocalized bonding σ-electrons (22). Although the carbon–carbon bonds of cyclopropane may have a p-component which plays an important role in protonation, Walsh's (18) treatment of cyclopropane also involves overlap between sp^2 orbitals, suggesting that the carbon–carbon bonds do have σ-bond character. Thus edge-protonated cyclopropane intermediates may be regarded as H-bridged nonclassical ions. Karabatsos and co-workers (23) have designated the formation of protonated cyclopropane from 1-propyl halides as the σ-route to a nonclassical carbonium ion.

Recently, Olah and Lucas (24) have attempted to observe protonated cyclopropane directly by NMR. In FSO_3H—SbF_5—SO_2ClF or HF—SbF_5—SO_2ClF solution above $-80°$, cyclopropane was protonated and cleaved to give acyclic products. Through various di- and trimerization processes and fragmentation, the products were, in varying amounts, the t-butyl and t-hexyl cations. At $-100°$, however, cyclopropane behaved differently. In FSO_3H—SbF_5—SO_2ClF solution, it formed a species the spectrum of which showed four lines, probably two doublets at -2.30 and -2.10 ppm, and a septuplet at -6.40 ppm, besides some ring-opened ions. On warming to $-80°$, the pair of doublets collapsed to a single doublet and this transformation was reversible. This NMR spectrum of the species formed at $-100°$ could be interpreted as that of protonated cyclopropane. The single proton, on the face of the cyclopropane ring or in an edge protonated but equilibrating system, could give rise to the septuplet at -6.40 ppm coupled to the six methylene hydrogens, which in turn appeared as a doublet. The presence of a pair of doublets at $-100°$ was suggested to mean the possibility that the methylene hydrogens above and below the plane of the ring could have different magnetic environments, both coupled to a single hydrogen, the geminal coupling being too small to be observed. It was pointed out, however, that the data obtained were insufficient and that much further work, such as decoupling experiments, study of deuterated derivatives, and investigation of the quenching products, will be required before it can be claimed that the observed spectrum was indeed that of the protonated cyclopropane.

III. PROTONATED CYCLOPROPANES FROM REACTIONS WITH CYCLOPROPANES

A. The Reaction of Sulfuric Acid with Cyclopropane

In the first of a series of three significant contributions dealing with protonated cyclopropane, Baird and Aboderin (25) reported that treatment of cyclopropane with D_2SO_4 for varying lengths of time resulted in the formation of mixtures of cyclopropane and monodeuteriocyclopropane plus small amounts of dideuteriocyclopropane. Solvolysis of the cyclopropane was also found to occur concurrently with exchange, the rate of solvolysis being about twice that of exchange. It was suggested that reversible formation of π-complex type of intermediates, c-C_3H_6—D^+, which could rearrange to c-C_3H_5D—H^+, may be responsible for the observed exchange. The second and more detailed contribution of Baird and Aboderin (11) concerns the solvolysis of cyclopropane in D_2SO_4 (the third paper in the series will be discussed in a later section). It was found that passage of cyclopropane, with stirring, through $8.43M$ D_2SO_4 at a flow rate of about 0.7 ml/ml of acid/min gave rise to 1-propyl hydrogen sulfate and 1-propanol. In some experiments, the acid sulfate was hydrolyzed by heating the reaction mixture at 50° for 36.5 hr before the alcohol was isolated. The 1-propanol was analyzed by NMR as its p-toluate ester, with the methyl group of the p-tolyl moiety serving as internal standard. The average deuterium distribution in the C-1, C-2, and C-3 positions of the alcohol was found to be 0.38, 0.17, and 0.46 deuterium atoms, respectively. This isotopic distribution was the same whether the 1-propanol was isolated before or after the acid sulfate was hydrolyzed.

The possibility that the observed deuterium distribution in the 1-propanol might be the result of equilibration prior to solvolysis (25) was ruled out under the conditions of rapid stirring and continuous cyclopropane passage, because the deuterated cyclopropane, once formed, would have essentially no probability of solvolyzing. Confirmation of this conclusion was indicated by the uptake of only one deuterium atom on the carbon skeleton of 1-propanol. Furthermore, it was also noted that the rates of solvolysis of cyclopropane in $8.43M$ D_2SO_4 and $8.41M$ H_2SO_4 were in the ratio of 0.66 : 1, which strongly suggested that cyclopropane was protonated slowly relative to its subsequent reactions. When $1,1$-d_2-1-

propanol was heated with 8.43M D$_2$SO$_4$ under conditions much more severe than those used in the isolation of solvolysis products, there was no loss of deuterium from C-1 nor any incorporation of deuterium into C-2 or C-3. These observations suggested that one could also eliminate the possibility of deuterium scrambling occurring after the formation of the acid sulfate or alcoholic products.

Baird and Aboderin (11) concluded that their results cannot be explained by a mechanism involving methyl-bridged ions, either alone or in equilibrium with isomeric methyl-bridged ions such as **12a** ⇄ **12b**, because this would give rise to equal amounts of deuterium at C-1 and C-2 in the

$$CH_2D \qquad\qquad CHD \qquad\qquad H_2C\text{-----}D$$
$$H_2C\!\!=\!\!=\!\!CH_2 \qquad H_2C\text{----}CH_3 \qquad H_2C\text{---}CH_2$$
$$\textbf{(12a)} \qquad\qquad \textbf{(12b)} \qquad\qquad \textbf{(13)}$$

resulting 1-propanol. Nor is it feasible to have a system of equilibrating 1-propyl cations via 1,3-hydride shifts, or an intermediate such as **13**, because this could not account for the presence of deuterium at C-2. It was pointed out that the observed isotopic distribution could be formally accounted for by a system of equilibrating primary carbonium ions, accompanied by some Wagner–Meerwein rearrangement of the methyl group. Such a mechanism was discarded because propene and 2-propyl derivatives, typical products from the 1-propyl cation, were essentially absent among the products of solvolysis of cyclopropane. The involvement of a π-complex type of face-protonated cyclopropane (**4**) was also considered to be unlikely in view of the association of π-complex intermediates with rapid and reversible proton exchange with solvent (26,27).

The mechanism suggested by Baird and Aboderin (11) involves the equilibration of hydrogen-bridged ions (edge-protonated cyclopropanes) via methyl-bridged ions as shown in Scheme I, the initial protonation could have proceeded via either route (*1*) or route (*2*). In order to explain the observed isotopic distribution, it was further assumed that solvolytic ring opening occurs primarily with the hydrogen-bridged ions **14a**, **14b**, and **14c** rather than with the carbon-bridged ions **12a** and **12b**. Since the formation of the 2-*d*-1-propyl product would require more extensive equilibration, it is not surprising that the smallest amount of deuterium is found at C-2. This mechanism, involving equilibration of edge-protonated cyclopropane intermediates, had a profound influence on subsequent investigations in this field and many workers have employed similar mechanisms in the interpretation of a variety of experimental results.

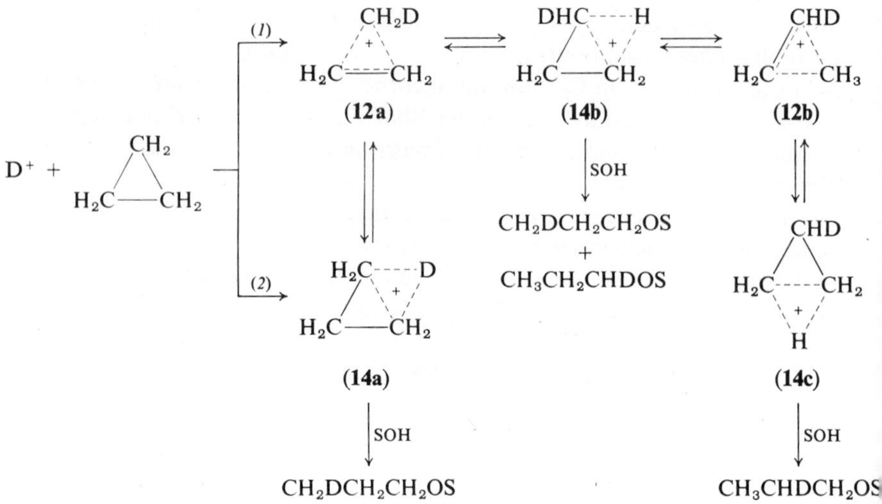

Scheme I

Deno and co-workers have attempted to confirm the results of Baird and Aboderin by directly monitoring the D-distribution in the 1-PrOSO$_3$D by NMR, thus avoiding the steps involving hydrolysis, extraction, VPC separation, preparation of p-toluate, and purification. Initially, it was reported (28) that the 1-propyl moiety of the product showed an equal distribution of deuterium atom at each of the three carbon positions. These results were later revised, and from experiments in 83–99% D$_2$SO$_4$, the D-distributions in the acid sulfate were found to be approximately 28:28:44%, respectively, at C-1, C-2, and C-3 (29).

Another attempt to confirm the findings of Baird and Aboderin was carried out by Lee and Gruber using tritiated sulfuric acid (30,31). Cyclopropane was passed through 9.2M or 13.8M t-H$_2$SO$_4$ at about the same flow rate as that used by Baird and Aboderin (but without stirring). The 1-propanol was recovered with the aid of inactive carrier either before or after the acid sulfate was hydrolyzed by heating the resulting reaction mixture at 50° for 30 hr. In the experiments with 9.2M t-H$_2$SO$_4$, the conditions of hydrolysis did not appreciably influence the isotopic distribution; the average t-distribution in the 1-propanol recovered with or without subsequent hydrolysis was 37, 26, and 37%, respectively, at C-1, C-2, and C-3. In the experiments with 13.8M acid, the subsequent heating of the reaction mixture to effect the hydrolysis of the acid sulfate also caused some additional isotopic scrambling. The average t-distribution at the C-1, C-2, and C-3 positions of the 1-propanol recovered without hydrolysis was 37, 27, and 36%, respectively. The observation that the smallest amount of t-label was found at C-2 was regarded as good confirmation of Baird's mechanism involving equilibrating protonated cyclopropane intermediates as depicted in Scheme I.

Summarized in Table I are the available data from the addition of labeled sulfuric acid to cyclopropane. As pointed out by Deno and coworkers (29), these data abundantly confirm the extensive scrambling of the label throughout the propyl moiety. It was suggested (29) that in the experiments with the more highly concentrated D_2SO_4, the c-$C_3H_6D^+$ ion could have a longer lifetime, and thus could give rise to an essentially statistical distribution of deuterium of $2:2:3$ ($28.5:28.5:43\%$) at C-1, C-2, and C-3. The polydeuterated product observed by Deno and coworkers (29) was regarded as arising from deuterated cyclopropanes (25) and this could be prevented by flushing the surface of the reaction mixture with nitrogen.

Of the various experiments summarized in Table I, the closest similarity in reaction conditions is found in experiment 1 with $8.43M$ D_2SO_4 and experiment 2 with $9.2M$ t-H_2SO_4. Lee and Gruber (30), however, did not stir the reaction mixture nor flushed its surface with nitrogen; thus the possibility that some of the products might have arisen from tritiated

TABLE I

Data from the Addition of Deuterated or Tritiated
Sulfuric Acid to Cyclopropane

Experiment	Reagent	Products	Total D introduced	Distribution of D or T, %		
				C-1	C-2	C-3
1	$8.43M$ (57%) D_2SO_4[a]	1-PrOD and 1-PrOSO$_3$D	1.00	38	17	46
2	$9.2M$ (60%) t-H_2SO_4[b]	1-PrOH and 1-PrOSO$_3$H	—	37	26	37
3	$13.8M$ (79%) t-H_2SO_4[b]	1-PrOH and 1-PrOSO$_3$H	—	37	27	36
4	83% D_2SO_4[c]	1-PrOSO$_3$D	—	∼28	∼28	∼44
5	92% D_2SO_4[c]	1-PrOSO$_3$D	1.5	∼28	∼28	∼44
6	92% D_2SO_4[c]	1-PrOSO$_3$D	1.0[d]	∼28	∼28	∼44
7	99% D_2SO_4[c]	1-PrOSO$_3$D	2.0	30	29	42

[a] Ref. 11.
[b] Ref. 30.
[c] Ref. 29.
[d] The surface of the reaction mixture was flushed with N_2 to remove deuterated cyclopropanes.

cyclopropanes was not eliminated. It has been pointed out (30) that if the difference in isotopic distributions between experiments 1 and 2 were really significant, a possible explanation could be given in terms of isotope effects. According to Scheme I, if protonation proceeds initially to give **12a** as the first intermediate (route *1*), there could be an isotope effect in the subsequent D-shift to **14a** and the H-shift to **14b**. If a T-label were involved the H—T isotope effect would be greater than the H—D isotope effect. Consequently, with D-labeling, the contribution to product formation from **14a** would be more and from **14b** and **14c** would be less than the analogous processes involving the T-label. The overall effect would fit qualitatively the difference in isotopic distributions from experiments 1 and 2. On the other hand, protonation might proceed directly to give **14a** (route *2*). Along with the H-shifts shown in Scheme I (Scheme I shows only the isotopically different protonated cyclopropane intermediates), D- or T-shifts could also occur to give rise to isotopically equivalent species, such as **15a–c**, all of which would lead to a product with the label at C-3. The difference in isotope effect from the use of D or T as label would result

| (15a) | (15b) | (15c) |

in a relatively smaller contribution of **15a–c** in the case of T-labeling. The net effect would again qualitatively account for the difference in isotopic distributions from experiments 1 and 2. As support for the probability that kinetic isotope effects may play some role in the equilibration of protonated cyclopropanes, it was pointed out that Nickon and Werstiuk (32) have observed a primary H—D isotope effect of about 2.1 in the 1,3-elimination of 6-*endo-d*-2-*exo*-norbornyl tosylate to give nortricyclene at "low base" concentrations which involves an ionic mechanism. Moreover, kinetic isotope effects apparently also come into play when 1-*t*-1-propanol is heated in $13.8M$ H_2SO_4 (30); the results from this study, however, will be deferred for discussion in a later section.

B. The Reaction of Hydrochloric and Other Acids with Cyclopropane

To study another possibility of protonation of cyclopropane followed by ring-opening product formation, Lee and co-workers (31,33) passed cyclopropane at room temperature through tritiated Lucas reagent (equimolar molar quantities of $12M$ *t*-HCl and $ZnCl_2$). The product, swept out by the cyclopropane and collected in cold traps, was found to be solely 1-chloropropane by VPC. The average t-distribution in this product was 38, 18, and 44%, respectively, at C-1, C-2, and C-3. It was also noted that under the experimental conditions employed, 1-chloropropane and tritiated Lucas reagent did not undergo H—T exchange. The rapid passage of the cyclopropane to sweep out the product was originally intended as a means of minimizing the length of contact between 1-chloropropane and Lucas reagent because Reutov and Shatkina (34,35) have reported the observation of various amounts of 1,3-hydride shift when 1-^{14}C-1-chloropropane was heated with HCl—ZnCl$_2$ for different lengths of time. Subsequently, however, it was found in this laboratory that the heating of 1-t-1-chloropropane with Lucas reagent at 50° under reflux for 100 hr resulted in no appreciable isotopic rearrangement in the recovered 1-t-1-chloropane (36).

Deno and co-workers (29) effected the addition of HCl and DCl to cyclopropane with FeCl$_3$ as a catalyst, the reaction being carried out at −34°, the boiling point of cyclopropane. The NMR spectrum from the HCl addition showed the exclusive formation of 1-chloropropane and that from the DCl addition showed the superposition of a doublet and a triplet for the C-1 and C-3 hydrogens, indicating the presence of CH$_2$ and CHD, but little CD$_2$, at C-2. The C-1:C-2:C-3 D-distribution in the 1-chloropropane was reported as 35:26:39%. Although their work on the addition of 83–99% D$_2$SO$_4$ to cyclopropane, described in the preceding section, gave only statistical distributions of deuterium in the 1-PrOSO$_3$D and thus did not confirm the reported findings of Baird and Aboderin (11) and of Lee and Gruber (29) that the smallest amount of isotopic label was found at C-2, the greater amount of deuterium at C-1 than C-2 observed in the addition of DCl to cyclopropane led Deno and co-workers (29) to support Baird's mechanism of equilibrating protonated cyclopropanes (Scheme I) involving predominant product formation from edge-protonated species rather than from methyl-bridged ions. Besides the work with D$_2$SO$_4$ and DCl, Deno and co-workers (29) also reported the addition to cyclopropane of 20% CH$_3$COOD–80% D$_2$SO$_4$ to give 1-propyl acetate, and of CF$_3$COOD to give 1-propyl trifluoroacetate. The C-1:C-2:C-3 D-distributions found for the acetate and the trifluoroacetate, respectively, were 28:34:38% and 28:28:43%.

A protonation mechanism has recently been suggested by Hightower and Hall (37,38) for the isomerization of cyclopropane and alkylcyclopropanes over silica–alumina to give olefins. In the earlier work with cyclopropane isomerization to propylene, the data could be explained by either a hydride transfer mechanism involving C$_3$H$_5$$^+$ or a proton addition mechanism involving $C_3H_7^+$ (39). However, the multiplicity of possible products from alkylcyclopropane isomerization made it possible for a choice in favor of proton addition. For example, the isomerization of methylcyclopropane gave 1-butene, *cis*-2-butene, and *trans*-2-butene in a ratio of about 1:1:2, respectively, with no isobutene nor cyclobutane. Arguments were presented to show that these products were incompatible with hydride transfer, but were compatible with proton addition, the required protons being furnished by the adsorbed carbonaceous residue on the catalyst (40). It was further suggested that in the case of cyclopropane, proton addition could give c-$C_3H_7^+$ similar to the nonclassical protonated cyclopropane of Baird and Aboderin (11), and this would explain various observations including the deuterium scrambling in the unisomerized cyclopropane from co-isomerization of cyclopropane-d_0 and cyclopropane-d_6 (37). From methycyclopropane, the c-$C_4H_9^+$ was believed to be probably different from the classical 2-butyl cation, since the product distribution obtained from methycyclopropane (37) was not that expected from the 2-butyl cation as observed previously from isomerization of n-butenes over the same catalyst (41). It was concluded, however, that the c-$C_4H_9^+$ ion involved in the interconversion of methylcyclopropane with the n-butenes would be of higher energy than the classical 2-butyl cation.

Recently, Cacace and co-workers (42) reported that the reaction of the He^3H^+ ions, formed from the decay of molecular tritium, with cyclopropane gave tritiated methane, cyclopropane, and ethylene as major products, together with smaller yields of tritiated ethane, propane, and propylene. It was suggested that the first step in the formation of the tritiated products likely was the protonation of the cyclopropane to give the excited species $(C_3H_7^*)_{exc}^+$ (the asterisk indicating a tritiated species). Part of the excited protonated ions would dissociate and the remainder would be stabilized by collision. The stabilized $C_3H_7^{*+}$ ions could then react with inactive cyclopropane by proton transfer to form the observed tritiated cyclopropane.

$$C_3H_7^{*+} + c\text{-}C_3H_6 \longrightarrow C_3H_6^* + C_3H_7^+$$

The fact that the $C_3H_6^*$ hydrocarbons contained a relatively larger amount of tritiated cyclopropane than tritiated propylene was regarded as evidence for a cyclic structure for the $C_3H_7^{*+}$ ion from the initial protonation.

C. The Reaction of Protic or Deuterated Acids with Substituted Cyclopropanes

On the basis of NMR data, Deno and co-workers (29) have reported that the addition of DCl to methylcyclopropane gave exclusively 2-

chlorobutane with the one deuterium exclusively at C-4. This is the expected product if D^+ were added to produce only the 2-butyl cation. The observation thus again reflects the greater stability of the 2-butyl cation over the possible protonated methycyclopropane. The treatment of cyclopropane-carboxylic acid (16) with 96% H_2SO_4 at 100° was found by Deno (29b) to give a 3:1 mixture of acetone and the acid sulfate of 4-hydroxybutanoic acid (17).

$$\triangleright\!-COOH \xrightarrow[100°]{96\% H_2SO_4} CH_3COCH_3 + HOSO_3(CH_2)_3COOH$$

(16)   (17)

However, Kushner (43) found that a similar treatment of cyclopropyl methyl ketone (18) with 96% H_2SO_4 gave only the acid sulfate of 5-hydroxy-2-pentanone (19).

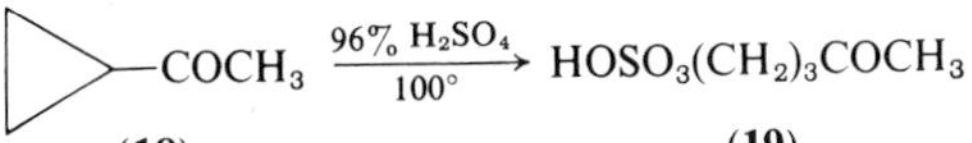

$$\triangleright\!-COCH_3 \xrightarrow[100°]{96\% H_2SO_4} HOSO_3(CH_2)_3COCH_3$$

(18)   (19)

The mechanism responsible for the decomposition of 16 to acetone is not clear. The formation of 17 and 19 from 16 and 18, respectively, possibly could proceed via either edge-protonated species such as 20 or classical ions such as 21.

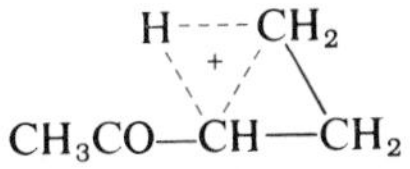

$$CH_3CO\!-\!\overset{\displaystyle H\text{----}CH_2}{\underset{(20)}{\underset{\big\backslash\;+\;\big/}{CH}\!-\!CH_2}}$$ $$CH_3COCH_2CH_2CH_2{}^+$$
                                                    (21)

Structure 20, if involved, apparently did not isomerize appreciably to 22, since 23, the product from 22, was not produced in detectable amounts (cf. acetylation of cyclopropane, Section III-D-1).

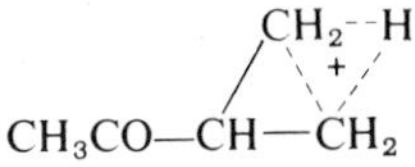

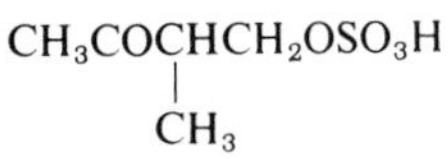

$$CH_3CO\!-\!\overset{\displaystyle CH_2\text{--}H}{\underset{(22)}{\underset{\big\backslash\;+\;\big/}{CH}\!-\!CH_2}}$$ $$CH_3COCHCH_2OSO_3H$$
                                            $$\underset{(23)}{\overset{\displaystyle |}{CH_3}}$$

The cleavage of the cyclopropane ring in nortricyclene (24) by protic or deuterated acids has been studied by Nickon and Hammons (44). Treatment of 24 with 0.08M H_2SO_4 in glacial acetic acid gave exo-norbornyl acetate, while reaction of 24 with hydrogen chloride in CH_2Cl_2 resulted in the formation of exo-norbornyl chloride. No endo-isomer was

detectable in either case by infrared analysis. Treatment of **24** in DOAc
with $0.14M$ D_2SO_4 gave *exo*-norbornyl acetate containing deuterium at
C-6 (**25**).

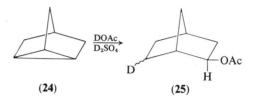

(**24**)   (**25**)

The mass spectrum of the norcamphor (**26**) derived from **25** indicated that
less than 3% of the molecules were multiply deuterated. Conversion of **26**
via the Wolff–Kishner reaction to deuterionorbornane (**27**) followed by
quantitative infrared analysis with the aid of known mixtures of *exo-d-*
and *endo-d*-norbornanes showed that the *exo*:*endo* ratio of the deuterium
in **27** was close to unity ($1:1.08 \pm 0.15$).

Cleavage of **24** with DCl in CH_2Cl_2 gave *exo*-norbornyl chloride,
whose mass spectrum revealed appreciable multiple deuteration (21.5%
d_0, 52% d_1, 22.5% d_2, 4% d_3; total of 1.09 D). The presence of d_2 and d_3
species indicated the intervention of reactions other than simple ring
cleavage. It was suggested that the extra deuterium could have entered by
exchange of cyclopropyl hydrogens prior to ring opening (25,29), or by the
formation of a deuterionorbornene followed by addition of DCl. Such a
deuterionorbornene possibly could arise by isomerization of notricyclene
or by elimination of HCl from the derived *exo*-norbornyl chloride.

In contrast to the reaction with DCl, formation of minor amounts of
multiply deuterated products from reaction with DOAc—D_2SO_4 indicated
that the deuterium entered during ring opening, and not prior or subsequent
to it. The complete selectivity of *exo*-acetate formation together with the
virtually equal *exo–endo* distribution of the deuterium at C-6 suggested
to Nickon and Hammons that **24** is converted to the carbon-bridged
norbornyl cation **28**. Whatever may be the steric course of any preceding
step, **28** would be expected to unite with nucleophile at C-1 and C-2 to
produce *exo*-acetates **25a** and **25b** in equal amounts, if isotope effects were
negligible (Scheme II).

Nickon and Hammons also pointed out that **28** apparently did not
return appreciably to the neutral tricyclic system because proton loss
would compete with deuteron loss, and thus would provide the means for
the eventual entry of more than one deuterium. Also, if reversible formation of edge-deutronated nortricyclene **29a** were to precede the generation
of **28**, then the conversion of **29a** to **28** must be essentially irreversible;

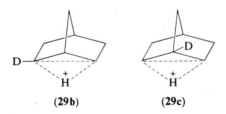

otherwise multiple deuteration would have resulted. The possibility of 6,2- and 6,1-hydride shifts in **28** was also discussed. It was pointed out, however, that the method used by these workers for the deuterium analysis was not sensitive enough to confirm, reliably, the possibility of such further scrambling of the D-label.

If product formation in the nortricyclene—DOAc—D_2SO_4 system were to result from reactions with protonated or deuteronated nortricyclenes by processes analogous to those suggested by Baird and Aboderin (11), the three hydrogen- and deuterium-bridged ions involved would be **29a**, **29b**, and **29c**. With isotope effects neglected, product formation

from these species would lead to twice as much 6-*endo*-deuterium as 6-*exo*-deuterium in the 2-*exo*-norbornyl acetate; in addition, the amounts of 2-*endo*-d-2-*exo*-norbornyl acetate and 6-*exo*-d-2-*exo*-norbornyl acetate would be equal. These expectations were not borne out by the experimentally observed formation of nearly equal amounts of 6-*endo-d*- and 6-*exo-d*-2-*exo*-norbornyl acetates, with possibly only minor amounts of 2-*endo-d*-2-*exo*-norbornyl acetate. Nickon and Hammons thus concluded that the principal acceptors of nucleophiles in the system studied were the carbon-bridged ions such as **28**, and not the hydrogen-bridged ions such as **29a–c**. In considering the mechanistic implications of protonated cyclopropanes, Lee and Kruger (45) also expressed a similar preference for equilibrating norbornonium ions as nucleophile acceptors in solvolytic reactions involving the norbornyl system.

In a study of the 2-norbornyl cation by NMR at low temperature, Jensen and Beck (46) obtained results confirming the earlier work of Schleyer and co-workers (47,48), but in addition fine structures in the spectra were observed. With 2-norbornyl bromide and gallium tribromide in SO_2 at $-80°$, the spectrum consisted of resonances at $\delta 5.2$, 3.1, and 2.1 ppm, downfield from TMS, with relative intensities of $4:1:6$ and with observed multiplicities of 7, 1, and 6, respectively. At $25°$, however, the spectrum was a single peak located at $\delta 3.3$ ppm. As concluded earlier (47,48), the spectra indicated exceedingly rapid exchange on the 1, 2, and 6 hydrogens ($\Delta F^{\ddagger} \leq 5.5$ kcal/mole), and slow 3,2-hydride shifts ($\Delta F^{\ddagger} \approx 11$ kcal/mole). Jensen and Beck also quoted a private communication from G. J. Karabatsos that in open-chain carbonium ions, α,β-secondary, secondary hydride shifts would occur with about the same activation energy as ion capture by solvent ($\Delta F^{\ddagger} \approx 2$ kcal/mole). Thus the observed slow 3,2-hydride shifts did not appear to be consistent with a classical ion being the stable species in the system used in the NMR studies, since the secondary, secondary hydride shifts would be expected to occur in this ion with a very low activation energy. It was estimated that the stable species would be $11 - 2 = 9$ kcal/mole (maximum) more stable than the classical ion. Jensen and Beck concluded that the observed spectra and rates of proton exchange can be adequately accounted for on the basis that carbon-bridged and hydrogen-bridged ions occur in the system as stable forms (Scheme III), or one of these two forms is a transition state and the other a stable species.* In reactions such as those studied by Nickon and Hammons (44) and in solvolyses (1,2,49–51), behaviors of the ions could be different from those under conditions of NMR studies (46). In these reactions, a major portion of product formation also could arise from equilibrating carbon-bridged and hydrogen-bridged ions (Scheme

* Very recently, G. A. Olah reported at the Twelfth Conference on Reaction Mechanisms, Brandeis University, June 19–22, 1968, that the Raman spectra of the norbornyl cation suggested that protonated nortricyclene may be a stable non-classical species.

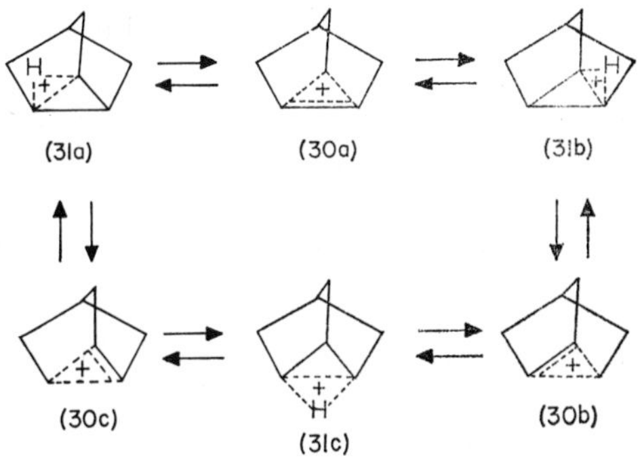

(31a)   (30a)   (31b)

(30c)   (31c)   (30b)

III). Probably, the carbon-bridged norbornonium ions **30a–c** might be the more stable species, with the edge-protonated cyclopropanes **31a–c** as transition states, although unequivocal differentiation between stable intermediates and transition states among these species would be very difficult.

D. Electrophilic Substitution Reactions with Cyclopropanes

1. Acylation

In 1957, Hart and Curtis (52) reported that cyclopropane was readily and exothermically absorbed by a solution of the 1:1 complex of acyl halides and $AlCl_3$ in chloroform at 0°. It was at first anticipated that the "normal" products would be the γ-chloroketones **32**.

$$R-\overset{O}{\overset{\|}{C}}-Cl + H_2C\overset{CH_2}{\overset{\triangle}{\underline{\quad\quad}}}CH_2 \xrightarrow{AlCl_3} R-\overset{O}{\overset{\|}{C}}-CH_2CH_2CH_2Cl$$

$$(32)$$

With $R = CH_3$, C_2H_5, $n\text{-}C_3H_7$, $i\text{-}C_3H_7$, $c\text{-}C_3H_5$, and C_6H_5, the "normal" products, **32**, were formed in lesser amounts than the "abnormal" products, the β-chloroketones **33**, or the corresponding olefin **34**, obtained

$$R-\overset{O}{\overset{\|}{C}}-\overset{}{\underset{\underset{CH_3}{|}}{CH}}-CH_2Cl \qquad\qquad R-\overset{O}{\overset{\|}{C}}-\overset{}{\underset{\underset{CH_3}{|}}{C}}=CH_2$$

(33)   (34)

after treatment with aqueous sodium bicarbonate, the latter treatment being applied in cases where the separation of the chloroketones in the initial products proved to be difficult. Similar results were obtained by

Hart and Levitt (53) when the work was extended to include acylation of substituted cyclopropanes:

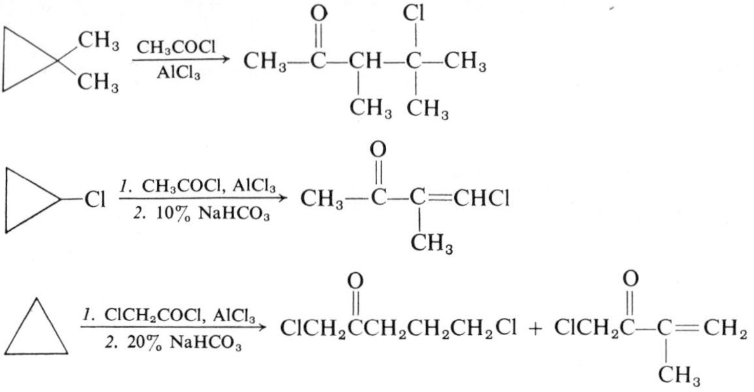

$$\triangleright\!\!<\!\!\begin{array}{c}CH_3\\CH_3\end{array}\quad\xrightarrow[AlCl_3]{CH_3COCl}\quad CH_3-\overset{\overset{\displaystyle O}{\|}}{C}-\underset{\underset{\displaystyle CH_3}{|}}{CH}-\underset{\underset{\displaystyle CH_3}{|}}{\overset{\overset{\displaystyle Cl}{|}}{C}}-CH_3$$

$$\triangleright\!\!-Cl\quad\xrightarrow[2.\ 10\%\ NaHCO_3]{1.\ CH_3COCl,\ AlCl_3}\quad CH_3-\overset{\overset{\displaystyle O}{\|}}{C}-\underset{\underset{\displaystyle CH_3}{|}}{C}=CHCl$$

$$\triangle\quad\xrightarrow[2.\ 20\%\ NaHCO_3]{1.\ ClCH_2COCl,\ AlCl_3}\quad ClCH_2\overset{\overset{\displaystyle O}{\|}}{C}CH_2CH_2CH_2Cl\ +\ ClCH_2\overset{\overset{\displaystyle O}{\|}}{C}-\underset{\underset{\displaystyle CH_3}{|}}{C}=CH_2$$

It was also noted that the reaction with chlorocyclopropane was slower and that the attempted acetylation of 1,1-dichlorocyclopropane resulted mainly in the recovery of unreacted material. With phenylcyclopropane, however, acetylation of the benzene ring predominated, giving rise to *p*-cyclopropylacetophenone as the major product.

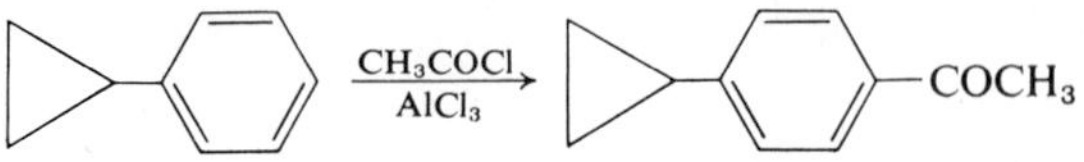

$$\triangleright\!\!-\!\!\bigcirc\quad\xrightarrow[AlCl_3]{CH_3COCl}\quad\triangleright\!\!-\!\!\bigcirc\!\!-COCH_3$$

A satisfactory explanation for the formation of both "normal" and "abnormal" acylation products of cyclopropanes remained elusive until more recently when protonated cyclopropanes were invoked as intermediates. In 1966, Hart and Schlosberg (54) reinvestigated the acetylation of cyclopropane and found by VPC analysis the presence of four products, the γ-chloroketone **35**, the β-chloroketone **36**, the olefin **37** and the newly observed α-chloroketone **38**. All four products were present when the reaction was carried out in a number of solvents, including CCl_4, CH_2Cl_2, CS_2, and nitrobenzene. From the NMR examination of the homogeneous

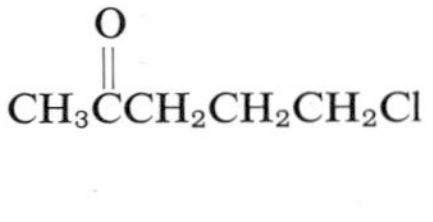

$$CH_3\overset{\overset{\displaystyle O}{\|}}{C}CH_2CH_2CH_2Cl$$

(35)

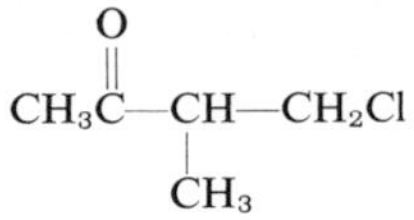

$$CH_3\overset{\overset{\displaystyle O}{\|}}{C}-\underset{\underset{\displaystyle CH_3}{|}}{CH}-CH_2Cl$$

(36)

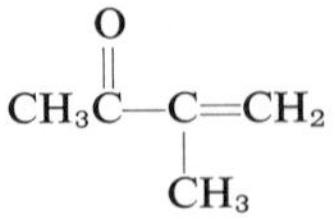

$$CH_3\overset{\overset{\displaystyle O}{\|}}{C}-\underset{\underset{\displaystyle CH_3}{|}}{C}=CH_2$$

(37)

$$CH_3\overset{\overset{\displaystyle O}{\|}}{C}-\underset{\underset{\displaystyle Cl}{|}}{CH}-CH_2CH_3$$

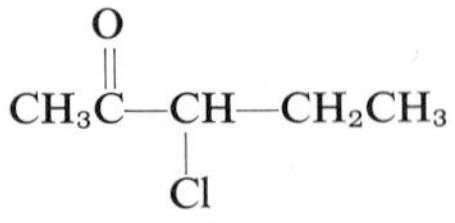

(38)

reaction mixture in CCl_4, it was shown that all four products were present before work-up; thus olefin 37 was formed directly, instead of by dehydrohalogenation during the work-up inferred in the earlier work (52,53). By carrying out the acetylation in the presence of added chloroketone 35, 36, or 38, or added acetylcyclopropane (39), it was demonstrated that the chloroketones did not interconvert during the reaction and that 39 did not suffer ring opening and thus could not be an intermediate in the acetylation of cyclopropane.

$$CH_3-\overset{\overset{\textstyle O}{\|}}{C}-HC\overset{\overset{\textstyle CH_2}{\triangle}}{}CH_2$$
(39)

$$CH_3\overset{\overset{\textstyle O}{\|}}{C}CH_2CH_2CH_2{}^+$$
(40)

Hart and Schlosberg (54) pointed out that any mechanism beginning with the formation of a classical carbonium ion, such as 40, would involve a series of irrational subsequent steps if such classical ions were to account for the observed products. Instead, they suggested that an acyl group could displace a proton from the cyclopropane, but that the proton remained associated with the cyclopropane ring as in 20. Structure 20 would react with a nucleophile (i.e., $AlCl_4{}^-$) to give 35 and 38, or would, by processes similar to those described by Baird and Aboderin (11), isomerize to 22. Reaction of 22 with nucleophile would give 36 and loss of the proton

$$CH_3CO-\overset{\overset{\textstyle H----CH_2}{\diagup\diagdown}}{CH}-CH_2$$
(20)

$$CH_3CO-\overset{\overset{\textstyle CH_2--H}{\diagup\diagup}}{CH}-CH_2$$
(22)

α to the carbonyl in 22 would give the olefin 37. It was further suggested that 22 should be more stable than 20 because the positive charge in 22 is farther away from the carbonyl group. One might then expect that if the nucleophile concentration were kept low to allow time for the conversion of 20 to 22, most of the product would be derived from 22. Hart and Schlosberg found that inverse addition gave results in accord with this hypothesis, the yields of 36 and 37 being increased at the expense of 35 and 38. For example, in CH_2Cl_2, the yields of 35, 36, 37, and 38 were changed by inverse addition from 24, 56, 16, and 4% to 4, 74, 20, and 2%, respectively. Because it was noted in the previous work (53) that electron-withdrawing substituents have a retarding effect on the rate of acylation of cyclopropane, Hart and Schlosberg regarded the formation of 20, or its immediate precursor, as the rate-determining step, and the subsequent steps as fast.

It is of interest to note that in the acetylation of nortricylene (24), Hart and Martin (55) obtained only the "normal" product, 6-acetyl-2-

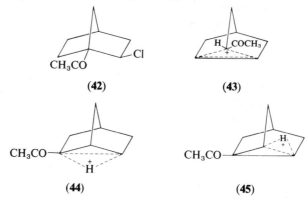

chloronorbornane (**41**). Although the stereochemical configurations of the substituents were not determined, from the work of Nickon and Hammons (44) it is probable that the 2-chloro group may be *exo* and the 6-acetyl group either *exo* or *endo*. The "abnormal" product, 1-acetyl-2-chloronorbornane (**42**) was not detected. Structure **41** could be derived from equilibrating classical ions, or by analogy with the conclusions of Nickon

and Hammon (44), from carbon-bridged ion **43**. If products were to be formed from edge-protonated species **44** and **45**, following the arguments of Hart and Schlosberg (54), one might expect **45** to be more stable than **44** and the major product should be **42**. The failure to detect **42** as a product is again in agreement with the suggestion that in the norbornyl system, product formation arises from reaction with carbon-bridged ions rather than with edge-protonated nortricyclenes.

2. Bromination

Deno and Lincoln (56) have subjected the addition of bromine to cyclopropane to careful examination. It was found that the reaction requires a Lewis acid ($FeBr_3$, $AlCl_3$, $AlBr_3$) as catalyst, and that 1,1-dibromopropane, 1,2-dibromopropane, 1,3-dibromopropane, and 1,1,2-tribromopropane were all produced (although under the conditions of some of the experiments the 1,1-dibromopropane was unstable; however, it did not give rise to the isomeric dibromopropanes on decomposition). Curiously, these workers also noted that chlorination of cyclopropane under similar conditions (but at a lower temperature), gave only 1,3-dichloropropane.

The production of all three dibromopropanes in the Lewis acid-catalyzed bromination of cyclopropane was interpreted to be the result of equilibration between protonated monobromocyclopropanes, c-$C_3H_6Br^+$. The mechanism was given by Deno and Lincoln as follows:

$$Br_2 + FeBr_3 \rightleftharpoons Br^+FeBr_4^-$$

$$Br^+ + c\text{-}C_3H_6 \rightleftharpoons \underset{(46)}{\overset{CH_2\ Br^+}{\underset{H_2C\text{——}CH_2}{\triangle}}} \rightleftharpoons \underset{(47)}{\overset{CHBr\ H^+}{\underset{H_2C\text{——}CH_2}{\triangle}}}$$

$$46 + Br^- \longrightarrow BrCH_2CH_2CH_2Br$$
$$47 + Br^- \longrightarrow Br_2CHCH_2CH_3 + BrCH_2BrCHCH_3$$

Structures **46** and **47** were meant to be noncommittal representations of the equilibrating c-$C_3H_6Br^+$ ions. With **46**, for example, it was suggested that either carbon-bridged or edge-attached geometries, **48**, or **49**, respectively, were attractive. At the time of the writing of that communication (September, 1966), Deno (28) was of the opinion that the evidence

$$\underset{(48)}{\overset{CH_2Br}{\underset{H_2C\text{——}CH_2}{\triangle^+}}} \qquad \underset{(49)}{\overset{CH_2\text{-}\text{-}\text{-}Br}{\underset{H_2C\text{——}CH_2}{\diagup^+}}}$$

then existing offered little basis for a choice between methyl-bridged ions (**12a,b**, Scheme I) and edge-protonated species (**14a–c**, Scheme I) in reactions involving protonated cyclopropanes. It was correctly pointed out (56) that the possible influence of isotope effects might invalidate the preference for predominant product formation from edge-protonated species, since this preference was based, to a large extent, on the observation of unequal amounts of deuterium at C-1 and C-2 (neglecting isotope effects) in the work of Baird and Aboderin (11). The possible consequence of isotope effects and the choice between methyl-bridged ions and edge-protonated cyclopropanes will be discussed more fully later in this review.

Deno and Lincoln (56) also pointed out that in textbooks a number of errors often appear regarding the bromination of cyclopropane. For example, the product is usually given as only 1,3-dibromopropane and that no catalyst is indicated, although in aromatic bromination reactions $FeBr_3$ is generally shown as the catalyst. In a competitive experiment between benzene and cyclopropane, $FeBr_3$ catalyzed bromination at

$-12°$ with a limited amount of bromine gave only bromobenzene and bromopropane (from HBr + cyclopropane), and no dibromopropane. Thus it was concluded that cyclopropane is much less reactive than benzene toward Br^+, and the Lewis acid catalyst is needed more in the bromination of cyclopropane than in the bromination of benzene. Subsequent to this work, a more detailed discussion on textbook errors regarding the halogenation of cyclopropane has been given by Gordon (57).

3. Friedel–Crafts Type of Reactions

It has been known for many years that cyclopropane, catalyzed by moist $AlCl_3$, can undergo Friedel–Crafts type of reactions in alkylating aromatic compounds (58–61). Recently, Deno and co-workers (29) have suggested that these reactions proceed with protonated cyclopropane, c-$C_3H_7^+$, acting as the electrophile in aromatic substitutions. It was found that the moist $AlCl_3$-catalyzed reaction between cyclopropane and C_6D_6 gave a mixture of 60% n-propylbenzene and 40% isopropylbenzene, one D being statistically incorporated into the n-propyl group of n-propylbenzene (28, 28, and 43% at C-1, C-2, and C-3, respectively) (29). The formation of the isopropyl product was interpreted as indicating that under the conditions of this reaction, the c-$C_3H_7^+$ ion had a relatively long lifetime and, therefore, was able to isomerize partly to the more stable 2-propyl cation before reacting with benzene. The statistical distribution of the D-label in the n-propyl group of n-propylbenzene was also regarded as indicative of a long lifetime for the c-$C_3H_6D^+$, and thus it was able to achieve complete equilibration before being captured.

Deno and co-workers (29) also found that in the alkylation of benzene derivatives with c-$C_3H_7^+$ (cyclopropane + moist $AlCl_3$), the percentage of isopropyl product increased as the aromatic became more resistant to alkylation. At 25°, the reaction with p-xylene, toluene, benzene, and chlorobenzene gave n-propyl to isopropyl product ratios of 92:8, 83:17, 70:30, and 31:69, respectively. These observations were interpreted again in terms of the lifetime of the protonated cyclopropane, c-$C_3H_7^+$. In reactions with aromatics that are more resistant to alkylation, such as chlorobenzene, the c-$C_3H_7^+$ was postulated to have a longer lifetime, and a greater opportunity to rearrange to the more stable 2-propyl cation. These results, together with the data indicating that the 1-propyl cation can give rise to protonated cyclopropane (to be discussed in Section IV) led Deno to conclude that the order of stability of $C_3H_7^+$ species is established as

$$\text{2-propyl}^+ > c\text{-}C_3H_7^+ > \text{1-propyl}^+$$

IV. PROTONATED CYCLOPROPANES FROM ALIPHATIC AND ALICYCLIC SYSTEMS NOT CONTAINING THE CYCLOPROPYL GROUP

A. The Deamination of 1-Propylamine

Roberts and Halmann (62) were the first workers to use isotopic tracers in the study of the nitrous acid deamination of 1-propylamine. From treatment of $1\text{-}^{14}C\text{-}1$-propylammonium perchlorate in 35% $HClO_4$ with a solution of $NaNO_2$ at 25°, both 1-propanol-^{14}C and 2-propanol-^{14}C were obtained. Product isolation was aided by adding inactive 1-propanol as carrier and the two alcohols were separated by fractionation through a Podbielniak Micro Column. The 1-propanol-^{14}C so obtained was further diluted with carrier and then degraded by oxidation to propionic acid followed by a Schmidt reaction to give ethylamine. It was found that the p-bromobenzenesulfonamide derivative of the ethylamine contained 8.5 $\pm$ 1% of the ^{14}C-activity. Roberts and Halmann suggested that this isotope position rearrangement of about 8.5% from C-1 to C-2 and C-3 was due to a 1,2-methyl shift, via the methyl-bridged ion **50**. This mechanism would place the 8.5% of the label at the C-2 position of the 1-propanol-^{14}C obtained in this reaction.

$$CH_3$$
$$H_2C\text{---}^{14}CH_2$$
(50)

A repetition of the work of Roberts and Halmann, but with a more complete degradation, was reported by Reutov and Shatkina in 1962 (10). In this reinvestigation, both inactive 1-propanol and 2-propanol were added as carriers during product isolation, and the alcohols were again separated by fractional distillation; the type of fractionating column employed, however, was not specified. The 1-propanol-^{14}C was degraded by oxidation to propionic acid and then to acetic acid, followed by the conversion of the latter via the Schmidt reaction to methylamine. In this way, the ^{14}C-activity at each of the three carbon positions of the 1-propanol-^{14}C was ascertained. It was reported that 8.0 $\pm$ 0.8% position rearrangement of the ^{14}C-label was found but the rearranged isotope was located entirely at C-3, no activity being observed at C-2. It was proposed that in the 1-propyl cation, either a one-stage 1,3-hydride shift or a two-stage successive 1,2-hydride shifts could account for these observations.

$$CH_3CH_2{}^{14}CH_2{}^+ \;\rightleftarrows\; {}^+CH_2CH_2{}^{14}CH_3$$
$$CH_3CH_2{}^{14}CH_2{}^+ \;\rightleftarrows\; CH_3\overset{+}{C}H{}^{14}CH_3 \;\rightleftarrows\; {}^+CH_2CH_2{}^{14}CH_3$$

Further results reported by Reutov at the 19th International Congress of Pure and Applied Chemistry (63) led to the conclusion that the 1,3-hydride shift was the preferred mechanism.

The apparent demonstration of the 1,3-hydride shift in the 1-propyl cation by Reutov has had a considerable influence on subsequent interpretations of related data obtained by other research workers. For example, as mentioned in the Introduction, Skell and Starer (7) observed in 1960 that the hydrocarbon products from the deamination of 1-propylamine and from the deoxideation reaction with 1-propanol both consisted of 90% propene and 10% cyclopropane. The symmetrical face-protonated cyclopropane, **4**, was suggested as being involved in the cyclopropane formation. If some 1-propanol were also derived from **4**, the work of Roberts and Halmann (62), and of Reutov and Shatkina (10) would have produced 1-propanol-^{14}C with the ^{14}C-label rearranged from C-1 to both C-2 as well as C-3. Since Reutov and Shatkina reported that the rearrangement was solely from C-1 to C-3, Skell and Starer in 1962 (9) considered that protonated cyclopropane **4** was excluded from playing any role in the deamination of 1-propylamine, and regarded the formation of cyclopropane in this reaction as arising from a "1,3-interaction," with the transition state formulated as **51**. In this communication (9), Skell and Starer also reported

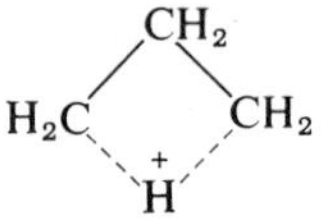

(51)

that the cyclopropane obtained from the deoxideation of 1-propanol-1,1-d_2 consisted of 94 ± 2% cyclopropane-d_2 and 5–6% cyclopropane-d_1. It was suggested that the likely route for the monodeuteriocyclopropane was a 1,3-interaction following the 1,3-hydride shift.

$$CH_3CH_2CD_2^+ \xrightarrow{-H^+} \text{propene} + \text{cyclopropane-}d_2$$

$$^+CH_2CH_2CD_2H \xrightarrow{-D^+} \text{propene} + \text{cyclopropane-}d_1$$

In an attempt to differentiate between 1,3-hydride shift and successive 1,2-hydride shifts as the explanation of the reported data of Reutov and Shatkina (10), Karabatsos and Orzech (64) studied the deamination reaction by treating 1-propylammonium-1,1,2,2-d_4 perchlorate in 35% HClO$_4$ with NaNO$_2$. It was anticipated that NMR examination of the 1-propanol-d_4 produced would provide a differentiation between **52** and **53**.

$$CH_3CD_2CD_2{}^+ \xrightarrow{1,3} {}^+CH_2CD_2CHD_2 \longrightarrow CHD_2CD_2CH_2OH$$
$$(52)$$

$$CH_3CD_2CD_2{}^+ \underset{\xleftarrow{}}{\overset{1,2}{\rightleftharpoons}} CH_3\overset{+}{C}DCD_3 \underset{\xleftarrow{}}{\overset{1,2}{\rightleftharpoons}} {}^+CH_2CHDCD_3 \longrightarrow CD_3CHDCH_2OH$$
$$(53)$$

Karabatsos and Orzech obtained a 36% yield of deuterated propanols consisting of 30% 1-propanol-d_4 and 70% 2-propanol-d_4. These were separated by VPC and examined by NMR. The spectrum of the 1-propanol-d_4 showed that besides the expected peaks for the hydroxyl and methyl protons, there was a broad singlet at 6.58τ assigned to the C-1 protons and a weak multiplet appearing in the C-2 proton region (8.5τ). The integrated area of the 8.5τ absorption was about 20% of the area of the signal at 6.58τ. It was pointed out, in a footnote, that the 20% value of the 8.5τ signal might be construed as indicating that the rearrangement could have occurred 60% via a 1,3-shift and 40% via successive 1,2-shifts; however, it was also considered that in such a combination of 1,3- and successive 1,2-shifts, the 6.58τ signal would not have been a clean singlet. These workers, therefore, concluded that the rearrangement was due mainly, if not exclusively, to a 1,3-hydride shift. The possibility of involving protonated cyclopropane to account for the presence of H at both C-1 and C-2 was not considered. The process via successive 1,2-hydride shifts was further eliminated by Karabatsos and Orzech (64) from the observation that the NMR spectrum of the 2-propanol-d_4 obtained in their studies showed no increase in proton absorption at C-2. Moreover, it was noted that the 2-propyl cation could not rearrange to the 1-propyl cation since the deamination of 2-propylamine gave only 2-propanol without a trace of 1-propanol. Similarly, the absence of 1-propanol from the deamination of 2-propylamine recorded in the earlier literature and the absence of any cyclopropane in the deoxideation with 2-propanol were cited by Skell and Starer (9) as evidence against successive 1,2-hydride shifts.

The preference for 1,3-hydride shift over protonated cyclopropane by Skell and Starer (9) and by Karabatsos and Orzech (64) were also in part due to the finding in their respective laboratories that the rearrangement from the neopentyl to the t-amyl cations did not involve a protonated cyclopropane. Karabatsos and Graham (65) reported that the treatment of 1-^{13}C-neopentyl alcohol with concentrated HBr gave 2-methyl-2-butene as a main product. If protonated cyclopropane 54 were involved, equal amounts of the ^{13}C-label would be located at C-3 and C-4 of the 2-methyl-2-butene (55 and 56), if isotope effects were neglected.

Such products would be distinguishable from each other by proton NMR spectroscopy because of the large difference in spin–spin coupling constants between ^{13}C and H attached to it (about 130 cps) and between ^{13}C and H on carbons removed from the ^{13}C by one or two bonds (4–6 cps). The spectrum of the 2-methyl-2-butene product showed ^{13}C side bands for the methyl protons with a $J_{^{13}C-CH_3}$ value of 4.8 cps, suggesting no detectable

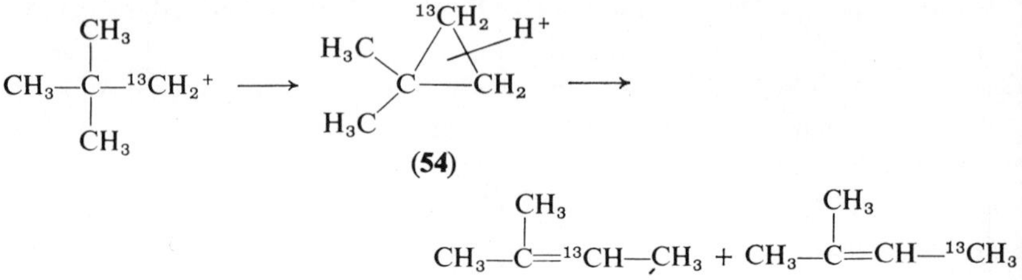

(54)

(55)   (56)

amount of **56** with ^{13}C at C-4. Moreover, analyses of the side band areas indicated that C-3 of the 2-methyl-2-butene contained as much excess ^{13}C as C-1 of the neopentyl alcohol. Thus the 2-methyl-2-butene product was **55**, with no detectable **56**, and it was concluded that a mechanism via protonated cyclopropane **54** would not be the main pathway in the rearrangement. Similarly, Skell, Starer, and Krapcho (66) studied the deoxideation with neopentyl-1,1-d_2 alcohol. From infrared and NMR examinations, the olefinic products, 2-methyl-1-butene and 2-methyl-2-butene, were found to contain D only at C-3, with no evidence of any D at C-4. These results again excluded protonated cyclopropane **57** as an intermediate in the neopentyl system, since **57** would give rise to olefins

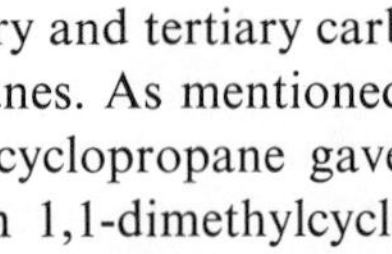

(57)

with D at both C-3 and C-4. More recent results of Deno and co-workers have indicated that secondary and tertiary carbonium ions are more stable than protonated cyclopropanes. As mentioned earlier (Section III-C), the addition of HCl to methylcyclopropane gave only 2-chlorobutane (29), while similar reactions with 1,1-dimethylcyclopropane gave only t-amyl products (29b). Thus the t-amyl cation most probably is much more stable than either edge- or face-protonated cyclopropane that might be derived from the neopentyl system. It is, therefore, not surprising that evidence for **54** or **57** was not detected in rearrangements of neopentyl to t-amyl cations. Up to about 1962, however, the work of Reutov, Karabatsos, and Skell appears to have firmly established the 1-3-hydride shift, and thus the 1,3-shift was strongly emphasized in a review on carbonium ions which appeared in an earlier volume in this series (67).

In discussing the equilibration of edge-protonated cyclopropane intermediates during the solvolysis of cyclopropane with D_2SO_4, Baird and Aboderin (11) suggested that edge-protonated cyclopropanes might also be responsible for the isotope position rearrangements in the 1-propanol from the deamination of 1-propylamine. It was pointed out that **58** and **59**, respectively, would be formed initially in the sytems studied by Reutov

$$
\begin{array}{cccc}
\underset{\text{(58)}}{\overset{\displaystyle H_2C\text{----}H}{H_2C\text{---}^{14}CH_2}} &
\underset{\text{(59)}}{\overset{\displaystyle H_2C\text{----}H}{D_2C\text{----}CD_2}} &
\underset{\text{(60)}}{\overset{\displaystyle H\text{----}CH_2}{H_2C\text{---}^{14}CH_2}} &
\underset{\text{(61)}}{\overset{\displaystyle D\text{----}CH_2}{D_2C\text{----}CDH}}
\end{array}
$$

and Shatkina (10) and by Karabatsos and Orzech (64). The 1-propanol derived from **58** or **59** would have the ^{14}C- or H-label located only at C-1 and C-3 as reported (10,64). To account for the reported absence of any label at C-2, it was assumed that in the deamination reaction the lifetime of the edge-protonated species was not long enough to allow for their equilibration with isomeric ions, such as **60** or **61**, that will lead to some ^{14}C or H located at the C-2 position of the 1-propanol.

In the last of the three papers by Baird and Aboderin on protonated cyclopropanes, the deamination of 1-propylamine-3,3,3-d_3 was reported to give a cyclopropane consisting of $43 \pm 1\%$ cyclopropane-d_2 and $57 \pm 1\%$ cyclopropane-d_3 (68). The relative amount of cyclopropane-d_3 observed was apparently too high to be readily accountable by the 1,3-hydride shift mechanism. Utilizing the data then existing, the following arguments may be presented (69). The formation of cyclopropane-d_2 and cyclopropane-d_3 following the 1,3-hydride shift could be depicted as follows:

$$
\underset{91.5\%}{\overset{\displaystyle CD_3}{H_2C\text{----}CH_2{}^+}}
\quad\rightleftharpoons\quad
\underset{8.5\%}{\overset{\displaystyle CD_2{}^+}{H_2C\text{----}CH_2D}}
$$

$$
\begin{array}{c}
-D^+ \Big\downarrow 3k_D \\
c\text{-}C_3H_4D_2 \\
\textbf{62-}d_2
\end{array}
\qquad\qquad
\begin{array}{cc}
-D^+ \Big\downarrow k_D & -H^+ \Big\downarrow 2k_H \\
c\text{-}C_3H_4D_2 & c\text{-}C_3H_3D_3 \\
\textbf{62-}d_2 & \textbf{62-}d_3
\end{array}
$$

Assuming $k_H = nk_D$, the percentage of cyclopropane-d_3 may be calculated from the following expression:

$$\% \; \textbf{62-}d_3 = \frac{(8.5)\,(2nk_D)}{(91.5)(3k_D) + (8.5)\,(k_D) + (8.5)\,(2nk_D)} \times 100$$

$$= \frac{17n}{283 + 17n} \times 100$$

If there were no isotope effect in the deprotonation ($n = 1$), one would predict no more than 6% **62-**d_3 according to this mechanism. With k_H/k_D as large as 7, the calculated relative amount of **62-**d_3 would be only about 30%, much less than the observed value of about 57%. Therefore, the idea that **62-**d_3 was derived from the 1-propyl cation after a 1,3-hydride shift appeared to be questionable, and the likelihood was enhanced in favor of cyclopropane formation from deprotonation of edge-protonated cyclopropane intermediates, as discussed in the preceding paragraph.

In order to conform to the reported overall 1,3-isotope position rearrangements, Aboderin and Baird (68) considered that in the deamination,

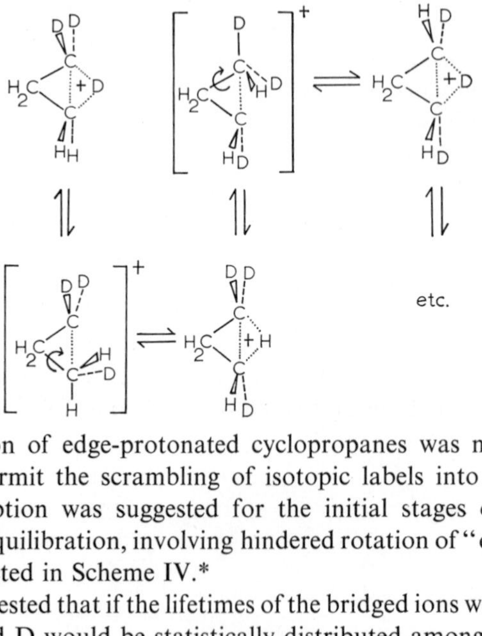

the equilibration of edge-protonated cyclopropanes was not sufficiently extensive to permit the scrambling of isotopic labels into C-2. A more detailed description was suggested for the initial stages of protonated cyclopropane equilibration, involving hindered rotation of "quasi-methyl" groups as depicted in Scheme IV.*

It was suggested that if the lifetimes of the bridged ions were sufficiently long, the H and D would be statistically distributed among the five positions involved. Loss of the bridging H or D would lead to cyclopropane-d_3

---

* This scheme was termed by Aboderin and Baird (68) as the Wiberg mechanism because it was suggested to these authors by Professor K. B. Wiberg.

or cyclopropane-d_2, respectively. By this mechanism, with the five hydrogen and deuterium atoms at C-1 and C-3 equivalent, an isotope effect for deprotonation of $k_H/k_D = 2.0$ would account for the observed result of $57 \pm 1\%$ cyclopropane-d_3. It was further pointed out that with still longer lifetimes for the bridged ions, as in solvolyses, subsequent rearrangements to involve the remaining two hydrogens at C-2 could take place via a transition state resembling the methyl-bridged ion.

With the demonstration in 1965 (12,13) that part of the deamination of 1-propylamine took place via essentially completely equilibrated protonated cyclopropanes which actually led to rearrangements of isotopic labels from C-1 equally to both C-2 and C-3 (see below), detailed differentiations between Scheme I and Scheme IV have become less essential. Lee and Kruger (45) have pointed out that with the three carbon positions (and all the hydrogen and deuterium atoms) in the protonated cyclopropane system derived from 1-propylamine-3,3,3-d_3 approaching complete equivalence, on purely statistical grounds, loss of a proton or a deuteron should give a $4:3$ ratio for cyclopropane-d_3 to cyclopropane-d_2, and this is almost exactly the same as the ratio of $57:43\%$ observed by Aboderin and Baird (68). The fact that it was not necessary to invoke isotope effects in explaining the results would indicate that the deprotonation is likely a fast process. In contrast to the deamination of 1-propylamine-3,3,3-d_3, the deoxideation of 1-propanol-1,1-d_2 gave only 5–6% cyclopropane-d_1, with $94 \pm 2\%$ cyclopropane-d_2 (9). It was suggested (45) that under the highly basic condition used in the deoxideation, the mixing of the hydrogen and deuterium atoms through equilibration of edge-protonated cyclopropanes likely would compete much less effectively with deprotonation, and this would result in a cyclopropane consisting predominantly of cyclopropane-d_2.

That protonated cyclopropane intermediates are likely responsible for the isotope position rearrangements observed in the deamination of labeled 1-propylamine was established from studies with 1-propylamine-2,2-d_2 (63-2-d_2), 1-propylamine-1-t (63-1-t) and 1-propylamine-1-^{14}C (63-1-^{14}C) reported from this laboratory, initially in 1965 (12,13) and in more detail in 1967 (45). In order to differentiate between the 1,3-hydride shift and successive 1,2-hydride shifts, it was anticipated that if only 1,3-shifts were involved, the NMR spectrum of the 1-propanol-d_2 obtained from the deamination of 63-2-d_2 would show no increase in C-2 proton absorption. The deamination of 63-2-d_2 was carried out under the conditions used by Roberts and Halmann (62) and the 1-propanol-d_2 and 2-propanol-d_2 products were recovered and purified by preparative VPC. The pertinent parts of the NMR spectra of these products and the starting material, **63**-2-d_2, as well as the spectra of the corresponding protio compounds are shown in Figure 1. From Figure 1*a* and *b*, it is seen that the 1-propanol-d_2 product appeared to show more C-2 proton absorption (ca. τ 8.45) than the original deuterated amine, thus suggesting that some of the C-2 deuterium has been rearranged, presumably to C-1 and/or C-3. Although the absorption at ca. τ 8.45 was too small to permit reliable integration, these results did suggest that 1,3-hydride shifts could not have been solely responsible for the isotope position rearrangements observed in the deamination of 1-propylamine.

If successive 1,2-hydride shifts were to play a role in the deamination of **63**-2-d_2, the 2-propanol-d_2 obtained should show more C-2 proton absorption than **63**-2-d_2. Figure 1*a* and *c*, however, appeared to indicate no significant increase in C-2 proton absorption (ca. τ 6.66) for the 2-propanol-d_2 product. This finding thus confirmed the earlier conclusions of Karabatsos and Orzech (64), and of Skell and Starer (9) that successive 1,2-hydride shifts could not be an important process in the deamination of 1-propylamine.

In order to obtain a more quantitative evaluation of the extents of isotopic scrambling suggested by the NMR studies with **63**-2-d_2, the deamination of **63**-1-*t* was carried out. The 1-propanol-*t* fraction from the product was purified by preparative VPC and degraded by oxidation, first to propionic acid and then to acetic acid. The activity of the propionic acid gave the *t*-content of C-2 and C-3 and the activity of the acetic acid showed the *t*-content of C-3. Each liquid sample was assayed after conversion to an appropriate solid derivative which had been repeatedly crystallized to constant specific activity to ensure radiochemical purity. The results showed that only about 3% of the *t*-label was rearranged from C-1 to C-2 and C-3. The rearranged isotope, however, was located at both C-2 and C-3 since only slightly more than one-half of the rearranged *t*-label was found at C-3. These findings, therefore, firmly established that the isotopic scrambling in the 1-propanol was not due solely to a 1,3-hydride shift.

To clarify the discrepancy between the above results and those reported by Reutov and Shatkina (10), Lee and Kruger (13,45) reinvestigated once more the deamination of **63**-1-[14]C. After the deamination reaction, the 1-propanol-^{14}C fraction from the product was purified by VPC before being degraded by oxidation to propionic acid and then to acetic acid, followed by the conversion of the latter to methylamine (10). Again, to ensure radiochemical purity, the samples were assayed as solid derivatives which had been repeatedly recrystallized until their specific activities were constant. The results showed that about 4% of the C-1 label

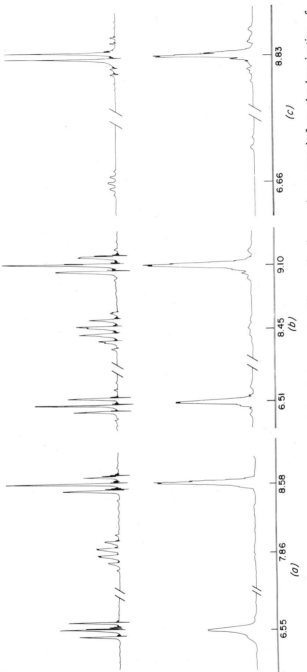

Fig. 1. NMR spectra of deuterio- (bottom row) and the corresponding protio- (top row) compounds from the deamination of 1-propylamine-2,2-d_2 (**63**-2-d_2). (*a*) Forty per cent solution of perchlorate salt of **63** and **63**-2-d_2 in D_2O. (*b*) 1-Propanol and 1-propanol-d_2. (*c*) 2-Propanol and 2-propanol-d_2.

was rearranged, and the rearranged ^{14}C was almost equally distributed at C-2 and C-3. The difference between these results and those reported earlier (10,62) might have been due largely to the purity of the compounds assayed for radioactivity. It was suggested (45) that, for example, if the 1-propanol-^{14}C were contaminated with some 2-propanol-^{14}C, and if the 2-propanol-^{14}C were oxidized to acetic acid during the degradation, a higher rearrangement with an apparent preponderance of activity at C-3 would result. Collins (70) has also pointed out to this reviewer that in the earlier work (10,62), the use of large amounts of inactive carrier, such as 1-propanol, might have diluted only the active 1-propanol without diluting the radioactive impurities, and some of these highly active impurities could have remained after the degradation. Inactive 1-propanol was also used as carrier by Lee and co-workers (12,13,45) in their studies with **63**-1-t and **63**-1-^{14}C; but in these instances, the compounds assayed were all rigorously purified by repeated crystallizations until the specific activities were constant.

The above results from **63**-2-d_2, **63**-1-t, and **63**-1-^{14}C clearly supported the involvement of protonated cyclopropane in giving rise to the isotopically scrambled 1-propanol. The almost equal isotopic distribution of the label at C-2 and C-3 from the work with **63**-1-t and **63**-1-^{14}C suggested that in the deamination of 1-propylamine, a portion of the reaction has proceeded through protonated cyclopropane intermediates in which the three carbon positions have become equivalent or were approaching complete equivalence. For the deamination reaction, therefore, it was not possible to differentiate between face-protonated cyclopropane, rapidly equilibrating edge-protonated cyclopropanes, or rapidly equilibrating methyl-bridged ions. Further discussions on the possibility of differentiating between these alternatives will be given in Section V. Since only 3–4% of the isotopic label has rearranged from C-1 to C-2 and C-3, only about 4–6% of the 1-propanol obtained in the deamination was derived from protonated cyclopropane intermediates. Lee and Kruger thus concluded that the 1-propyl cation from the deamination of 1-propylamine could react with solvent to give 1-propanol without rearrangement, deprotonate to give propene, undergo irreversible 1,2-hydride shift which will lead to 2-propanol and propene, and cyclize to protonated cyclopropane intermediates which, in turn, could either deprotonate to give cyclopropane or react with solvent to give isotopically scrambled 1-propanol.

The conclusion that the isotopic scrambling in the 1-propanol from the deamination of 1-propylamine resulted from protonated cyclopropane intermediates was independently arrived at in 1965 by Karabatsos, Orzech, and Meyerson (14). These workers studied the deamination of

1-propylamine-1,1-d_2 (**63**-1-d_2) and 1-propylamine-2,2-d_2 (**63**-2-d_2) and examined the resulting 1-propanol-d_2 by mass spectrometry. The parentless-ethyl ions from the trimethylsilyl ether of the 1-propanol were used as measures of the isotopic composition of the C-1 methylene group of the 1-propanol. After appropriate corrections, the net results were summarized as follows:

$$CH_3CH_2CD_2NH_2 \xrightarrow{0°} C_2H_5-CD_2OH + C_2H_4D-CHDOH + C_2H_3D_2-CH_2OH$$
$$\begin{array}{llll} \text{63-1-}d_2 \\ 100\%\ d_2 & \quad 97.8\% & \quad 0.6\% & \quad 1.6\% \end{array}$$

$$CH_3CH_2CD_2NH_2 \xrightarrow{40°} C_2H_5-CD_2OH + C_2H_4D-CHDOH + C_2H_3D_2-CH_2OH$$
$$\begin{array}{llll} \text{63-1-}d_2 & \quad \textbf{(64)} & \quad \textbf{(65)} & \quad \textbf{(66)} \\ 100\%\ d_2 & \quad 95.7\%;\ 96.0\% & \quad 1.0\%;\ 0.8\% & \quad 3.3\%;\ 3.2\% \end{array}$$

$$CH_3CD_2CH_2NH_2 \xrightarrow{40°} C_2H_5-CD_2OH + C_2H_4D-CHDOH + C_2H_3D_2-CH_2OH$$
$$\begin{array}{llll} \text{63-2-}d_2 & \quad \textbf{(64a)} & \quad \textbf{(65a)} & \quad \textbf{(66a)} \\ 100\%\ d_2 & \quad 1.2\% & \quad 0.9\% & \quad 97.9\% \end{array}$$

The above data indicated once more that the 1-propanol arose mainly from a path leading to isotopically unrearranged alcohol and partly from a path leading to extensively rearranged alcohol. The higher concentration of **64a** than **65a** from **63**-2-d_2 was regarded as evidence against reversible 1,2-hydride shifts since a simple set of reversible 1,2-hydride shifts beginning with $CH_3CD_2CH_2{}^+$ could lead to **65a**, whereas the formation of **64a** would require the occurrence of several 1,2-shifts before reaction with solvent.

The finding that the deamination at 40° of either **63**-1-d_2 or **63**-2-d_2 gave the same concentration of $C_2H_4D-CHDOH$ ([**65**] = [**65a**]) was regarded as highly significant since this indicated that **63**-1-d_2 and **63**-2-d_2 must have given rise to the same isotope position intermediate, or its equivalent, prior to the formation of the isotopically scrambled 1-propanol. It was concluded that this intermediate could be best formulated as some protonated cyclopropane. The contribution of the protonated cyclopropane path to the formation of 1-propanol was estimated to be about 3% at 0° and about 5% at 40°; this is in good agreement with the estimated contribution of 4–6% at 25° from the studies with **63**-1-t and **63**-1-^{14}C (12,13). The earlier conclusion of Karabatsos and Orzech (64) that the isotopically rearranged 1-propanol from the deamination of 1-propylamine-1,1,2,2-d_4 resulted "mainly, if not exclusively" from 1,3-hydride shifts was regarded as in error chiefly because of the limitations involved in assaying minor amounts of proton absorption by NMR. It was pointed out, for example, that in the work with 1-propylamine-1,1,2,2-d_4, a major source of error was the presence of some 1-propylamine-1,2,2-d_3 which was not detected by NMR.

In formulating the protonated cyclopropane pathway, Karabatsos, Orzech, and Meyerson (14) considered the mechanism shown in Scheme V.

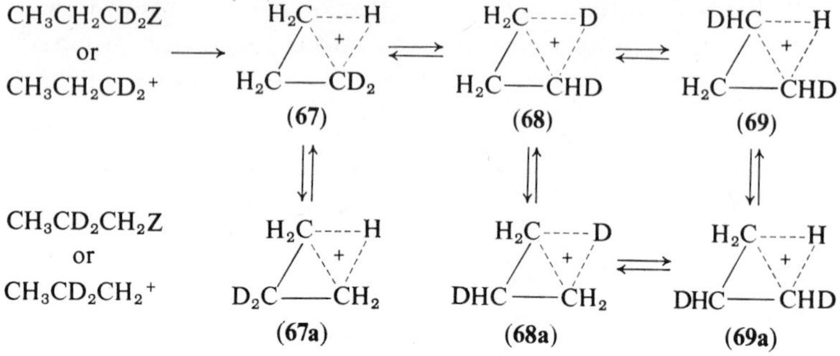

$CH_3CH_2CD_2Z$
or $\longrightarrow$
$CH_3CH_2CD_2{}^+$

(diagram of bridged species 67, 68, 69 in top row; 67a, 68a, 69a in bottom row)

$CH_3CD_2CH_2Z$
or
$CH_3CD_2CH_2{}^+$

Scheme V

In order to account for the finding that the concentration of C_2H_4D—CHDOH was the same from either 63-1-d_2 or 63-2-d_2 ([65] = [65a]), it was suggested that, according to Scheme V, 67 $\rightleftarrows$ 67a would have to be much faster than the reaction of these ions with solvent to form 1-propanol, and this would be indistinguishable from a symmetrical protonated cyclopropane. To explain the finding that from 63-2-d_2, the concentration of C_2H_5—CD_2OH (64a), obtainable from reaction of 67, was greater than that of C_2H_4D—CHDOH (65a), obtainable from reaction of 68, it was suggested that the mechanism as given in Scheme V would require 67 $\rightleftarrows$ 68 to be slower than 67 $\rightleftarrows$ 67a. In the work of Baird and Aboderin (11) on the solvolysis of cyclopropane with D_2SO_4, to account for the presence of the smallest amount of deuterium at the C-2 position of the 1-propyl product, it was concluded that in the mechanism shown in Scheme I (p. 136), 14a $\rightleftarrows$ 14b (analogous to 67 $\rightleftarrows$ 68) was faster than 14b $\rightleftarrows$ 14c (analogous to 67 $\rightleftarrows$ 67a), and this would be opposite to the suggested required condition to account for the concentration of 64a being greater than the concentration of 65a. Hence Karabatsos and co-workers (14) ruled out Scheme V as a valid mechanism. As alternatives, these workers proposed processes A and B, given in Scheme VI. For path A to be correct, the initially formed methyl-bridged ion, 70, would have to proceed to 67 and 67a as fast or faster than it could react with solvent to give 1-propanol. For path B (direct formation of edge-protonated species without the intervention of methyl-bridged ions) to be correct, it was suggested that 67 and 67a would have to be formed in the same ratio from either labeled species 63-1-d_2 or 63-2-d_2.

From the above discussion based on the arguments advanced by Karabatsos and co-workers (14), it would appear that in the deamination

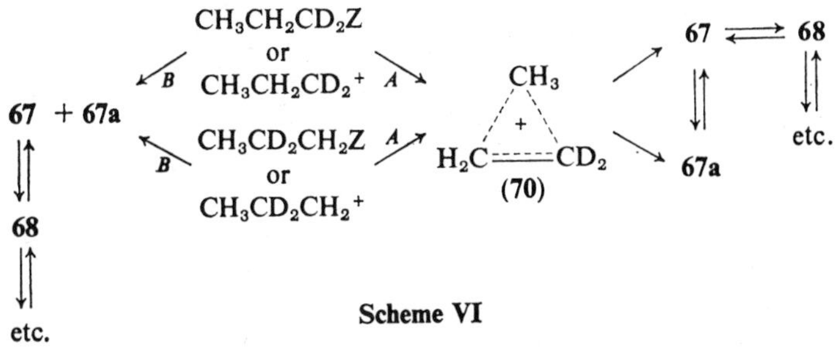

$$CH_3CH_2CD_2Z$$

or

$$67 \rightleftharpoons 68$$

Scheme VI

of 1-propylamine, initial formation of a methyl-bridged ion would be possible (Scheme VI, path *A*), but the direct formation of an edge-protonated cyclopropane would require some restrictions (Scheme VI, path *B*, but not Scheme V). However, the differentiation between Scheme V and path *B*, Scheme VI may not be necessary. Unlike the behavior observed during sulfuric acid solvolysis of cyclopropane where complete equivalence of all three carbon positions was not attained (11,30,31), the deamination of **63**-1-*t* and **63**-1-^{14}C (12,13) indicated that the resulting isotopically scrambled 1-propanol was derived from protonated cyclopropane intermediates in which all three carbon positions were essentially equivalent. The greater energy contents of the ions derived from deamination than from solvolysis likely may have played a part in accounting for this difference. In terms of equilibrating protonated cyclopropanes, for the ions derived from the deamination, all equilibrations shown in Scheme V, including **67** ⇌ **67a** and **67** ⇌ **68**, would be fast relative to reaction with solvent. With **67** ⇌ **67a** and other equilibrations being fast processes, Scheme V and Scheme VI, path *B*, would become indistinguishable. To account for the greater concentration of **64a** over **65a** from **63**-2-d_2, instead of **67** ⇌ **68** being slower than **67** ⇌ **67a**, there could be an isotope effect favoring the formation of **64a** from **67**, which involved the breaking of a partial C—H bond, over the formation of **65a** from **68**, which involved the breaking of a partial C—D bond. Thus the protonated cyclopropane pathway to 1-propanol in the deamination of 1-propylamine could be formulated by equilibrating edge-protonated cyclopropanes, with or without the intervention of methyl-bridged ions.

B. Formation of Cyclopropanes in Carbonium Ion Reactions

1. Deaminations

As mentioned in the preceding section, the hydrocarbon products from the nitrous acid deamination of 1-propylamine were found to consist of 10% cyclopropane and 90% propene (7) and that this cyclopropane likely resulted from deprotonation of protonated cyclopropane intermediates. In a number of other deaminations of primary amines, cyclopropanes have also been reported, and protonated cyclopropanes have probably played an important part in their formation. Among the earliest of such observations was that of Silver (71,72), who reported initially in 1960 (71), that the deamination of 3-methyl-2-butylamine in acetic acid gave among the products about 15% 1,2-dimethylcyclopropane. A similar reaction with isoamylamine gave about 2% 1,2-dimethylcyclopropane, and no 1,1-dimethylcyclopropane. From both reactions, the *cis:trans* ratio of the 1,2-dimethylcyclopropane was about 1:2. After rejecting the possibility of a carbene mechanism, it was suggested that loss of H^+ from methyl-bridged ion **71** might be responsible for the 1,2-dimethylcyclopropane formation. This mechanism is essentially that of deprotonation of a protonated cyclopropane.

$$CH_3-HC \stackrel{\displaystyle \overset{CH_3}{\diagup \; \overset{+}{} \; \diagdown}}{=\!=\!=\!=} CH-CH_3$$
(71)

A considerable amount of work has been done on the deamination of isobutylamine (**72**) and *n*-butylamine (**73**) as these systems could provide data on the effects of a methyl substituent at the C-2 or C-3 position of 1-propylamine. Friedman and co-workers (73) have studied the diazotization of **72** and **73** in aprotic media (solvents not proton donors) and found that the products were markedly different from those obtained in protic media. Some of the results on the hydrocarbons produced are given in Table II. It is seen that in aprotic media, hydrocarbon yields tended to be enhanced, while skeletal rearrangements and double bond migration were minimized and cyclopropane formation was significantly increased. It was suggested that in aqueous systems, the cationic species would be stabilized by solvation and thus rearrangement to thermodynamically more stable intermediates would occur. On the other hand, under aprotic conditions, products would be derived from kinetic rather than thermodynamic factors. In addition to other carbonium ion processes such as rearrangement, elimination, and substitution, it was proposed that poorly solvated (unstabilized) cations formed in aprotic media might yield, by neighboring group participation (internal solvation) the more stable protonated cyclopropane intermediate, and cyclopropanes could then result by simple loss of proton.

With *sec*-butylamine, Friedman and co-workers (73) found that in the

TABLE II

Hydrocarbon Products from Diazotization of Isobutylamine and n-Butylamine[a]

Acid	Solvent	Percent yield of hydrocarbons	Methyl-cyclopropane	Iso-butene	trans-2-Butene	cis-2-Butene	1-Butene
						Composition	
Isobutylamine							
HOAc	CHCl₃	33	14	73	4.2	2.1	6.9
HCl	CHCl₃	8	15	57	5.0	3.7	19
HOAc	Glyme	35	10	76	4.1	2.2	7.7
HCl	Glyme	20	9.7	68	5.6	3.2	13
HOAc	HOAc	20	4.5	62	14	7.1	12
HOAc	50% aq. HOAc	12	2.5	40	25	11	
n-Butylamine							
HOAc	CHCl₃	8	4.3		4.5	2	89
HCl	CHCl₃	9	3.1		3.1	2	92
HOAc	Glyme	18	2.9		6.6	3	88
HCl	Glyme	12	3.5		6.6	3	87
HOAc	HOAc—NaNO₂	5	0.7		25	11	63
HOAc	50% HOAc—NaNO₂	11	0.3		29	14	56

[a] Amine (0.005 mole), acid (0.005 mole), and alkyl nitrite (0.0055 mole) in 10 ml of solvent at reflux.

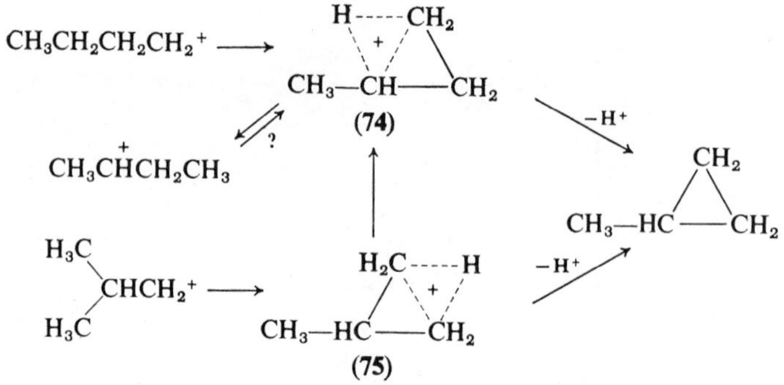

$$CH_3CH_2CH_2CH_2{}^+ \longrightarrow$$

$$\underset{(74)}{\overset{H\text{-----}CH_2}{\underset{CH_3-CH\text{----}CH_2}{\diagdown\,\overset{+}{\diagup}}}}$$

$$\underset{CH_3\overset{+}{CH}CH_2CH_3}{}$$

$$\underset{H_3C}{\overset{H_3C}{\diagdown}}CHCH_2{}^+ \longrightarrow$$

$$\underset{(75)}{\overset{H_2C\text{-----}H}{\underset{CH_3-HC\text{----}CH_2}{\diagdown\,\overset{+}{\diagup}}}}$$

$$-H^+$$

$$\overset{CH_2}{\underset{CH_3-HC\text{----}CH_2}{\diagup\diagdown}}$$

hydrocarbon products, the 2-butenes predominated when diazotization was carried out in aqueous media, while aprotic diazotization gave mainly 1-butene. About 4% of methylcyclopropane was also observed. Since the *n*-butyl and *sec*-butyl cations did not give any isobutyl or *t*-butyl derivatives (no *t*-butyl derivative could be derived from either 74 or 75), it was concluded that edge-protonated methylcyclopropane 74 rather than 75 could best describe the intermediate.

To investigate the possibility that carbenic and cationic processes might be occurring concurrently during diazotization, Jurewicz and Friedman (74) studied the deamination of isobutylamine-1,1-d_2 (72-1-d_2). Some pertinent processes are given below.

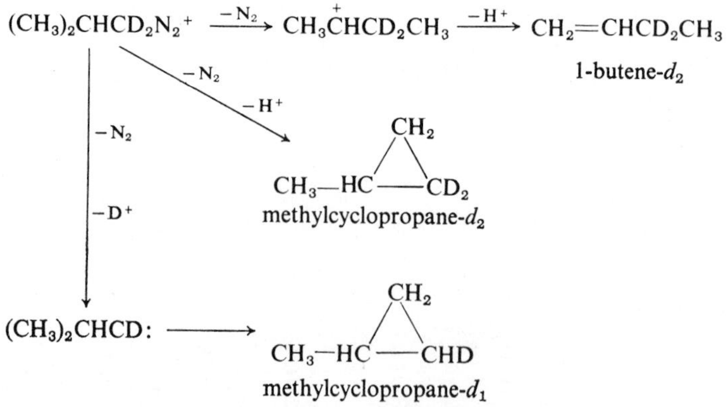

$$(CH_3)_2CHCD_2N_2{}^+ \xrightarrow{-N_2} CH_3\overset{+}{C}HCD_2CH_3 \xrightarrow{-H^+} CH_2{=}CHCD_2CH_3$$

1-butene-d_2

$$-N_2$$
$$-H^+$$

$$\overset{CH_2}{\underset{CH_3-HC\text{----}CD_2}{\diagup\diagdown}}$$
methylcyclopropane-d_2

$$-N_2$$
$$-D^+$$

$$(CH_3)_2CHCD: \longrightarrow \underset{CH_3-HC\text{----}CHD}{\overset{CH_2}{\diagup\diagdown}}$$
methylcyclopropane-d_1

Because significant amounts of hydrogen–deuterium exchange were found to take place during aprotic diazotization (75), an absolute deuterium determination of the methylcyclopropane would not differentiate between the cationic and carbenic mechanisms. However, since 1-butene could arise only from a cationic process, the difference in absolute deuterium content of the 1-butene and methylcyclopropane from 72-1-d_2

would be a measure of the carbenic process. It was found by low-voltage mass spectrometry that from the diazotization of **72**-1-d_2 in chloroform or in DOAc–D$_2$O, the deuterium contents in the 1-butene and methylcyclopropane were essentially the same, indicating that carbenic intermediates were not involved under either aprotic or protic conditions. There was, however, a minor difference of about 1–2% less deuterium in the methylcyclopropane, and this was attributed to a limited involvement of equilibrating deuteronated (or protonated) methylcyclopropane, which could lose a portion of its deuterium in its conversion to methylcyclopropane.

Karabatsos and co-workers also studied the deamination of isobutylamine-1,1-d_2 (**72**-1-d_2) (76) and of 1-butylamine-1,1-d_2, 2,2-d_2, and 3,3-d_2 (**73**-1-d_2, 2-d_2, and 3-d_2) (77) and analyzed the products by mass spectrometry and NMR. The deamination of **72**-1-d_2, carried out in aqueous HClO$_4$, gave a 70% yield of an alcohol mixture composed of 71.5% *t*-butyl, 18.0% *sec*-butyl, and 10.5% isobutyl alcohols. The isobutyl alcohol was found to be the unrearranged (CH$_3$)$_2$CHCD$_2$OH, thus eliminating the involvement of **76** and **77** in its formation. The *sec*-butyl alcohol consisted

$$
\begin{array}{cc}
\text{H}_2\text{C}\text{-----}\text{H} & \text{H}_2\text{C}\text{-----}\text{D} \\
\diagup \;\; + \;\; \diagdown & \diagup \;\; + \;\; \diagdown \\
\text{CH}_3\text{—HC}\text{——}\text{CD}_2 & \text{CH}_3\text{—HC}\text{——}\text{CHD} \\
\textbf{(76)} & \textbf{(77)}
\end{array}
$$

of about 93% **78** and 7% **79**, the latter being attributed to interconversion between 2-butyl cations. From these results, Karabatsos and co-workers (76) concluded that in the deamination of **72**, protonated methylcyclopropane could be detected only by the formation of methylcyclopropane. Apparently protonated methylcyclopropane could rearrange to the stable secondary 2-butyl cation much faster than the equilibration between isomeric protonated methylcyclopropanes, but not fast enough to prevent the formation of methylcyclopropane. Another factor which might further

$$
\begin{array}{c}
(\text{CH}_3)_2\text{CHCD}_2{}^+ \longrightarrow \quad \overset{\text{CH}_3}{\underset{\text{CH}_3\text{HC}\!\!=\!\!=\!\!\text{CD}_2}{\diagup\;+\;\diagdown}} \\[2em]
\downarrow \qquad \swarrow \\[1em]
\overset{+}{\text{CH}_3\text{CHCD}_2\text{CH}_3} \;\rightleftharpoons\; \overset{+}{\text{CH}_3\text{CHDCDCH}_3} \\[1em]
\downarrow \qquad\qquad\qquad \downarrow \\[1em]
\underset{\text{OH}}{\text{CH}_3\text{CHCD}_2\text{CH}_3} \qquad \underset{\text{OH}}{\text{CH}_3\text{CHDCDCH}_3} \\[0.5em]
\textbf{(78)} \qquad\qquad\qquad \textbf{(79)}
\end{array}
$$

hinder such equilibrations was suggested to be the 1,2-eclipsing interaction illustrated in **80**. The lack of equilibration between protonated methyl-cyclopropanes in the deamination of **72** noted by Karabatsos is somewhat at variance with the findings of Friedman and co-workers (73,74), who concluded that there was a *limited* involvement of equilibrating protonated methylcyclopropanes in the deamination of this amine; the reaction media used by these two groups of workers, however, were different.

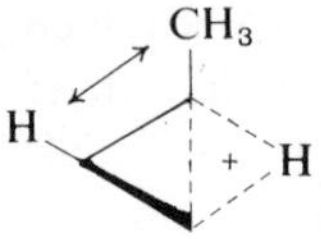

(80)

The deamination of any one of the three deuterated 1-butylamines, **73**-1-d_2, 2-d_2, or 3-d_2 (**77**) gave a 1-butanol fraction which was less than 0.1% isotopically rearranged. Thus the carbon-bridged species such as **81** and

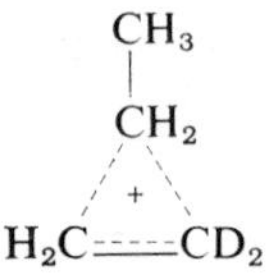

(81) 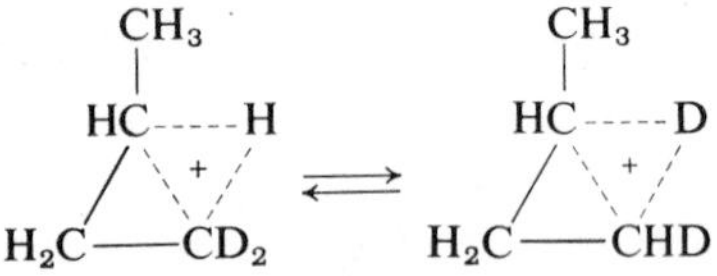   **(82)**

H—D mixing processes such as **82** were excluded from the paths leading to 1-butanol from **73**-1-d_2, 2d_2, or 3-d_2. The results from analyses of the 2-butanol fractions are summarized as follows:

$$CH_3CH_2CH_2CD_2NH_2 \longrightarrow CH_3CH_2\underset{OH}{CH}CHD_2 + CH_3\underset{OH}{CH}CH_2CHD_2$$

100% d_2 74%   26%

$$CH_3CH_2CD_2CH_2NH_2 \longrightarrow$$
$$CH_3CH_2\underset{OH}{CD}CH_2D + CH_3\underset{OH}{CH}CHDCH_2D + CH_3\underset{OH}{CH}CD_2CH_3$$

100% d_2 75%   25%

$$CH_3CD_2CH_2CH_2NH_2 \longrightarrow CH_3CD_2\underset{OH}{CH}CH_3 + CH_3\underset{OH}{CD}CHDCH_3$$

100% d_2 83%   17%

No nominally 1,3-hydride shift was observed, thus no 2-butanol was formed from edge-protonated methylcyclopropane **74**. All the 2-butanol products could be accounted for by 1,2-hydride shifts, with 2-butyl cation ⇆ 2-butyl cation playing a part (the two alcohols, $CH_3CH(OH)$-$CHDCH_2D$ and $CH_3CH(OH)CD_2CH_3$, constituting 25% of the 2-

butanol from **73**-2-d_2 could not be distinguished by the mass spectral data). From these results, Karabatsos and co-workers (77) again concluded that the only experimental evidence for protonated methylcyclopropane in the deamination of **73** was the formation of methylcyclopropane. Repulsive interaction depicted by **80** again was suggested as playing a role in minimizing further equilibrations.

In contrast to the formation of some methylcyclopropane in the deamination of **72** and **73**, no 1,1-dimethylcyclopropane was found in the deamination of neopentylamine (**83**) in protic media (7,71,78,79); Friedman and co-workers (73), however, have stated in a footnote that about 1% dimethylcyclopropane was formed by aprotic diazotization of **83**. Using **83** labeled at C-1 with ^{13}C or D, Karabatsos and co-workers (79) found that the *t*-amyl alcohol obtained contained the label only at the C-3 position, thus ruling out the intervention of 1,3-hydride shifts, protonated cyclopropanes, or H-bridged ions in the rearrangement from the neopentyl to the *t*-amyl systems. The stability of the *t*-amyl cation obviously was of controlling importance.

It may be worthwhile to mention at this point that in 1956, Fort and Roberts (80) attempted to find, but did not observe, any 1,3-shifts in the nitrous acid deamination of 3-phenyl-1-^{14}C-1-propylamine and of 3-(*p*-methoxyphenyl)-1-^{14}C-1-propylamine. The products obtained were the corresponding 3-aryl-1-propanol, 1-aryl-2-propanol (from a 1,2-hydride shift) and 3-arylpropene, with negligible amounts of isotope position rearrangement in all cases. These results apparently ruled out for these reactions the intermediacy of protonated arylcyclopropanes such as **84** and **85**. Since no 1-aryl-1-propanol was detected, even though the 1-aryl-1-propyl cation would be more stable than the 1-aryl-2-propyl cation,

possibly in these systems, repulsive 1,2-eclipsing interaction between Ar and H in **84** and **85**, analogous to that depicted in **80** (76), may have assumed greater importance. It would, however, be of interest to ascertain whether any arylcyclopropanes were produced in these reactions, since Fort and Roberts apparently did not look for these products.

The highest yields of cyclopropane derivatives from deamination reactions were reported by Dauben and Laug (81,82). These workers found that when *trans*-8-hydrindanylcarbinylamine (**86**) was allowed to react with HNO$_2$, tricyclo[4.3.1.0^{1,6}]decane (**87**) was formed in 74% yield.

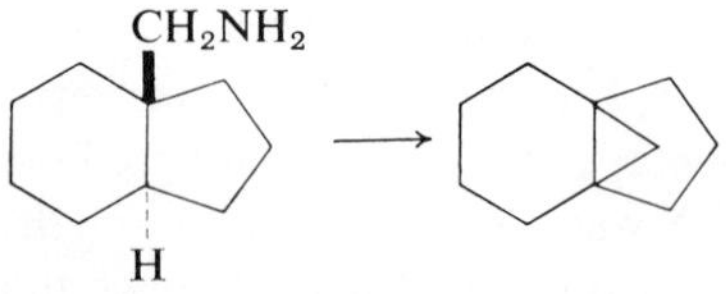

(86)   (87)

Similarly, the deamination of 18-amino-5α-pregnane-3β,20-diol **(88)** with
HNO₂ in HOAc gave 14β,18-cyclo-5α-pregnane-3β,20-diol **(89)** in 35–45%
yield. Since **86** and **88** are really substituted neopentylamines, the high yields
of products containing the cyclopropane ring are in sharp contrast to the
absence of any cyclopropane derivative in similar deaminations of neopentylamine itself. Dauben and Laug suggested that the first step in these

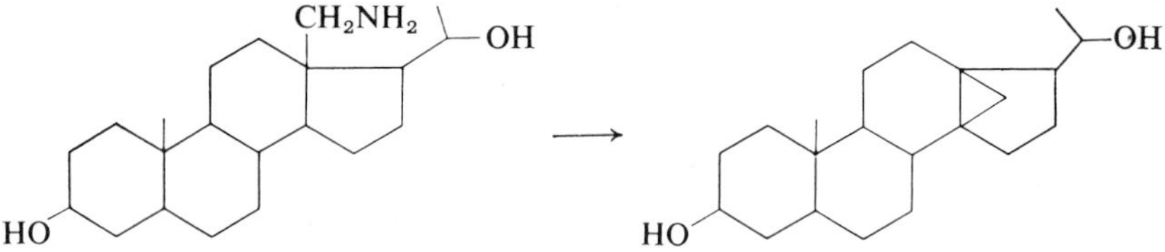

(88)   (89)

reactions is likely a Wagner–Meerwein rearrangement to give the bridgehead carbonium ion **90**. Collapse of such an ion to form an olefin would
be retarded since the resulting double bond would be highly strained by
being at a bridgehead. Intramolecular alkylation of the carbonium ion
by the other bridgehead carbon atom with concomitant loss of a proton
was proposed as the path giving rise to the cyclopropane ring system, the
alkylation being facilitated by the spatial proximity of the two carbon

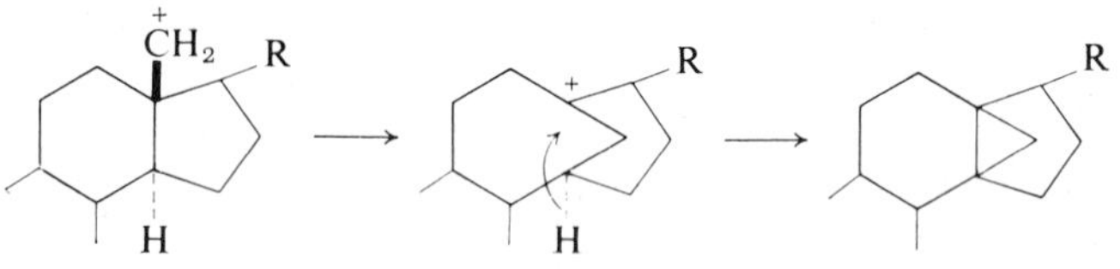

(90)

atoms involved. Whether a protonated cyclopropane would or would not
play a part in the intramolecular alkylation has not been clarified; however, it is conceivable that this alkylation could proceed via a protonated
cyclopropane, followed by loss of a proton to give the final product.

A significant contribution to the chemistry of protonated cyclopropanes is the work of Edwards and Lesage (83) who studied the deamination of certain alicyclic α-aminoketones and found the formation of
cyclopropanes as one of the main components among a variety of prod-

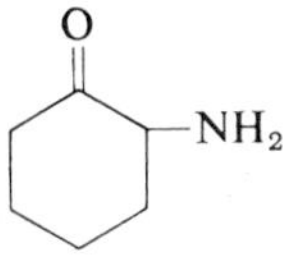

(91) 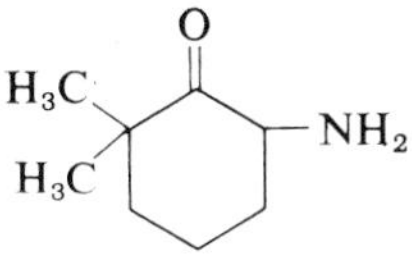 (92) 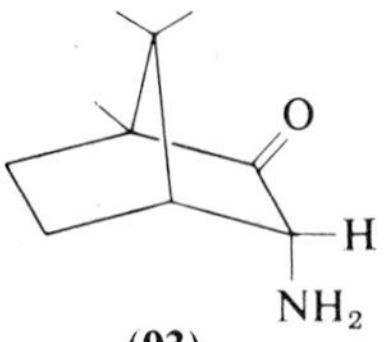 (93)

ucts. Specifically, the compounds investigated were 2-aminocyclohexanone (91), 2-amino-6,6-dimethylcyclohexanone (92), and 3-*endo*-aminocamphor (93). The cyclopropane derivatives obtained were, respectively, bicyclo-[3.1.0]hexan-2-one (94), 3,3-dimethylbicyclo[3.1.0]hexan-2-one (95), and cyclocamphanone (96). As an illustration of the variety of products and

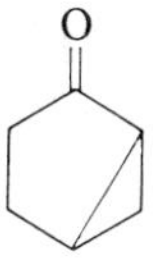

(94) 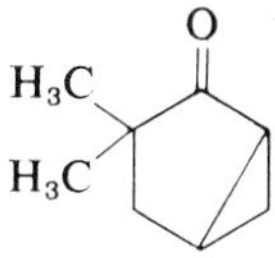 (95) 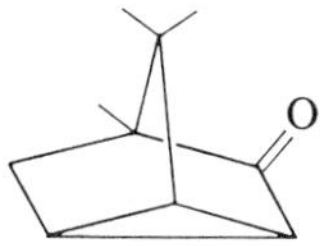 (96)

their possible mechanistic implications, the results from 91 are presented below in more detail.

Solutions of the hydrochloride of 91 in aqueous H_2SO_4 or $HClO_4$ were treated with $NaNO_2$. The products were separated into readily extractable neutral and acid fractions and very water-soluble materials. The latter proved to be mainly 3-hydroxy-2-methylcyclopentanone (97) (7–10%) and adipic acid (small quantities were recovered, presumably from the oxidation of α-hydroxycyclohexanone). The main product recovered was cyclopentanecarboxylic acid, the maximum yield obtained being about 50%. The extractable neutral fraction contained at least eight components, the two major ones being 2-methyl-2-cyclopenten-1-one (98) (about 16%) and the bicyclic ketone 94 (6.5–9%). The infrared spectrum, indicated that the conjugated cyclohexenone (99) was essentially absent. The variety of these products is in sharp contrast to the relatively simple product composition obtained from a reexamination of the deamination of cyclohexylamine by Edwards and Lesage (averaging about 80% cyclo-hexanol, 14% cyclohexyl nitrite, 4% cyclohexene, and 2% bicyclo-[3.1.0]hexane).

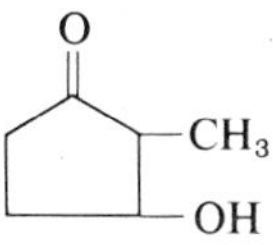

(97) 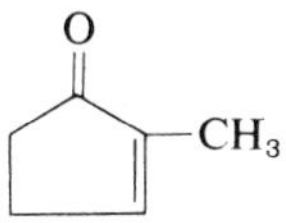 (98) 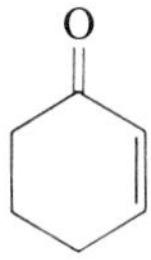 (99)

In discussing the above observations, Edwards and Lesage suggested that the generation of a carbonium ion adjacent to a carbonyl would be energetically unfavorable. One would then expect that such an ion would be delocalized, if possible, by participation of nearby C—C and C—H bonds, and this was borne out by the high yields of unusual products in the deamination of α-aminoketones. Of pertinence to protonated cyclopropanes is the proposal that the major neutral products from **91**, namely **94**, **97**, and **98**, were all derived from a common set of bridged ion intermediates. The first intermediate was formulated as **100** derived from neighboring C—C bond participation. Structure **100** then could be converted, via H-bridged ion **101**, to methyl-bridged ion **102**. Classical ion **103**, which would give rise to **97** and **98**, could be derived from **102**, and deprotonation of **101** would give **94**. Interconversion between **100** and **102** via **101** is really the equilibration between carbon-bridged ions via an

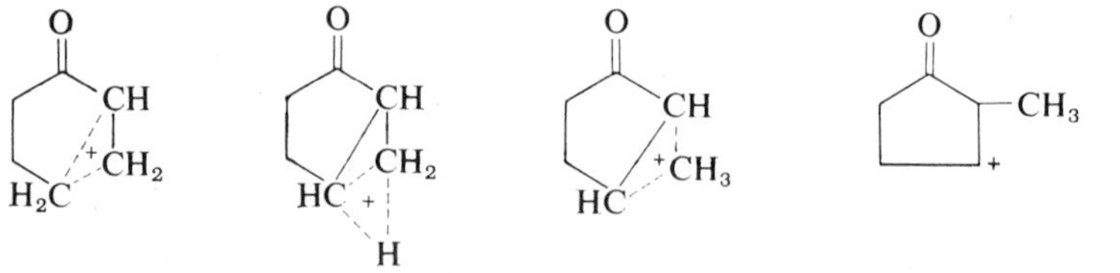

(100)   (101)   (102)   (103)

edge-protonated cyclopropane. It is noteworthy that Edwards and Lesage reported these ideas some six months prior to the publication of the classical work of Baird and Aboderin (11) on equilibrating protonated cyclopropanes. Edwards and Lesage also stated, in footnote 11 of their paper, that although the work of Karabatsos and Graham (65) and of Skell, Starer, and Krapcho (66) (discussed on p. 153 of the present review) constituted good evidence that symmetrically protonated cyclopropanes are not intermediates in methyl migration in carbonium ion reactions, "edge-protonated" cyclopropanes have not been excluded, and they expressed the belief that such intermediates would give rise to cyclopropanes in high yields. Again, it is worthwhile to note that these ideas were expressed in 1963, when the concept of protonated cyclopropanes had not yet gained wide acceptance.

From the above discussion, it is seen that in a number of deamination reactions, cyclopropanes, presumably arising from deprotonation of protonated cyclopropanes, can be found among the reaction products, generally as minor constituents. Probably the most important factors controlling the extents of cyclopropane formation are the relative stabilities of the intermediates involved in the competing reaction pathways. With simple aliphatic systems, protonated cyclopropanes appear to be more stable than primary carbonium ions, while secondary and tertiary carbonium ions are much more stable than protonated cyclopropanes. That relatively more cyclopropanes can be obtained under conditions of kinetic control has been demonstrated by the work of Friedman and co-workers on aprotic diazotizations. If the initially generated carbonium ion is especially unstable energetically, such as in the deamination of α-amino-ketones, neighboring group participations, including cyclopropane formation, have been found to be enhanced.

2. Other Reactions

The formation of cyclopropanes from deoxideation of alkoxides with bromoform in strong base has been reported in a number of systems by Skell and co-workers. The fact that the deoxideation with 1-propanol and the deamination of 1-propylamine both gave a hydrocarbon product consisting of 10% cyclopropane and 90% propene was one of the criteria which led Skell and Starer (7,8) to suggest a carbonium ion mechanism for deoxideations. These workers (7,8) also noted that in the deoxideation with iso-, n-, and sec-butyl alcohols, methylcyclopropane was formed in 4, 2, and less than 0.5%, respectively, among the C_4H_8 products. These results are fairly similar to those observed in the deamination of the butyl-amines in protic media (73), and the source of methylcyclopropane may be the deprotonation of protonated methylcyclopropane suggested for the deamination reactions. No cyclopropane derivatives were detected from the deoxideation with isopropyl, t-butyl, t-amyl, and neopentyl alcohols (7). In the deoxideation with 2-methyl-1-butanol, Skell and Maxwell (84) found 2.1% ethylcyclopropane and 2.0% trans-1,2-dimethyl-cyclopropane among the hydrocarbon products. No analogous study on the deamination of 2-methyl-1-butylamine has been reported.

Friedman, Shechter, and co-workers (85) have reported that deoxidea-tion studies with cyclopropylcarbinol yielded cyclobutene, methylene-cyclopropane, bicyclo[1.1.0]butane, 1,3-butadiene, ethylene, and acetylene. The source of the bicyclobutane has not been clearly established. These authors suggested that the reaction apparently involved highly energetic intramolecular cationic paths including carbon–hydrogen insertion, carbon skeleton rearrangement, and fragmentation. These same workers also reported that the aprotic diazotization of cyclopropylcarbinylamine resulted in extensive intramolecular insertion in which bicyclobutane was the principal product. Although the possible role of protonated cyclo-propanes was discussed in previous studies on aprotic diazotization of the butylamines (73,74), protonated cyclopropane was not invoked in discussing the formation of bicyclobutane. The similarity in the intramolecular products was noted for the aprotic diazotization of cyclopropylcarbinylamine and for the cationic decomposition of sodium cyclopropanecarboxaldehyde p-tosylhydrazone in proton-donor solvents (86), and it was suggested (85) that these products appeared to be derived from poorly solvated cyclopropylmethyldiazonium ion intermediates which were highly energized (for a suggested mechanism for bicyclobutane formation from the decomposition of the p-tosylhydrazone, see ref. 87).

In a study on the electrolysis of some simple aliphatic acids at carbon anodes, Koehl (88) has observed the formation of products derived from carbonium ion-like intermediates. Thus, for example, anodic oxidation of butanoic acid at a carbon anode in the presence of an excess of the acid gave among the products cyclopropane and propene in a molar ratio of 1:2. Similarly, the electrolysis of pentanoic acid, 3-methylbutanoic acid, and 2-methylbutanoic acid gave, respectively, 8–13%, 7–10%, and 3–5% methylcyclopropane among the C_4H_8 hydrocarbons. The proportions of cyclopropane or methylcyclopropane formed in these electrolyses were higher than those found in the corresponding deoxideation or deamination reactions carried out in aqueous media. However, it might be worth noting that the amounts of methylcyclopropane obtained were fairly similar to those observed in the aprotic diazotization of the butylamines (73). In discussing the mechanism of these electrolyses, Koehl suggested that at the anode, an alkyl radical could be generated via the usual Kolbe sequence and then oxidized to a carbonium ion of high energy. These highly energetic carbonium ions would then yield products by elimination to give cyclopropanes and part of the olefins and by skeletal rearrangement to more stable intermediates which, in turn, would yield olefins and substitution products. Such an explanation of the formation of relatively large proportions of cyclopropanes is not dissimilar from the views of Edwards and Lesage (83) on the stabilization of highly energetic carbonium ions by the protonated cyclopropane pathway, nor from the ideas of Friedman and co-workers (73) on aprotic diazotization to give rise to poorly solvated and thus less stable carbonium ions which led to more cyclopropane formation.

In discussing cyclopropane formation as evidence for protonated cyclopropanes, it should also be mentioned that in the extensively investigated acetolysis of exo-2-norbornyl brosylate, more recently, Winstein and co-workers (89) have reported the formation of about 4% elimination product, the composition of which was 98% nortricyclene and 2% norbornene. In view of the recent conclusion of Olah favoring protonated nortricyclene as a stable intermediate (see footnote on p. 144 and ref. 90), it is quite conceivable that the nortricyclene obtained in the acetolysis could have been derived from deprotonation of protonated nortricyclene. In a study of a related system, Benjamin, Ponder, and Collins (91) have found that the hydrolysis of the dideuterated tosylate **104** in aqueous acetone (containing sodium carbonate) led to the quantitative production of **105**, **106**, and **107** in yields of 25, 60, and 15%, respectively. These results were rationalized on the basis of a stereospecific

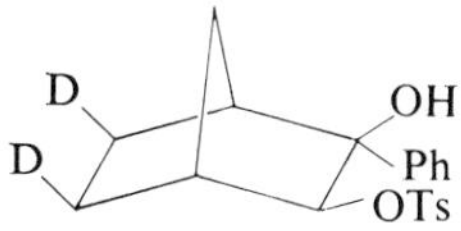

(104) 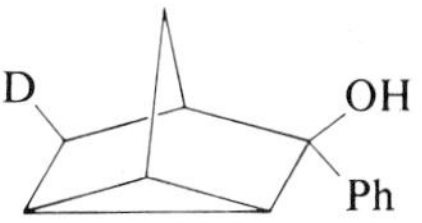   (105)

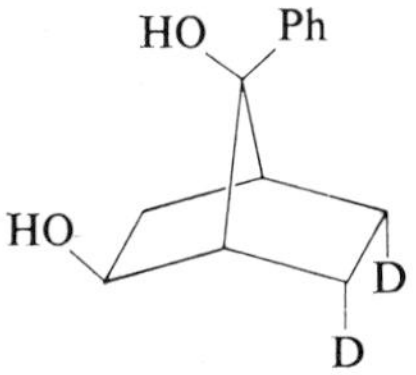

(106) 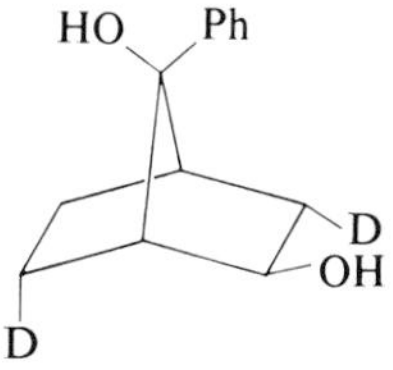   (107)

deuterium migration via deuteronated cyclopropane **109**, with the loss of a deuteron from **109** as the suggested source of the phenylnortricyclanol **105**.

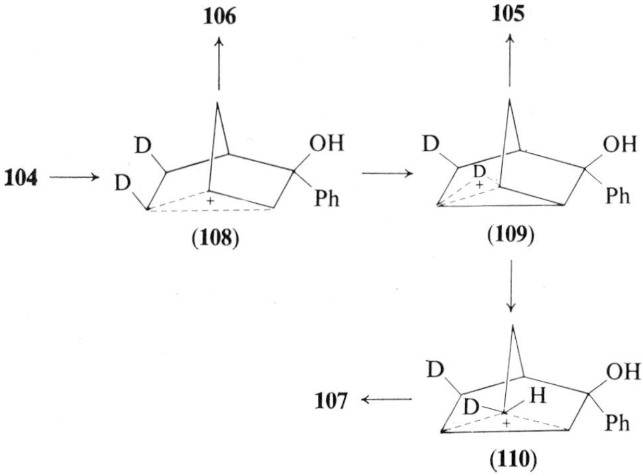

106   105

104 ⟶ (108)   (109)

107 ⟵ (110)

C. Isotopic Scrambling Studies with the 1-Propyl System

Besides the work on labeled 1-propylamine discussed in Section IV-A, a number of other reactions involving labeled 1-propyl derivatives have been investigated (23,30,33,92). Lee and Kruger (92) reported that the acetolysis of $1\text{-}^{14}\text{C}$-1-propyl tosylate (111) gave chiefly the unrearranged 1-propyl acetate, with less than 1% 2-propyl acetate, and there was no isotopic scrambling in the 1-propyl ester. The formolysis of 111 also gave mainly 1-propyl formate together with $1\text{--}2\%$ 2-propyl formate. A small amount of isotope position rearrangement was found in the 1-propyl formate, and by the use of a sample of 111 of high specific activity (about 4×10^6 cpm/mmole), it was ascertained that the total rearrangement of the ^{14}C-label from C-1 was 0.83%, with 0.68% found at C-3 and 0.15% at C-2. This small amount of isotopic scrambling was attributed to a very small portion of the reaction proceeding through equilibrating edge-protonated cyclopropanes 112a–c. It was suggested that more product

$$
\begin{array}{ccc}
\overset{14}{\text{H}}\text{---}\text{CH}_2 & \overset{14}{\text{H}_2\text{C}}\text{----}\text{H} & ^{14}\text{CH}_2 \\
\diagdown \; \overset{+}{} \; \diagup & \diagup \; \overset{+}{} \; \diagdown & \diagup \qquad \diagdown \\
\text{H}_2\text{C}\text{-----}\text{CH}_2 & \text{H}_2\text{C}\text{-----}\text{CH}_2 & \text{H}_2\text{C}\text{----}\text{CH}_2 \\
& & \overset{+}{} \\
& & \text{H} \\
(112a) & (112b) & (112c)
\end{array}
$$

was derived from 112a and 112b, formed in the earlier stages of the equilibration, than from 112c, thus accounting for the presence of a greater amount of the rearranged label at C-3 than C-2.

In contrast to the above results (92), other workers have failed to detect any significant amount of isotope position rearrangement in the solvolysis of labeled 1-propyl tosylate in formic acid. In their study on the acetolysis and formolysis of 111, Reutov and co-workers (93) found, in agreement with Lee and Kruger, no isotopic scrambling in the acetolysis. With the formolysis, however, although the degradation of the 1-propyl formate obtained gave a sample of active acetic acid suggesting the possibility of a total of about 0.9% rearrangement, the products of further degradation of the acetic acid showed no detectable activity. It was thus concluded that the formolysis of 111 also resulted in an insignificant degree of isotopic scrambling. These workers (93) further suggested that a possible factor contributing to the difference between their results and those of Lee and Kruger might be the difference in reaction time (12 hr in refluxing formic acid compared to 18 hr used by Lee and Kruger). Similarly, Karabatsos, Fry, and Meyerson (23) have reported that the "hydrolysis" of $CH_3CH_2CD_2OTs$ in 99% formic acid at 75° gave 1-propanol with no measurable isotope position rearrangement on the basis of mass spectral data. It was stated in a footnote that under their reaction conditions, the alcohol fraction consisted of 94% 1-propanol and 6% 2-propanol. Thus in this case, presumably it was the alcohol from the hydrolysis, and not the formate ester, that was analyzed. Pointing out that Lee and Kruger obtained less than 1% rearrangement at reflux temperature and that the reaction mechanism was not quite limiting under these conditions (94), Karabatsos and co-workers concluded that their failure to detect isotopic scrambling in this reaction was not surprising.

In conjunction with the study on the protonation of cyclopropane with tritiated Lucas reagent, discussed earlier in this review (Section III-B), the reaction between ordinary Lucas reagent and 1-t-1-propanol was also investigated (33). The product obtained was chiefly 1-chloropropane, with 4–6% 2-chloropropane. Degradation of the 1-chloropropane showed about 1% isotope position rearrangement from C-1 to C-2 and C-3, and the rearranged t-label was more or less equally distributed at C-2 and C-3. These small amounts of rearrangements were felt to be real because experiments with 1-t-propanol of widely different specific activities all gave similar results. The extent of rearrangement was of the same order of magnitude as that observed in the formolysis of 111 (92); however, the nearly equal distribution of the rearranged label at C-2 and C-3 was different from that observed in the formolysis, which resulted in more of the C-1 label rearranging to C-3 than C-2. This comparison would appear to suggest that of that small portion of reaction proceeding through equilibrating protonated cyclopropanes, there was more extensive equilibration before product formation in the reaction between 1-propanol and Lucas reagent than in the formolysis of 1-propyl tosylate.

In connection with the study on the reaction between H_2SO_4-t and cyclopropane (Section III-A), because the heating of the reaction mixture in $13.8M$ H_2SO_4 caused additional rearrangements in the resulting product, some experiments were carried out to determine the effect of $13.8M$ H_2SO_4 on 1-t-1-propanol (30). When 1-t-1-propanol was heated in $13.8M$ H_2SO_4 at $50 \pm 2°$ for 30 hr, a 30% recovery of an approximately 1:2 mixture of 1-propanol and 2-propanol was obtained. Degradation of the recovered 1-propanol showed a total of 5.2 and 6.9% isotopic rearrangement in duplicate runs, the rearranged t-label being about equally distributed at C-2 and C-3. These rearrangements again may be attributed to equilibrating protonated cyclopropanes. It was also found that the specific activity of the recovered 1-propanol, isolated with or without dilution by inactive carrier, was some 15–19% greater than that of the original alcohol. Since some 70% of the 1-propanol employed in each experiment was lost, presumably largely by elimination processes, it was suggested (30) that the enrichment in t-activity in the recovered 1-propanol might be due to an isotope effect, whereby H was lost more readily than T. Since a simple El process involving only the classical 1-propyl cation would not require the removal of the t-label, the explanation of the enrichment based on isotope effects would require the shifting or change in position of the t-label from the original C-1 location before the loss of H or T. Such shifting of the label could be by the equilibration of protonated cyclopropanes as well as by a 1,2-hydride shift to the 2-propyl cation before elimination.

The most extensive isotopic scrambling observed in reactions involving the open chain 1-propyl system reported so far is the work of Karabatsos, Fry, and Meyerson (23) on the partial isomerization of 1-bromopropane (**113**) to 2- bromopropane (**114**) with AlBr$_3$. Using different amounts of the catalyst and contact times of 5 or 6 min at 0°, various

TABLE III

Isotopic Distributions from Reactions between
1-Bromopropane and AlBr$_3$

			D-Distribution in recovered 1-PrBr, %		
Run	Reactant	Conversion to 2-PrBr, %	C$_2$H$_5$—CD$_2$Br	C$_2$H$_4$D—CHDBr	C$_2$H$_3$D$_2$—CH$_2$Br
1	CH$_3$CH$_2$CD$_2$Br	45	97.1	0.6	2.3
2	CH$_3$CH$_2$CD$_2$Br	79	84.5	3.3	12.2
3	CH$_3$CH$_2$CD$_2$Br	80	79.8	5.0	15.2
4	CH$_3$CD$_2$CH$_2$Br	58	1.5	3.0	95.5
5	CH$_3$CD$_2$CH$_2$Br	65	2.1	5.8	92.1

extents of conversion to **114** were observed. Starting with 1,1-d_2- or 2,2-d_2-1-bromopropane (**113**-1-d_2 or **113**-2-d_2), the recovered **113**-d_2 and **114**-d_2 were converted to the corresponding alcohols and analyzed as the trimethylsilyl ethers by mass spectrometry. The isotopic distributions found in the recovered **113**-d_2 are summarized in Table III. Similarly, NMR and mass spectral analysis of the 1-bromopropane recovered from the reaction of 1-^{13}C-1-bromopropane (**113**-1-^{13}C) with AlBr$_3$, after 80% conversion to 2-bromopropane, gave the following data:

$$CH_3CH_2{}^{13}CH_2Br \longrightarrow CH_3CH_2{}^{13}CH_2Br + CH_3{}^{13}CH_2CH_2Br + {}^{13}CH_3CH_2CH_2Br$$
$$100\%\ {}^{13}C \qquad 85.7 \pm 0.2\% \qquad 3.7 \pm 0.9\% \qquad 10.6 \pm 0.6\%$$

These scramblings in the 1-bromopropane recovered from studies with **113**-1-d_2, **113**-2-d_2 and **113**-1-^{13}C were interpreted as arising from edge-protonated cyclopropane intermediates.

Karabatsos, Fry, and Meyerson (23) also reported that analyses of the samples of 2-bromopropane, **114**-d_2, obtained from runs 1–3 listed in Table III, showed that these bromides were better than 97% isotope-position unrearranged, in contrast to the 15–20% isotope position rearrangements in the 1-bromopropane, **113**-d_2, recovered in runs 2 and 3. Similarly, in the study with **113**-1-^{13}C, analysis of the 2-bromopropane obtained revealed no ^{13}C at C-2, thus indicating to these authors that the 2-bromopropane has undergone essentially no rearrangement. It was concluded that the 2-bromopropane arises from a nominally irreversible 1,2-hydride shift. Deno (29b), however, has pointed out to the present reviewer that in the partial isomerization of labeled 1-bromopropane to 2-bromopropane, since some isotope position rearrangement has occurred in the 1-bromopropane, during the course of the reaction, some 2-bromopropane must have been derived from isotopically scrambled 1-bromopropane. It is, therefore, not logical to have a 2-bromopropane product *without any* isotope position rearrangement. Most probably, since the net amount of rearrangement to C-2 of the 2-bromopropane would likely be small, because of the limitations of the methods of analysis, the results of Karabatsos and co-workers from analysis of the various samples of 2-bromopropane indicated that under their conditions, little or no *detectable* isotopic change could be observed at the C-2 position.

In connection with a study on the Friedel–Crafts alkylation of benzene with labeled 1-chloropropane, presently in progress in this laboratory, Woodcock (36) has investigated the effect of AlCl$_3$ on 1-*t*-1-chloropropane. Using a 1-PrCl:AlCl$_3$ ratio of 17:1 and contact times of 5 and 20 min at 0°, the conversions to 2-PrCl were, respectively, about 70 and 80%. The recovered 1-PrCl was degraded and the *t*-label was found to have rearranged from C-1 to both C-2 and C-3. These findings, as to be expected, were quite similar to those of Karabatsos and co-workers (23) in the partial isomerization of 1-PrBr to 2-PrBr with AlBr$_3$. With a contact time of 5 min, a total of about 9% of the label was rearranged, and the rearranged *t*-activity was located approximately equally at C-2 and C-3. After a contact time of 20 min, the total rearrangement was about 18%, with somewhat more of the rearranged label found at C-3 than C-2. These results appeared to suggest that with the longer contact time, equilibration among the protonated cyclopropane intermediates were more extensive, with the C-2:C-3 activity ratio beginning to pass 1:1 on its approach to the statistical distribution of 2:3.

V. GENERAL DISCUSSION AND CONCLUSIONS

In the present review, the experimental evidence has been summarized to support the conclusion that protonated cyclopropane intermediates could play significant roles in various cationic reactions. When the substrate is cyclopropane itself, such as in the reaction of cyclopropane with H_2SO_4 or with Lucas reagent, all the product could be regarded as being derived from equilibrating protonated cyclopropane intermediates. On the other hand, in reactions with simple aliphatic systems which could give rise to protonated cyclopropanes, this route generally constitutes only a minor part of the overall reaction because of more favorable competing processes such as direct displacements leading to no rearrangement and 1,2-shifts leading to the more stable secondary or tertiary cations. As evidence accumulated in favor of protonated cyclopropanes, there has also been a wider usage of such intermediates in the interpretation of experimental results. For example, Olah and co-workers (95) have found that starting from either 1-methylcyclopentyl or cyclohexyl precursors at $-60°$, only the 1-methylcyclopentyl cation was observed by NMR. It was suggested that a rearrangement mechanism involving a protonated cyclopropane intermediate should merit consideration.

Although protonated cyclopropane intermediates have been discussed chiefly in terms of edge-protonation, much of the experimental results in support of protonated cyclopropane could be rationalized by either equilibrating edge-protonated species or equilibrating methyl-bridged ions. For example, in the work of Baird and Aboderin (11) on the reaction of cyclopropane with D_2SO_4, it was concluded that the isotopic distribution in the resulting 1-propanol would not fit a mechanism involving equilibrating methyl-bridged ions, such as 12a ⇌ 12b, because this would necessarily give a product with equal amounts of D at C-1 and C-2, which was not observed. Such a conclusion was arrived at without considering any possible influence of isotope effects. From Scheme VII, if k' were not equal to k (and k'_H may not equal to k_H), the D contents at the C-1 and C-2 positions of the 1-propanol would not necessarily have to be equal.

Collins (70) has carried out calculations which could show that using secondary kinetic isotope effects of reasonable magnitudes, most isotopic scrambling data from D- or T- labelling work on protonated cyclopro-

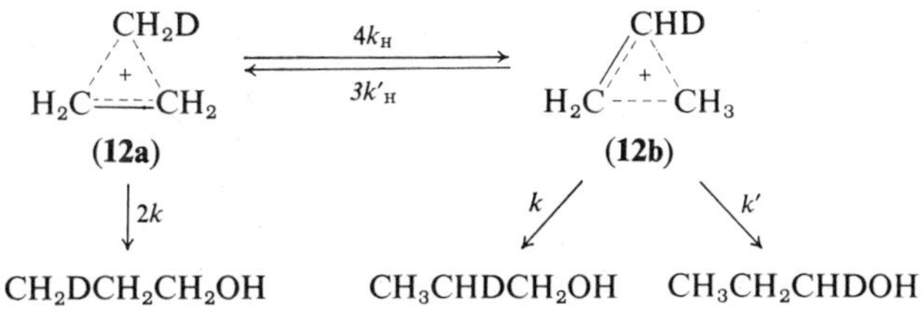

$$\begin{array}{ccc}
\underset{\text{(12a)}}{\overset{\displaystyle \text{CH}_2\text{D}}{\triangle}\text{H}_2\text{C}\overset{+}{=\!=\!=}\text{CH}_2} & \underset{3k'_{\text{H}}}{\overset{4k_{\text{H}}}{\rightleftharpoons}} & \underset{\text{(12b)}}{\overset{\displaystyle \text{CHD}}{\triangle}\text{H}_2\text{C}\overset{+}{-\!-\!-\!-}\text{CH}_3}
\end{array}$$

Scheme VII structure: 12a → 2k → CH$_2$DCH$_2$CH$_2$OH; 12b → k → CH$_3$CHDCH$_2$OH and k′ → CH$_3$CH$_2$CHDOH

Scheme VII

panes might be rationalized equally well by either equilibrating edge-protonated cyclopropanes or equilibrating methyl-bridged ions. In these calculations, however, it was not possible to predict the direction of the isotope effect. In Scheme VII, for example, in order to give the observed result of more D at the C-1 than the C-2 position of the product, k' would have to be greater than k. It is uncertain whether or not one should reasonably expect k' to be greater than k in reactions with a system such as **12b**. Furthermore, to account for Baird's data, k'/k would have to be $0.38/0.17 = 2.2$, which appears to be too high.

Collins (70) has suggested that the only clear-cut pieces of evidence in favor of edge-protonated intermediates are those obtained from the formolysis of $1\text{-}^{14}\text{C}\text{-}1$-propyl tosylate (**111**) reported by Lee and Kruger in 1966 (92) and from the reaction of $1\text{-}^{13}\text{C}\text{-}1$-bromopropane with AlBr_3 reported by Karabatsos, Fry, and Meyerson in 1967 (23). In these studies with labelling by isotopic carbon, secondary kinetic isotope effects are negligible. Calculations based on mechanisms A and B, given below, would show that in the isotope position rearrangements from C-1 to C-2

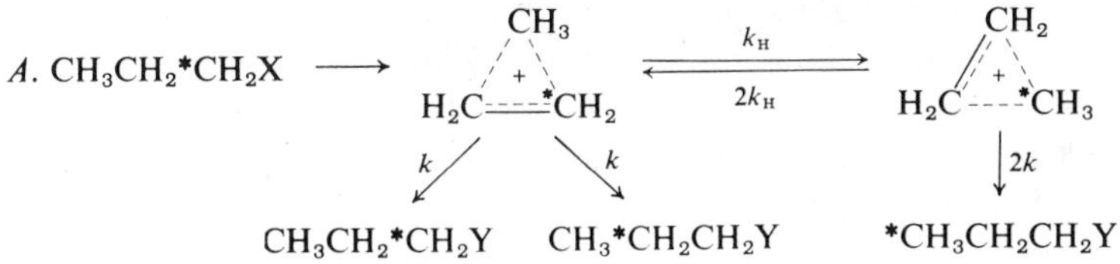

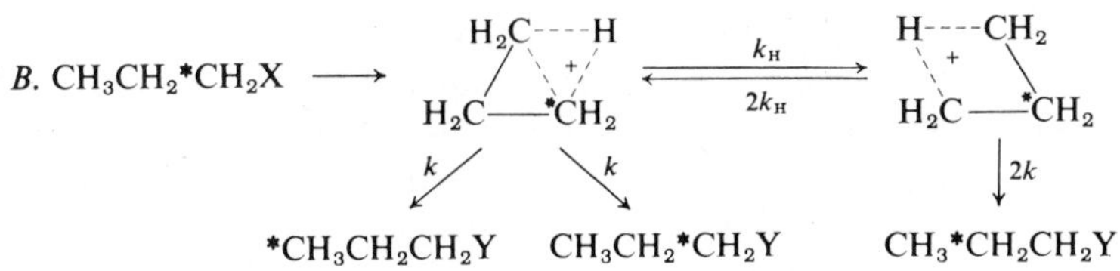

$A.$ $\text{CH}_3\text{CH}_2{}^{*}\text{CH}_2\text{X} \longrightarrow$

$$\underset{\text{CH}_3\text{CH}_2{}^{*}\text{CH}_2\text{Y} \quad \text{CH}_3{}^{*}\text{CH}_2\text{CH}_2\text{Y}}{\overset{\displaystyle \text{CH}_3}{\triangle}\text{H}_2\text{C}\overset{+}{=\!=\!=}{}^{*}\text{CH}_2} \underset{2k_{\text{H}}}{\overset{k_{\text{H}}}{\rightleftharpoons}} \underset{{}^{*}\text{CH}_3\text{CH}_2\text{CH}_2\text{Y}}{\overset{\displaystyle \text{CH}_2}{\triangle}\text{H}_2\text{C}\overset{+}{-\!-\!-\!-}{}^{*}\text{CH}_3}$$

$B.$ $\text{CH}_3\text{CH}_2{}^{*}\text{CH}_2\text{X} \longrightarrow$

$$\underset{{}^{*}\text{CH}_3\text{CH}_2\text{CH}_2\text{Y} \quad \text{CH}_3\text{CH}_2{}^{*}\text{CH}_2\text{Y}}{\overset{\displaystyle \text{H}_2\text{C}\text{-}\text{-}\text{-}\text{-}\text{H}}{\triangle}\text{H}_2\text{C}\text{—}{}^{*}\text{CH}_2} \underset{2k_{\text{H}}}{\overset{k_{\text{H}}}{\rightleftharpoons}} \underset{\text{CH}_3{}^{*}\text{CH}_2\text{CH}_2\text{Y}}{\overset{\displaystyle \text{H}\text{-}\text{-}\text{-}\text{-}\text{CH}_2}{\triangle}\text{H}_2\text{C}\text{—}{}^{*}\text{CH}_2}$$

and C-3, mechanism A, involving methyl-bridged ions, could not give rise to a product with more label at C-3 than C-2. On the other hand, with mechanism B, involving edge-protonation, more ${}^{*}\text{CH}_3\text{CH}_2\text{CH}_2\text{Y}$ than

$CH_3{}^*CH_2CH_2Y$ could be obtained. Since the results from the above ^{14}C- and ^{13}C- labelling studies showed more of the label rearranging to C-3 than C-2, edge-protonation is favored over methyl-bridged ions. The same conclusion favoring edge-protonation was also arrived at by Lee and Kruger (45) from a qualitative consideration of the finding that a greater amount of the rearranged label was found at C-3 than C-2 in the formolysis of **111**. In view of the positive evidence for edge-protonated cyclopropane from the work with ^{14}C- and ^{13}C- labelling, and since theoretical considerations also favor edge-protonation (Section II), it is, therefore, quite reasonable to represent protonated cyclopropanes as edge-protonated species.

The equilibration between edge-protonated cyclopropanes could proceed via methyl-bridged ions as originally proposed by Baird and Aboderin (11) (Scheme I). If products were derived predominantly from the edge-protonated species, the methyl-bridged ion could be regarded as the transition state. Alternatively, the methyl-bridged ion might also be an intermediate, but of somewhat higher energy, and the interconversion would take place via an unsymmetrical transition state structure in between the methyl-bridged ion and the edge-protonated species along the reaction coordinate. With D- or T- labelling, although Scheme I adequately describes all the isotopically distinguishable intermediates, a more extensive equilibration is shown in Scheme VIII, which includes stereochemically different, but isotopically indistinguishable species. Actually, a still more complete scheme would include two additional series of equilibration, beginning with **12c** and **12c′**, analogous to **12a** ⇆ **14b** ⇆ **12b** ⇆ **14c** ⇆ **12b′** ⇆ **14b′** ⇆ **12a**. Thus it is not surprising that under conditions which could give a long lifetime to the protonated cyclopropane intermediates, a completely statistical distribution of the D-label in the ratio of 2:2:3 for the C-1, C-2, and C-3 positions would be attained in the 1-propyl product, as has been observed by Deno and co-workers (29).

At the beginning of this review, it was stated that the nortricyclonium ion proposed in 1951 (1) was really the first suggested protonated cyclopropane intermediate. Mention was also made of the controversy on whether the 2-norbornyl cation is a nonclassical ion or a pair of rapidly equilibrating classical ions. The communication of Olah, Commeyras, and Liu (90) has just appeared as the writing of the present review was nearing completion. Pointing out that in the earlier NMR studies (46–48), no clear differentiation on the NMR time scale between a rapidly equilibrating pair of ions or a bridged, symmetrical ion was possible, these workers (90) selected Raman spectroscopy as a means of making this differentiation, assuming that vibrational transition rates would be faster than hydride shifts or Wagner–Meerwein rearrangement. The cation studied (**115**) was

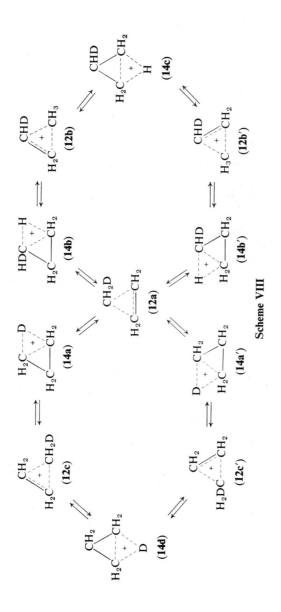

Scheme VIII

generated from 2-*exo*-norbornyl chloride in FSO_3H—SbF_5—SO_2 at $-80°$. Its Raman spectrum was found to be similar to those of nortricyclene and 3-bromonortricyclene, but different from those of norbornane and 2-*exo*-norbornyl halides. The same spectrum was observed when nortricyclene was protonated in FSO_3H—SbF_5—SO_2 at $-80°$. Based on the presumption that in Raman spectroscopy the time scale of chemical shifts of equilibration cannot affect the observed structure, it was concluded that the structure of ion **115** is not a rapidly equilibrating 2-norbornyl system, but is that of protonated nortricyclene. Although a face-protonated nortricyclene would conform with the experimental facts and would have C_{3v} symmetry, because MO considerations favor edge-protonation, it was suggested that protonated nortricyclene is an edge-protonated species. Such a structure would not have C_{3v} symmetry; but it was felt that it would still be in agreement with the Raman spectroscopic observations because the close similarity between the Raman spectra of nortricyclene and 3-bromonortricyclene, as well as between those of norbornane and 2-*exo*-halonorbornanes, clearly indicate that skeletal similarities and not overall point group symmetries are reflected in the main spectral similarities. The equivalence of the four protons at C-1, C-2, and C-6 observed by NMR was explained as resulting from proton scrambling around all three edges of the cyclopropane ring via what amounts to be a "corner-protonated" transition state or intermediate (with a plane of symmetry). Strong interaction was suggested as being responsible in keeping the C-2 to C-6 distance similar to that in nortricyclene. In view of these conclusions, perhaps some of the very extensive results obtained in studies of the norbornyl system might merit reassessment with the possibility of giving more importance to protonated nortricyclene, rather than a subsidiary role to the carbon-bridge norbornonium ion to which it has been assigned in many discussions in the literature.

Acknowledgment

Sincere appreciation is extended to Professor N. C. Deno for making available to the writer numerous experimental findings before their publication, to Professor C. J. Collins for isotope-effect calculations and suggestions relating to the differentiation between edge-protonation and methyl-bridged ions, and to both of these gentlemen for additional comments and discussions which were valuable and stimulating to the writer of this review.

References

1. J. D. Roberts and C. C. Lee, *J. Am. Chem. Soc.*, **73**, 5009 (1951).
2. J. D. Roberts, C. C. Lee, and W. H. Saunders, Jr., *J. Am. Chem. Soc.*, **76**, 4501 (1954).

3. S. Winstein and D. Trifan, *J. Am. Chem. Soc.*, *74*, 1154 (1952).
4. G. E. Gream, *Rev. Pure Appl. Chem.*, *16*, 25 (1966).
5. G. D. Sargent, *Quart. Rev.*, *20*, 301 (1966).
6. H. C. Brown, *Chem. Eng. News*, *45*, No. 7, 87 (1967); *Chem. Brit.*, *2*, No. 5, 199 (1966).
7. P. S. Skell and I. Starer, *J. Am. Chem. Soc.*, *82*, 2971 (1960).
8. P. S. Skell and I. Starer, *J. Am. Chem. Soc.*, *81*, 4117 (1959).
9. P. S. Skell and I. Starer, *J. Am. Chem. Soc.*, *84*, 3962 (1962).
10. O. A. Reutov and T. N. Shatkina, *Tetrahedron*, *18*, 237 (1962).
11. R. L. Baird and A. A. Aboderin, *J. Am. Chem. Soc.*, *86*, 252 (1964).
12. C. C. Lee, J. E. Kruger, and E. W. C. Wong, *J. Am. Chem. Soc.*, *87*, 3985 (1965).
13. C. C. Lee and J. E. Kruger, *J. Am. Chem. Soc.*, *87*, 3986 (1965).
14. G. J. Karabatsos, C. E. Orzech, Jr., and S. Meyerson, *J. Am. Chem. Soc.*, *87*, 4394 (1965).
15. A. Colter, E. C. Friedrich, N. J. Holness, and S. Winstein, *J. Am. Chem. Soc.*, *87*, 378 (1965).
16. J. A. Berson and P. W. Grubb, *J. Am. Chem. Soc.*, *87*, 4016 (1965).
17. R. Hoffmann, *J. Chem. Phys.*, *40*, 2480 (1964).
18. A. D. Walsh, *Trans. Faraday Soc.*, *49*, 179 (1949).
19. M. Hanack and H.-J. Schneider, *Angew. Chem. Intern. Ed.*, *6*, 666 (1967).
20. W. A. Bernett, *J. Chem. Educ.*, *44*, 17 (1967).
20a. J. D. Petke and J. L. Whitten, *J. Am. Chem. Soc.*, *90*, 338 (1968).
21. L. Joris, P. v. R. Schleyer, and R. Gleiter, *J. Am. Chem. Soc.*, *90*, 327 (1968).
22. P. D. Bartlett, *Nonclassical Ions*, W. A. Benjamin, New York, 1965, p. v.
23. G. J. Karabatsos, J. L. Fry, and S. Meyerson, *Tetrahedron Letters*, *1967*, 3735.
24. G. A. Olah and J. Lukas, *J. Am. Chem. Soc.*, *90*, 933 (1968).
25. R. L. Baird and A. A. Aboderin, *Tetrahedron Letters*, *1963*, 235.
26. E. L. Purlee and R. W. Taft, Jr., *J. Am. Chem. Soc.*, *78*, 5807 (1956).
27. L. G. Cannell and R. W. Taft, Jr., *J. Am. Chem. Soc.*, *78*, 5812 (1956).
28. N. C. Deno, Lecture given at the 152nd National A.C.S. Meeting, New York, N.Y., September 12–16, 1966.
29. (a) N. C. Deno, D. LaVietes, J. Mockus, and P. C. Scholl., *J. Am. Chem. Soc.*, *90*, 6457 (1968); (b) N. C. Deno, private communications.
30. C. C. Lee and L. Gruber, *J. Am. Chem. Soc.*, *90*, 3775 (1968).
31. C. C. Lee, L. Gruber, and K. M. Wan, *Tetrahedron Letters*, *1968*, 2587.
32. A. Nickon and N. H. Werstiuk, *J. Am. Chem. Soc.*, *89*, 3915, 3917 (1967).
33. C. C. Lee, W. K. Y. Chwang, and K. M. Wan, *J. Am. Chem. Soc.*, *90*, 3778 (1968).
34. O. A. Reutov and T. N. Shatkina, *Bull. Acad. Sci. U.S.S.R. Div. Chem. Sci. (English Transl.)*, *1963*, No. 1, 180.
35. O. A. Reutov, *Pure Appl. Chem.*, *7*, 203 (1963).
36. D. J. Woodcock, to be published.
37. J. W. Hightower and W. K. Hall, *J. Am. Chem. Soc.*, *90*, 851 (1968).
38. H. R. Gerberich, J. W. Hightower, and W. K. Hall, *J. Catalysis*, *8*, 391 (1967).
39. J. G. Larson, H. R. Gerberich, and W. K. Hall, *J. Am. Chem. Soc.*, *87*, 1880 (1965).
40. J. W. Hightower and W. K. Hall, *J. Am. Chem. Soc.*, *89*, 778 (1967).
41. J. W. Hightower and W. K. Hall, *J. Phys. Chem.*, *71*, 1014 (1967).
42. F. Cacace, M. Caroselli, R. Cipollini, and G. Ciranni, *J. Am. Chem. Soc.*, *90*, 2222 (1968).

43. A. S. Kushner, Ph.D. thesis, The Pennsylvania State University, University Park, Pa., 1966, p. 117.
44. A. Nickon and J. H. Hammons, *J. Am. Chem. Soc.*, *86*, 3322 (1964).
45. C. C. Lee and J. E. Kruger, *Tetrahedron*, *23*, 2539 (1967).
46. F. R. Jensen and B. H. Beck, *Tetrahedron Letters*, *1966*, 4287.
47. P. v. R. Schleyer, W. E. Watts, R. C. Fort, Jr., M. B. Comisarow, and G. A. Olah, *J. Am. Chem. Soc.*, *96*, 5679 (1964).
48. M. Sanders, P. v. R. Schleyer, and G. A. Olah, *J. Am. Chem. Soc.*, *86*, 5680 (1964).
49. C. C. Lee and L. K. M. Lam, *J. Am. Chem. Soc.*, *88*, 2831 (1966).
50. C. C. Lee and L. K. M. Lam, *J. Am. Chem. Soc.*, *88*, 2834 (1966).
51. C. C. Lee and L. K. M. Lam, *J. Am. Chem. Soc.*, *88*, 5355 (1966).
52. H. Hart and O. E. Curtis, Jr., *J. Am. Chem. Soc.*, *79*, 931 (1957).
53. H. Hart and G. Levitt, *J. Org. Chem.*, *24*, 1261 (1959).
54. H. Hart and R. H. Schlosberg, *J. Am. Chem. Soc.*, *88*, 5030 (1966); *90*, 5189 (1968).
55. H. Hart and R. A. Martin, *J. Org. Chem.*, *24*, 1267 (1959).
56. N. C. Deno and D. N. Lincoln, *J. Am. Chem. Soc.*, *88*, 5357 (1966).
57. A. J. Gordon, *J. Chem. Educ.*, *44*, 461 (1967).
58. V. N. Ipatieff, H. Pines, and B. B. Corson, *J. Am. Chem. Soc.*, *60*, 577 (1938).
59. A. V. Grosse and V. N. Ipatieff, *J. Org. Chem.*, *2*, 477 (1937).
60. V. N. Ipatieff, H. Pines, and L. Schmerling, *J. Org. Chem.*, *5*, 253 (1940).
61. L. Schmerling, *Ind. Eng. Chem.*, *40*, 2072 (1948).
62. J. D. Roberts and M. Halmann, *J. Am. Chem. Soc.*, *75*, 5759 (1953).
63. O. A. Reutov, *Congress Lectures, 19th International Congress of Pure and Applied Chemistry*, Butterworths, London, 1963, pp. 203–227.
64. G. J. Karabatsos and C. E. Orzech, Jr., *J. Am. Chem. Soc.*, *84*, 2838 (1962).
65. G. J. Karabatsos and J. D. Graham, *J. Am. Chem. Soc.*, *82*, 5250 (1960).
66. P. S. Skell, I. Starer, and A. P. Krapcho, *J. Am. Chem. Soc.*, *82*, 5257 (1960).
67. N. C. Deno, *Progress in Physical Organic Chemistry*, Vol. 2, S. Cohen, A. Streitwieser, Jr., and R. W. Taft, Eds., Interscience, New York, 1964, p. 129.
68. A. A. Aboderin and R. L. Baird, *J. Am. Chem. Soc.*, *86*, 2300 (1964).
69. R. L. Baird, private communications.
70. C. J. Collins, private communications.
71. M. S. Silver, *J. Am. Chem. Soc.*, *82*, 2971 (1960).
72. M. S. Silver, *J. Org. Chem.*, *28*, 1686 (1963).
73. J. H. Bayless, F. D. Mendicino, and L. Friedman, *J. Am. Chem. Soc.*, *87*, 5790 (1965).
74. A. T. Jurewicz and L. Friedman, *J. Am. Chem. Soc.*, *89*, 149 (1967).
75. J. H. Bayless and L. Friedman, *J. Am. Chem. Soc.*, *89*, 147 (1967).
76. G. J. Karabatsos, N. Hsi, and S. Meyerson, *J. Am. Chem. Soc.*, *88*, 5649 (1966).
77. G. J. Karabatsos, R. A. Mount, D. O. Rickter, and S. Meyerson, *J. Am. Chem. Soc.*, *88*, 5651 (1966).
78. M. S. Silver, *J. Am. Chem. Soc.*, *83*, 3482 (1961).
79. G. J. Karabatsos, C. E. Orzech, Jr., and S. Meyerson, *J. Am. Chem. Soc.*, *86*, (1964).
80. A. W. Fort and J. D. Roberts, *J. Am. Chem. Soc.*, *78*, 584 (1956).
81. W. G. Dauben and P. Laug, *Tetrahedron Letters*, *1962*, 453.
82. W. G. Dauben and P. Laug, *Tetrahedron*, *20*, 1259 (1964).
83. O. E. Edwards and M. Lesage, *Can. J. Chem.*, *41*, 1592 (1963).
84. P. S. Skell and R. J. Maxwell, *J. Am. Chem. Soc.*, *84*, 3963 (1962).

85. J. Bayless, L. Friedman, J. A. Smith, F. B. Cook, and H. Shechter, *J. Am. Chem. Soc.*, *87*, 661 (1965).
86. J. A. Smith, H. Shechter, J. Bayless, and L. Friedman, *J. Am. Chem. Soc.*, *87*, 659 (1965).
87. K. B. Wiberg and J. M.Lavanish, *J. Am. Chem. Soc.*, *88*, 365 (1966).
88. W. J. Koehl, Jr., *J. Am. Chem. Soc.*, *86*, 4686 (1964).
89. S. Winstein, E. Clippinger, R. Howe, and E. Vogelfanger, *J. Am. Chem. Soc.*, *87*, 376 (1965).
90. G. A. Olah, A. Commeyras, and C. Liu, *J. Am. Chem. Soc.*, *90*, 3882 (1968).
91. B. M. Benjamin, B. W. Ponder, and C. J. Collins, *J. Am. Chem. Soc.*, *88*, 1558 (1966).
92. C. C. Lee and J. E. Kruger, *Can. J. Chem.*, *44*, 2343 (1966).
93. T. N. Shatkina, A. N. Lovtsova, and O. A. Reutov, *Izv. Akad. Nauk. S.S.S.R., Ser. Khim.*, *1967*, 2748.
94. S. Winstein and H. Marshall, *J. Am. Chem. Soc.*, *74*, 1120 (1952).
95. G. A. Olah, J. M. Bollinger, C. A. Cupas, and J. Lukas, *J. Am. Chem. Soc.*, *89*, 2692 (1967).

Electrolytic Reductive Coupling: Synthetic and Mechanistic Aspects

By Manuel M. Baizer and John P. Petrovich

Central Research Department, Monsanto Company,
St. Louis, Missouri

CONTENTS

I. INTRODUCTION

Organic electrolytic coupling reactions are about as old as electro-organic chemistry itself. They have presented synthetic opportunities and mechanistic challenges for over 100 years. The progress of research in both categories has been sporadic. Synthetic chemists have apparently been deterred by the facts that the modus operandi is more complex than that required for ordinary organic preparations—though not more cumbersome than the more familiar operations of high-pressure hydrogenation—and that mixtures of products are obtained which are often accessible by other, simpler routes. Physical organic chemists have presumably been deterred from studying reactions which by their very nature are mechanistically very complex: they simultaneously involve elements of adsorption, hetero-geneous catalysis, reactions within an electric field, concentration and pH gradients, and bulk reactions in heterogeneous or homogeneous liquid media.

Within the last decade, however, there has been ample demonstration that electrolytic coupling reactions often involve unique features of convenience, directness, or economy. Beginning even earlier, the development of new techniques and instrumentation, such as potential control, polarography and its variants (including cyclic voltammetry), ESR spectroscopy, etc., has progressed to the extent that coupling reactions and their mechanisms may be studied rationally.

Electrolytic oxidative coupling reactions have recently been reviewed (1). Electrolytic reductive coupling reactions *inter alia* have been reviewed by Knunyants and Gambaryan (2), by Swann (3), by Allen (4), and by Schultz and Popp (5). The synthetic promise of the electrochemical dimerization method was presented in a 1964 paper by Fioshin and Tomilov (6). The synthetic aspects of electrolytic reductive coupling of activated olefins were summarized in 1967 (7).

It is the purpose of this chapter to present representative examples of the several types of electrolytic reductive couplings that have been reported and to examine critically the mechanisms or hypotheses that have been advanced to explain the results obtained. In this discussion electrolytic reductive coupling is construed to encompass reactions in which two or more organic moieties are joined following a step of electron transfer from an electrode. As will be seen later, the coupling may be the result of a single electrochemical (E) event followed by chemical (C) combination of intermediates or may be a consequence of combinations of successive E and C reactions. The coupling may be *inter-* or *intra*molecular.

For classification purposes this survey is organized on the basis of the type of electroreducible starting materials from which the reductively coupled products are prepared. Tabulations of data available elsewhere (3) will not be presented here.

II. GENERAL SYNTHETIC CONSIDERATIONS

The general procedures for assembling suitable cells, choosing and preparing electrodes, and carrying out electrolytic reactions are described by Swann (3) and by Allen (4). In general the types of compounds (e.g., nitro compounds, carbonyl compounds, activated olefins) which can be cathodically reductively dimerized (e.g., to azo compounds, pinacols, and 1,4-disubstituted butanes, respectively) can also under other conditions be further electroreduced. To obtain optimum yields of the reduced coupled products it is necessary to control reaction parameters, such as: pH, concentration of substrate, concentration and nature of supporting electrolyte, cathode potential, overpotential of the electrode as well as its physical state, and temperature. The cathode potential may be controlled *extrinsically* by manual or automatic means (potentiostat) or *intrinsically*, as in the older literature, by the current density and by choosing an electrode which by its nature limits (through hydrogen discharge, for example) the negative voltage obtainable. The "state of the art" has advanced to the extent that the effect upon the distribution of products of systematically varying most of the parameters can be rationalized. In some cases (e.g., the physical state of the cathode) the data are still largely empirical in nature.

III. GENERAL MECHANISTIC CONSIDERATIONS

A complete understanding of the mechanism of an electrochemical reductive-coupling reaction requires elucidation of (*a*) the chemical step, if any, preceding the electron transfer, (*b*) the electron-transfer step, (*c*) the nature of the reduced intermediate, (*d*) the reaction of the reduced intermediate at the electrode or within the electric field, (*e*) the reaction of the intermediate or its product from step (*d*) in the bulk of the solution.

Several tools are available for the investigation of electrochemical mechanisms. Of these, the most widely used are polarography [for (*a*) and (*b*)], coulometry [for (*b*)], cyclic voltammetry [for (*a*), (*b*), and (*c*)], electron spin resonance spectroscopy [for (*c*)] and controlled potential electrolysis [for (*d*) and (*e*)].

Polarography involves current–voltage curves most commonly using a dropping mercury electrode. A typical stepwise two-electron reduction is illustrated in Figure 1. This method yields the reduction potential of the

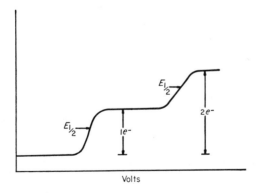

Figure 1

depolarizer as well as an estimate of the number of electrons involved in each step. Valuable information regarding the mechanism of organic reduction has been obtained from polarographic studies conducted as a function of solvent [protic (8) and aprotic (9)], hydrogen ion concentration (10) and substituents (11). Detailed treatment of the interpretation of polarographic data is given in the references cited (12).

Coulometry is the measurement of the total current required to consume all of a depolarizer at a fixed potential. If the electrode product is stable or if the electrode product is unstable but does not form an electroactive species at the fixed potential, then the coulometric measurement gives the exact number of electrons involved in a particular polarographic wave. Recently developed, thin-layer coulometric techniques give a rapid and fairly accurate method for this analysis (13).

The most useful electrochemical technique for the investigation of electrochemical mechanisms is cyclic voltammetry. Cyclic voltammetry is an extension of stationary-electrode polarography (14) or "peak polarography." The technique involves current–voltage curves, but the voltage is swept at a faster rate than in simple polarography. The result of this rapid voltage scan is a maximum in the current–voltage curve. From a single cyclic scan one can obtain valuable information about the reversibility of the electrochemical system (15), stability of electrogenerated intermediates, and possibly the existence of electroactive products. A typical cyclic polarogram for a reversible system is given in Figure 2b, and an irreversible system is shown in Figure 2a. By examining the curves obtained from a series of voltage scans at various scan rates (16) one obtains information as to the type of electrochemical mechanism operative under these conditions, e.g., EC or ECE. The rate constant for the decomposition of the electrochemically generated intermediate can also be obtained by this technique.

With the advent of electrolysis in an ESR cavity (17) several interesting

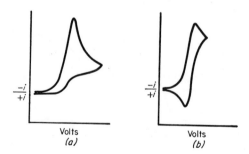

Figure 2

and informative studies have been conducted on the structure of electrochemically generated intermediates. With present techniques the method has been limited to species with lifetimes of at least 1 sec for completely resolved spectra and 0.01 sec for detection (18).

Any mechanistic study requires an unequivocal identification of the reaction products. A controlled potential electrolysis on a scale sufficiently large to allow identification of reaction products is imperative. Without this tool mechanistic studies are largely speculative; with it interpretations must be cautious.

In all mechanistic schemes it must be remembered that the electron transfer occurs at an interface. In many cases the first follow-up chemical reaction occurs at or near the electrode surface and, therefore, must be influenced by the electric field (19). The structure of the electrode–solution interface (the double layer) probably influences the chemical reaction. There are many pitfalls in attempting to elucidate the reactions of electrogenerated charged species (e.g., anion radicals) without simultaneously considering the nature of the counterion and the solvent involved. In a negative sense, very little is known about the structure of the double layer

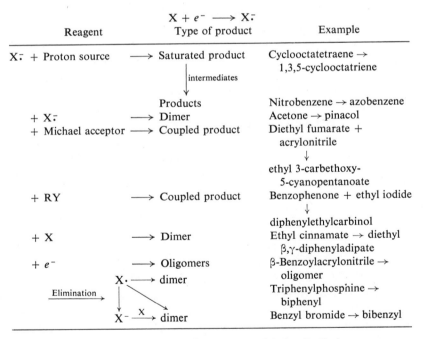

Chart I. Reactions of Electrogenerated Anion Radicals

$$Z^+ \xrightarrow{e^-} Z\cdot$$

Z· + electrode $\longrightarrow$ Z$_x$M		Dimethylcyanoethylsulfonium bromide
		$\downarrow$
		bis-cyanoethyl mercury
Z· + e^- $\longrightarrow$ Z$^-$		Triphenylcyanomethylphosphonium bromide
		$\downarrow$
		triphenylcyanomethylenephosphorane

Chart II. Reactions of Electrogenerated Free Radicals

in organic systems. A complete understanding of organic electrode processes cannot be realized before the questions regarding adsorption and the structure of the double layer are answered.

Electrolytic reductive coupling reactions involve the chemistry of electrolytically generated anion radicals from electrically neutral starting materials and of radicals from onium compounds. A broad generalization of this chemistry is given in Charts I and II. In general, the chemistry of anion radicals is quite similar to the chemistry of simple anions. The addition of anion radicals to activated olefins (19) as well as their participation in nucleophilic displacement (20,21) reactions have been observed. The addition of a proton to an anion radical yields a free radical. Several reaction paths have been observed for these intermediate free radicals. They can dimerize, react with the metallic electrode, or be reduced to the corresponding anion. In the latter case typical anionic reactions are subsequently observed. The specific mode of reaction of the intermediates from particular depolarizers will be discussed in detail later in this chapter.

IV. REDUCTIVE COUPLING REACTIONS

A. Saturated Carbonyl Compounds

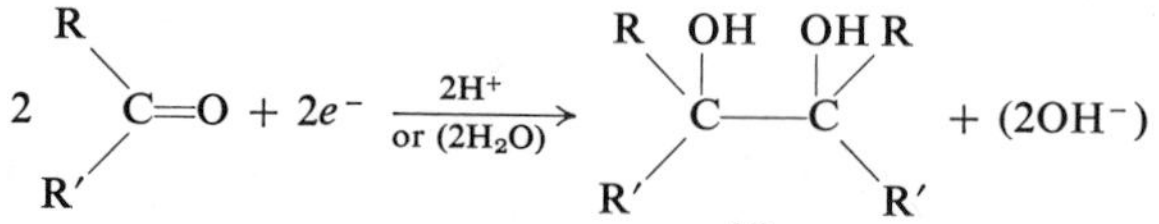

$$2 \underset{R'}{\overset{R}{\diagdown}}C{=}O + 2e^- \xrightarrow[\text{or } (2H_2O)]{2H^+} \underset{R'}{\overset{R}{\diagdown}}\underset{|}{\overset{OH}{C}}{-}\underset{R'}{\overset{OH}{\underset{|}{C}}}\overset{R}{\diagup} + (2OH^-)$$

(1)

R is aliphatic or aromatic; R′ = R or H

Aliphatic and aromatic aldehydes and ketones have been electro-reduced to *vic*-diols (1). A wide variety of conditions (low and high pH's, low and high overvoltage cathodes, etc.) has been reported (3,4) so that one must agree with the comment in a recent paper (22) that coherent organization of the data is very difficult.

In general the aromatic carbonyl compounds give better results than the aliphatic. It is obvious that, if the reduction is carried out in alkaline medium, base-catalyzed condensation reactions of any available α-methylene groups will detract from the chemical yield of **1**. The *current efficiency* may not be impaired (1 mole of **1** requires 2 F). Too low a pH or too high a temperature may catalyze the rearrangement of **1** to the corresponding pinacolone (e.g., 23).

Nuclear substituents which are not themselves electroreducible at the cathode voltage required for the carbonyl group reduction do not interfere with the course of the pinacol reduction. *p*-Hydroxy- (24) and *p*-aminoacetophenone (25) give their respective pinacols in good yield as do *p*-hydroxy- (26) and *p*-dimethylaminobenzaldehyde. In the last case mentioned the ratio of diastereoisomeric isomers of 4,4'-bisdimethylaminohydrobenzoins obtained was 1.48:1. The pinacol from *p*-aminoacetophenone could not be prepared chemically but was successfully prepared electrochemically. Escherlich and Moest (27) had reduced Michler's ketone to the pinacol at copper but not at mercury or lead; Elofsen found (28) that pinacol yields were high at mercury when controlled potential (ca. -0.95 V) and rapid stirring were used. Furfural (29) and 3-acetylpyridine (24) behave like other "aromatic" carbonyl compounds; the former yields its pinacols in 63% yield. Mannich bases have been successfully reduced to their pinacols (30).

Mixed (or unsymmetrical) pinacols have been obtained in moderate yield by electrolysis of a mixture of *p*-dimethylaminoacetophenone and *p*-methoxyacetophenone (31). The yield of mixed pinacol was 20–30%; among the byproducts were the pinacols formed from each of the individual ketones.

Diketones $RCO(CH_2)_nCOR$ can form inter- and intramolecular reductive coupling products. Acetylacetone is an interesting starting material

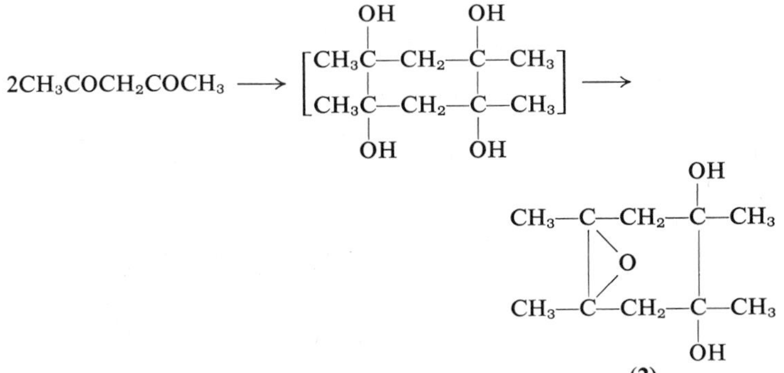

$$(2)$$

since it can form coupled products both by electrooxidation and by electro-reduction. The former category was studied by Johnston and Stride (32) who obtained an 11% yield of tetracetylethane. The latter was investigated by Jadot (33) who, electrolyzing in 40% sulfuric acid at 15° and 6 amp, obtained a product $C_{10}H_{18}O_3$ to which he assigned structure **2**.

Casals and Wiemann (34), however, in reducing acetylacetone with magnesium–acetic acid or zinc–acetic acid obtained the tricyclic compound $C_{10}H_{16}O_3$ which presumably arose via coupling of two radicals **3** followed by loss of water:

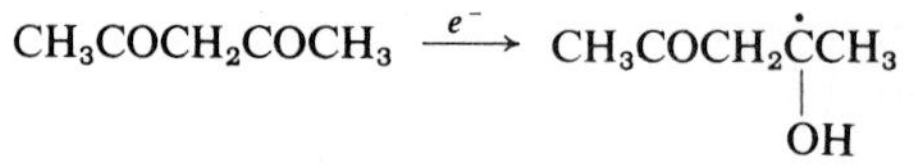

$$CH_3COCH_2COCH_3 \xrightarrow{e^-} CH_3COCH_2\overset{\cdot}{C}CH_3$$
$$\underset{OH}{|}$$

(3)

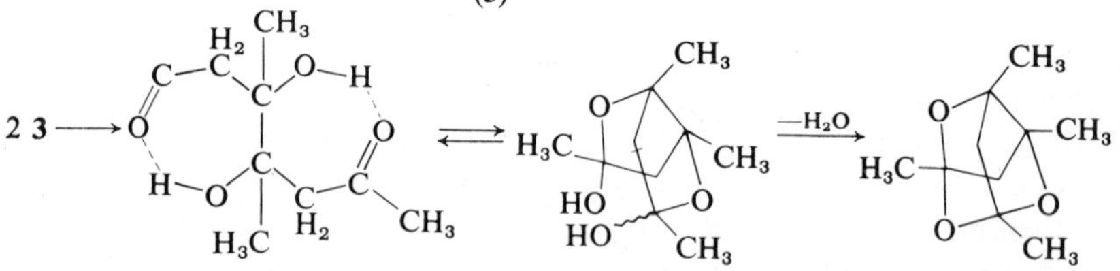

They did not find $C_{10}H_{18}H_3$ among the metal–acid reduction products.

1,3-Diphenyl-1,3-propanedione (35) showed complex polarographic behavior. Bulk electrolysis at pH 4.2 (citrate buffer) at the first reduction wave yielded the glycol **4**; reduction at the second wave yielded mixtures considered to be probably isomers of **5**.

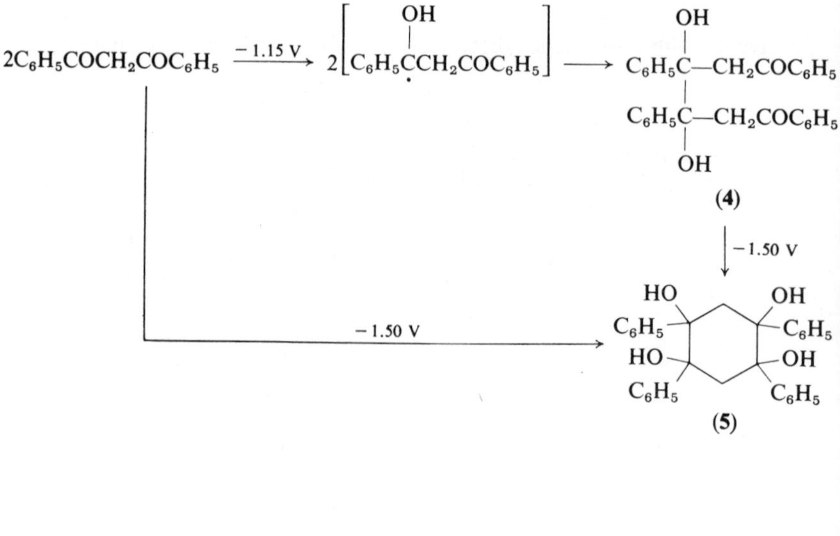

Steric effects as well as the stability of the radical are important factors in pinacol formation from the reduction of carbonyls. The reduction of 2-phenylindandione-1,3 in acid solution yields the stable diradical **6** which has been observed by ESR. No pinacol was formed in this system (36).

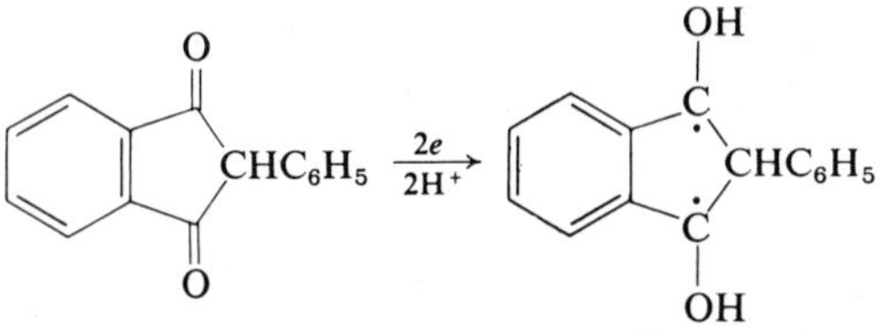

(6)

The reduction of 1-acetylnaphthalene in ethanolic potassium hydroxide at mercury (-1.8 V vs. SCE) in a divided cell has been reported (37) to take an unexpected course. Instead of the anticipated pinacol, the oxepin **(7)** was obtained.

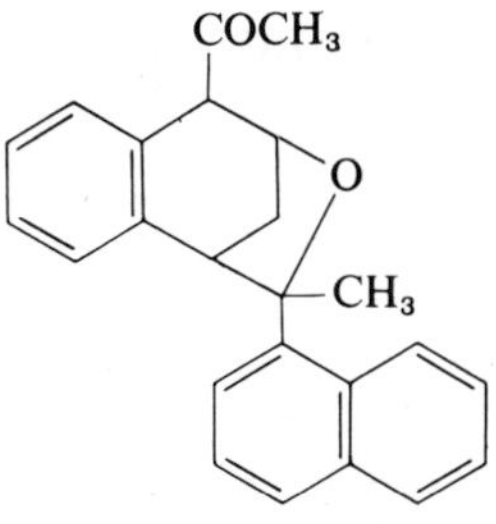

(7)

The authors rationalize the formation of this product as follows: One-electron reduction of starting material leads to a radical for which at least the two canonical forms **8** and **9** may be written:

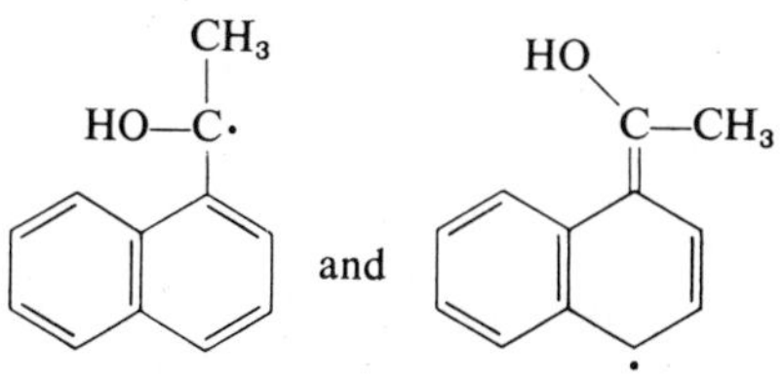

Coupling of **8** and **9** yields **10**; ketonization of **10** to **11** and intramolecular addition of the hydroxy group of **11** to the olefinic bond yields **7**.

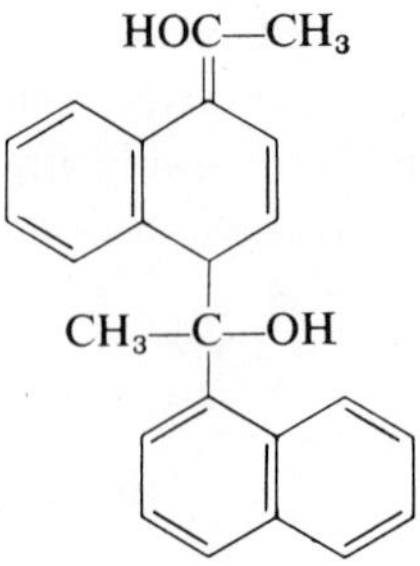

(10) 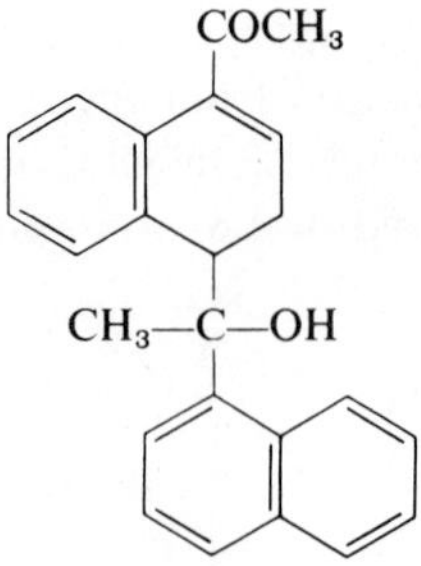   (11)

The structure of **7** is deduced from NMR and mass spectroscopic data as well as the fact that **7** is converted by acetic-sulfuric acid to **12**.

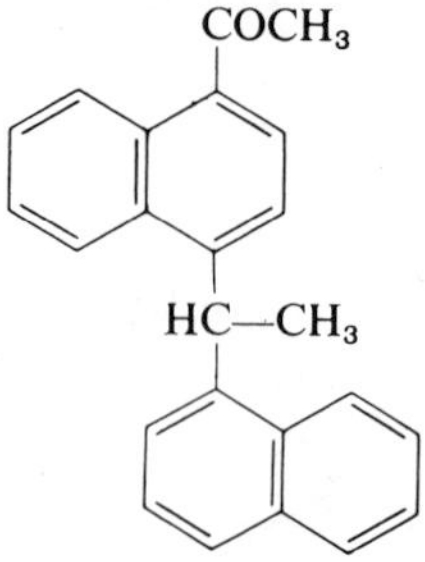

(12)

Phenylmagnesium bromide and methylmagnesium iodide each in the presence of magnesium react (38) with 9-benzoylanthracene and 9-acetylanthracene to yield reductive products (**13**) in which coupling has occurred through the 10,10′ positions.

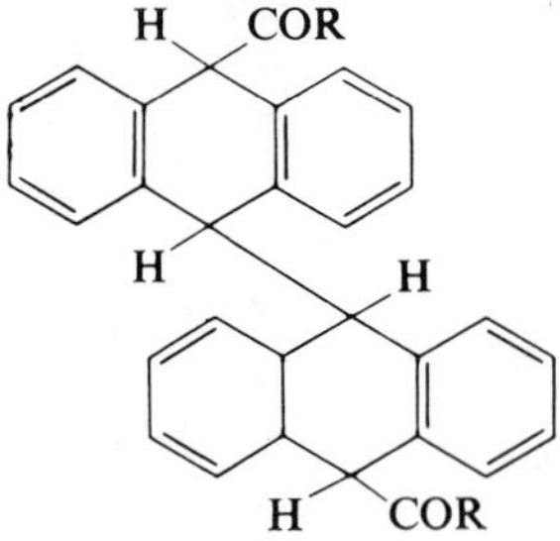

R = acetyl or benzoyl

(13)

Analogous electrochemical reductions have not been reported.

Gourley and Grimshaw (39) studied the formation of cyclic glycols from **14**, **15**, and **16**, in electrolysis at mercury of aqueous or ethanolic potassium or sodium hydroxide solutions of the ketones. The runs were potentiostated to suppress discharge of the alkali metal.

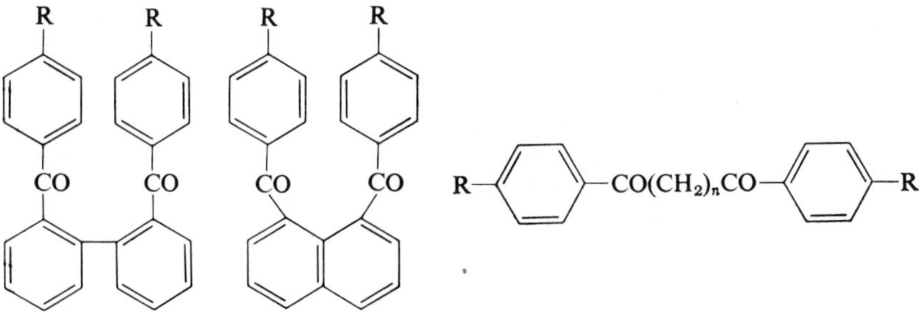

(14)   (15)   (16)

R = OH or OMe n = 3,4,5 when R = H

Intramolecular reductive coupling of **14** led to moderate yields of *cis* and *trans* cyclic glycols. The cyclic glycols were not obtained from **15** but presumably were intermediates to the pinacolone actually isolated. Cyclic compounds (glycols and pinacolones) were obtained from **16** when $n = 3$; polymers were obtained when n was 4 or 5

It is interesting that the reduction of *o*-dibenzoylbenzene by 1 atom-equivalent of sodium (40) yields the reduced dimer **19** presumably via **17** and **18**:

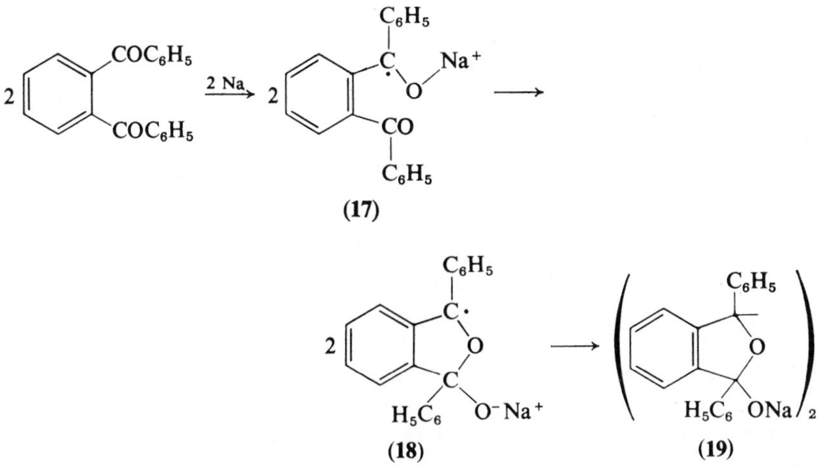

(17)

(18)   (19)

The electrochemical equivalent of this intramolecular attack upon carbonyl of an oxygen anion has not been reported.

Mixed reductive coupling between saturated carbonyl compounds and other species has not been extensively investigated. Nelson and Collins reported (20) the formation of a small quantity of the ethyl ether of phenyl-ethylcarbinol during the electrolysis of benzaldehyde. The supporting electrolyte was tetramethylammonium iodide. The coupling at both the

$$C_6H_5CHO + 2C_2H_5I \xrightarrow{2e^-} C_6H_5CH(OC_2H_5)C_2H_5 + 2I^-$$

carbon and the oxygen of the carbonyl group is unusual. In a related experiment Wawzonek and Gundersen (21) found that diphenylethyl-carbinol was formed upon electrolysis of benzophenone in dimethyl formamide containing potassium iodide as electrolyte:

$$(C_6H_5)_2CO + EtI \longrightarrow (C_6H_5)_2C(C_2H_5)OH$$

The mechanism for product formation from the electrochemical reduction of carbonyl functions is very complex. It depends importantly on the structure of the depolarizer, the cation of the electrolyte, the solvent, and the concentration and nature of the proton source. Examples of all of the paths outlined in Chart III have been proposed and many of them have been demonstrated.

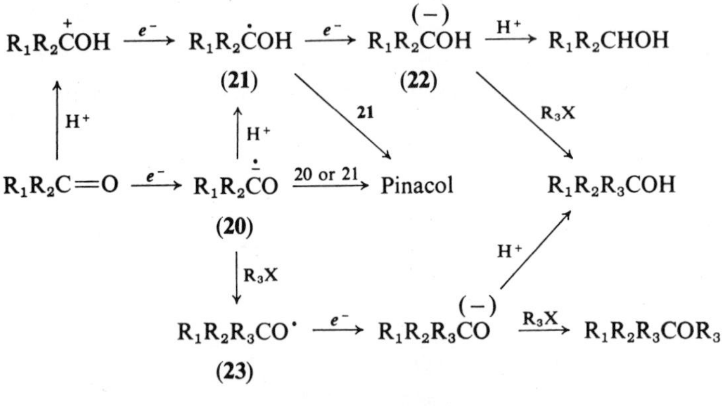

Chart III. Paths to Product Formation from Electrolysis of Carbonyls. R_1 = aryl, alkyl; R_2 = aryl, alkyl or H; and R_3X = alkyl halide or activated olefin

Polarographic studies have shown that the electrochemical reduction of carbonyl functions involves two one-electron steps. The potential of the first step shifts anodically with decreasing pH, while the second step is essentially independent of pH. Therefore, in acidic media, two clearly separated polarographic waves are observed, while in basic media a single two-electron wave is obtained. With weak acids, in the intermediate pH range, three polarographic waves can be seen in certain instances (41).

The products from this reduction depend on the proton-donating ability of the media. In acidic media, the initially formed anion radical is protonated rapidly in almost a concerted process. The resulting radical is stable to reduction at the reduction potential of the ketone, and, therefore, a sufficient concentration of these radicals can be obtained to facilitate dimerization. In contrast, in basic media the initially formed anion radical can be protonated by water, but the resulting radical being unstable toward reduction can be reduced, leading to the formation of considerable amounts of alcoholic products. Pinacols can be formed, however, in basic media, probably by the reaction of a radical (21) with an anion radical (20), rather than the dimerization of two radicals, since the diastereo ratio, i.e., *dl/meso*, is considerably different in acidic and basic media (42).

$$2\phi COCH_3 \xrightarrow[2H^+]{2e^-} \phi-\underset{\underset{H}{|}}{\overset{\overset{OH}{|}}{C}}-\underset{\underset{H}{|}}{\overset{\overset{OH}{|}}{C}}-\phi \qquad \begin{array}{cc} dl/meso & pH \\ 0.9/1.4 & >7 \\ 2.5/3.2 & <7 \end{array}$$

There is some indication that the oxygen atom of the anion radical may be associated with the electrode surface in basic media although probably not in acidic media. This association and/or solvation has been used to rationalize the formation of exclusively *erythro* 1,2-diphenylpropanol from α-methyldesoxybenzoin at a pH of 8.

The intermediates from the electrochemical reduction of carbonyl functions have been trapped. The ethyl ether of ethylphenylcarbinol was obtained from the electrolysis of benzaldehyde in the presence of ethyl iodide. Under these strictly anhydrous conditions the benzaldehyde anion radical (20) reacts with ethyl iodide forming the oxy radical (23) which can be reduced and reacts further with ethyl iodide forming the observed product. Similarly, diphenylethylcarbinol is formed from the reduction of benzophenone in dimethylformamide as solvent in the presence of ethyl iodide. The trapping of benzaldehyde anion radical with carbon dioxide has also been reported.

Sugino and Nonaka (43) have reported very interesting cross-couplings between acetone and acrylonitrile and acetone and maleic acid at mercury cathodes in 20% sulfuric acid:

$$(CH_3)_2CO + CH_2{=}CHCN + 2e^- + 2H^+ \xrightarrow[ca.\ 70\%]{} (CH_3)_2\underset{\overset{|}{OH}}{C}-CH_2CH_2CN$$
$$\text{+ lactone} \quad (1)$$

$$(CH_3)_2CO + \underset{CHCOOH}{\overset{CHCOOH}{\|}} + 2e^- + 2H^+ \xrightarrow[-H_2O]{} (CH_3)_2C\underset{\underset{}{\overset{|}{O}}}{\quad\quad}CH\underset{\overset{|}{CH_2}}{\quad}-COOH \quad (2)$$
$$\underset{\overset{\|}{O}}{\overset{\diagdown C \diagup}{}}$$

In an extension of the Sugino and Nonaka work (43) Nicolas and Pallaud (44) reduced acetone at mercury in acidic solution in the presence of styrene, α-methylstyrene, indene, cyclohexene, 1-heptene, and 1,3-butadiene. In each case they obtained low yields (1–24% based on current) of reductively coupled products which were *tert*-alcohols,

$$(CH_3)_2CO + \quad \overset{\diagdown}{}C{=}C\overset{\diagup}{} + 2H^+ + 2e^- \longrightarrow (CH_3)_2C(OH)\overset{|}{C}{-}\overset{|}{C}H$$

or di-*tert*-alcohols:

$$2(CH_3)_2CO + 2 \quad \overset{\diagdown}{}C{=}C\overset{\diagup}{} + 2H^+ + 2e^- \longrightarrow$$

$$(CH_3)_2C(OH){-}\overset{|}{C}{-}\overset{|}{C}{-}\overset{|}{C}{-}\overset{|}{C}{-}\overset{|}{C}(OH)(CH_3)_2$$

The reaction of eq. (1) is a particularly interesting and informative model system for pointing up the difficulty in making a definitive choice of reaction paths. The various paths are outlined in Chart IV.

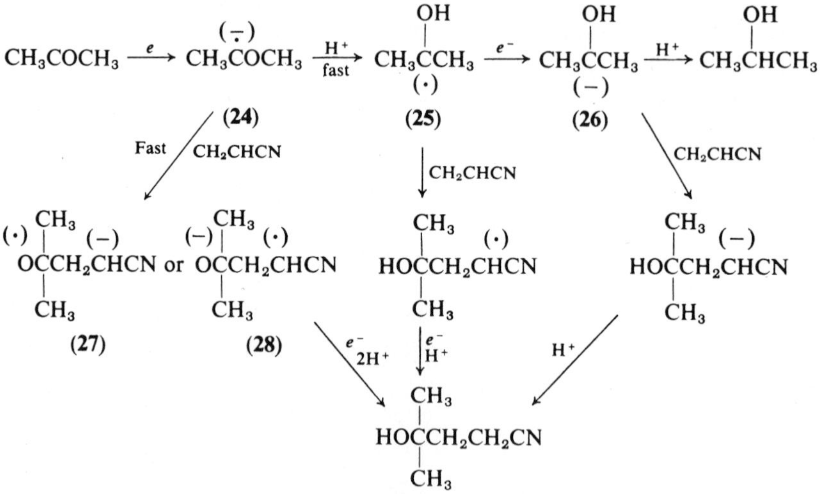

Chart IV. Possible Mechanisms for the Reductive Coupling of Acetone and Acrylonitrile

The original proposal by Sugino, based on careful potentiostatically controlled macroelectrolyses, was that the anion **26**, which has also been postulated to participate in pinacol and organometallic formation (45,46), was the precursor of the coupled product. Since the main product (ca. 95%) obtained from the reduction of acetone alone in sulfuric acid at a mercury cathode is isopropanol, presumably via **26**, this proposal was reasonable.

Brown and Lister have conducted an admirable study of the sulfuric acid-acetone-AN system by microelectrolytic techniques (47). From polarization curves it is clear that acrylonitrile enters the reaction prior to the second electron-transfer step. This evidence eliminates the anion **26** as an important intermediate. The mechanism proposed by Brown and Lister involves the radical **25**. The evidence shows one electron transfer and two fast chemical reactions followed by the final electron transfer. The order of the chemical reactions is difficult to determine. A kinetic model assuming protonation as the first step (**24** → **25**) gave a reasonable fit for the experimental data. However, substitution of acrylonitrile for the protonation in a fast chemical step (**24** → **28**) would be expected to give an indistinguishable result. In addition, during the electrolysis, the area near the electrode surface would show a higher pH than the bulk of solution due to the protonation reaction. This, in effect, could alter the course of the reaction. A further complication arises if the anion radical **24** reacts with acrylonitrile. Would the attack by **24** be anionic or radical in nature yielding **27** or **28**, respectively? If the reaction involves the radical **25**, as suggested by Brown, or **24** leading to the formation of **28**, polymeric products would be expected but are not observed. As pointed out by Brown, this radical could reduce to the anion in the reducing atmosphere of the electrode, thereby avoiding polymerization. All of these considerations make the definitive choice of the critical intermediate very difficult.

The major byproducts in the pinacol synthesis are the carbinol, the hydrocarbon, and the organometallic derivative. By changing the conditions the byproduct may often be made the major product. In the reduction of carbonyl compounds perhaps more than in any other area of electrolytic reductive coupling is our scanty fundamental knowledge of the role of the cathode material revealed (46). Metals of similar overvoltage may give in one case the carbinol and in another the pinacol (4). At a *smooth* zinc cathode acetone is reduced in aqueous alkaline solution to a mixture of pinacol and isopropyl alcohol. The surface changed during a prolonged electrolysis. To obtain a constant 30% yield of pinacol, Khomjakov et al. (48) added first zinc salts and then cyanide to the catholyte. Swann's classic studies of cathodes (3) have, among other results, led to procedures for reducing ketones to hydrocarbons.

B. α-β-Unsaturated Carbonyl Compounds

Reductive couplings, chemical and electrochemical, of carbonyl compounds conjugated with olefinic groups have been a continuously active area of investigation for many years. As indicated in a recent general review paper (49), J. Wiemann and his students have devoted over 25 years to an elucidation of the complexities encountered in this field. Some of the older work was rationalized by Pasternak (50); a more recent organization of the complex data has been presented by Zuman (41).

In this series reductive coupling may in principle occur (*a*) via the β-carbon atoms to yield 1,6-diketones and their subsequent transformation products (eq. (3)), (*b*) via the carbonyl groups to yield divinylglycols (eq. (4)), and (*c*) via the β-carbon atom of one molecule and the carbonyl group of a second molecule (eq. (5)) to give a "mixed" γ-hydroxy-carbonyl intermediate* and then its ultimate product:

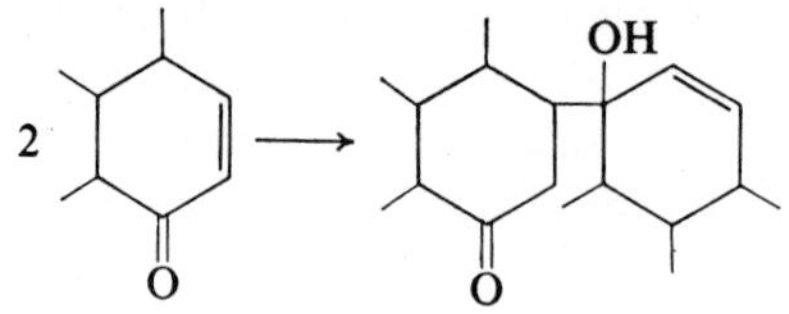

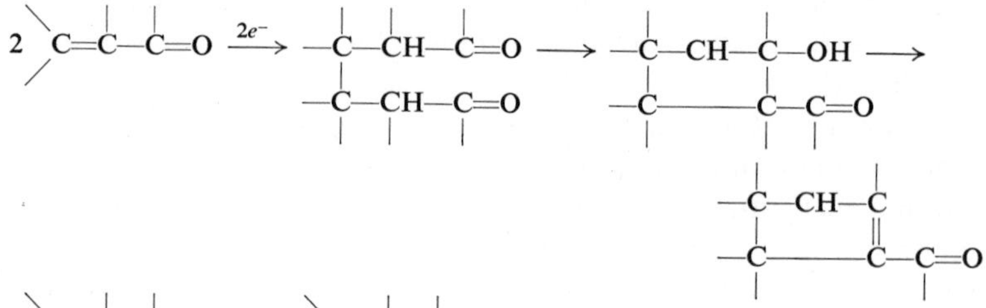

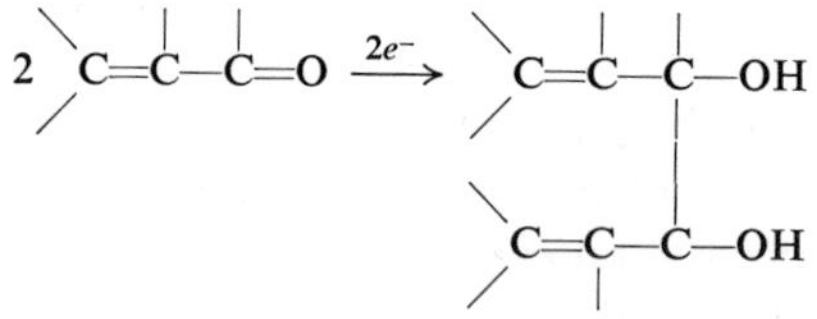

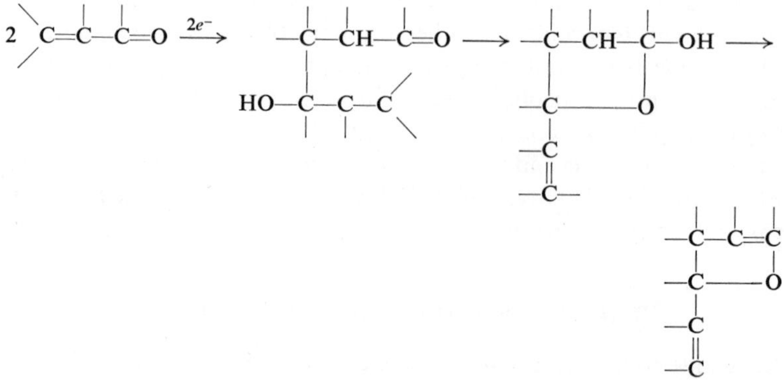

* L. J. Sargent and U. Weiss (51), report unsymmetrical reductive coupling of 14-hydroxycodeinone by Zn + HOAc or Zn–Cu + HCOOH.

For example, by reducing methyl vinyl ketone at mercury in an undivided cell, Wiemann and Bouguerra (52) obtained several products depending upon the conditions of electrolysis:

a. At -1.35 to 1.43 V in a solution containing acetic acid-sodium acetate-ethanol 2,7-octanedione (eq. (3)), 1,2-dimethyl-1,2-divinylglycol (eq. (4)), and polymer in the ratio of 65:13:17, respectively. Evidence was also obtained for the mercurial byproduct $Hg(CH_2CH_2COCH_3)_2$ which had been previously reported (53).

b. At -1.42 V in a solution containing tetraethylammonium *p*-toluene-sulfonate and aqueous acetonitrile or tetrahydrofuran the products were the octanedione (eq. (3)) and 2-methyl-3-acetyl-1-cyclopentene formed from the dione by cyclization and dehydration (eq. (5)). Analogous results were found in the reduction of mesityl oxide (54).

Misono and co-workers (55) obtained cyclopentenealdehyde by reductive coupling of acrolein presumably via the intermediate adipaldehyde (eq. (3)).

The mechanism for the reduction and for electrolytic reductive coupling of α,β-unsaturated carbonyls depends importantly on the availability of protons. In acid solution pinacol formation occurs through a free radical intermediate. Depending on the strength of the acid, the protonated form of the carbonyl may be reduced directly to the radical or the carbonyl may be reduced to the anion radical which is protonated in a fast step. Either of these paths yield allylic radicals which can dimerize head-to-head, head-to-tail, or tail-to-tail, yielding pinacol, carbonyl, alcohol, or dicarbonyl, respectively. In neutral or basic media the α,β-unsaturated carbonyl

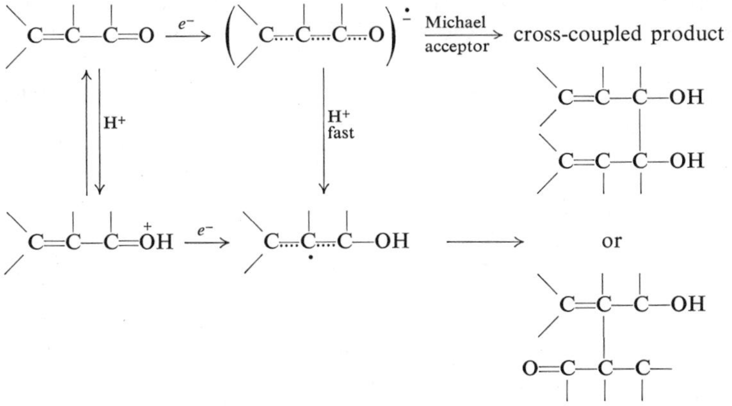

Chart V. Paths for Electrolytic Reductive Coupling of α,β-Unsaturated Carbonyls in Acid Media

is reduced to an anion radical which can react with starting material to yield the dimer ketyl. This type of mechanism will be discussed in more detail in the section on activated olefins.

Baizer and Anderson (56) reported mixed reductive couplings between methyl vinyl ketone and diethyl fumarate (eq. (6)), benzalacetone and acrylonitrile (eq. (7)), mesityl oxide and acrylonitrile (eq. (8)), and methyl vinyl ketone and 4-vinyl pyridine (eq. (9)). In all cases electrolyses were at a mercury cathode and in neutral or slightly alkaline aqueous medium containing tetraalkylammonium alkarylsulfonates.

$$CH_3COCH{=}CH_2 + \mathit{trans}\ \begin{matrix}CHCOOC_2H_5\\ \|\\ CHCOOC_2H_5\end{matrix} \xrightarrow[2H_2O]{2e^-}\ \begin{matrix}CH_2COOC_2H_5\\ |\\ CHCOOC_2H_5 + 2OH^-\\ |\\ CH_2CH_2COCH_3\end{matrix} \qquad (6)$$

$$C_6H_5CH{=}CHCOCH_3 + CH_2{=}CHCN \xrightarrow[2H_2O]{2e^-}\ \begin{matrix}C_6H_5CHCH_2COCH_3\\ |\\ CH_2CH_2CN\end{matrix} + 2OH^- \qquad (7)$$

$$(CH_3)_2C{=}CHCOCH_3 + CH_2{=}CHCN \xrightarrow[2H_2O]{2e^-}\ \begin{matrix}(CH_3)_2CCH_2COCH_3\\ |\\ CH_2CH_2CN\end{matrix} + 2OH^- \qquad (8)$$

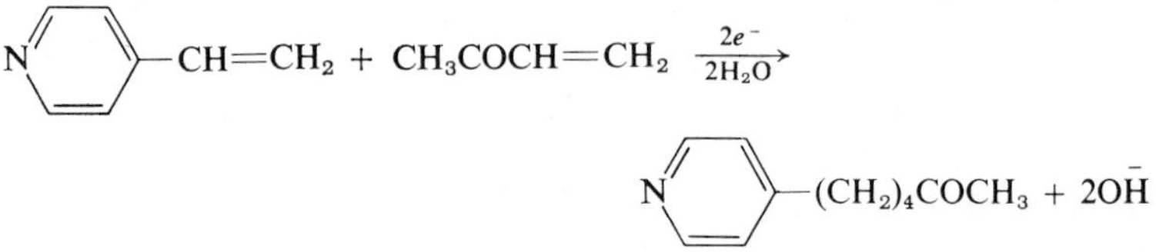

$$\textbf{(29)}$$

$$\xrightarrow[2H_2O]{2e^-}$$

$$-(CH_2)_4COCH_3 + 2O\bar{H} \quad (9)$$

Wiemann and Bouguerra (57) likewise obtained **29** according to eq. (8), although in smaller yield, together with self-coupling products of mesityl oxide. From electrolysis of mesityl oxide and ethyl acrylate in acetic acid–sodium acetate they obtained a small yield of **30** contaminated by its cyclodehydration product **31**.

$$\begin{matrix}(CH_3)_2C{-}CH_2COCH_3\\ |\\ CH_2CH_2COOC_2H_5\end{matrix}$$

(30)

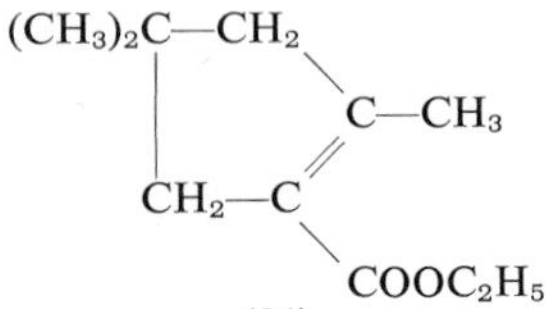

(31)

C. α,β-Unsaturated Acids

C. L. Wilson and co-workers (58) prepared β,γ-diphenyladipic acid in 55% yield (ca. 1:1 *meso/dl*) by electroreduction of cinnamic acid in alcohol-aqueous sulfuric acid at a mercury cathode. [Lead gave no bi-molecular acids but had been reported previously (Inoue) to yield 3,4-diphenyl-1,6-hexanediol]. Other organic water-miscible solvents could be used. The reductive coupling worked well for o-chloro-, m-hydroxy-, and p-methoxy-cinnamic acids, poorly for o-cyanocinnamic acid and not at all for the three nitro- and p-aminocinnamic acids (58).

In earlier work Wilson (59) had also reductively dimerized sorbic acid under similar conditions. From this work it is clear that dimeric products are formed through free radical dimerization. The electrode reaction probably involves a direct electron transfer to the organic depolarizer rather than the reduction of hydrogen. The organic species may be associated with a proton when electron transfer occurs.

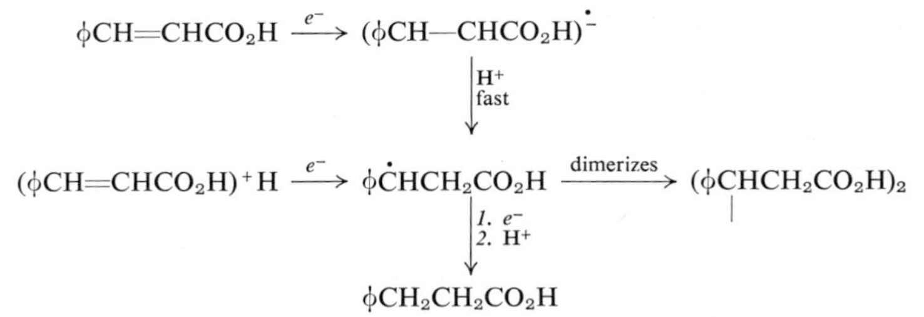

$$\phi CH=CHCO_2H \xrightarrow{e^-} (\phi CH\text{---}\overset{\bullet}{C}HCO_2H)^-$$

$$\downarrow \begin{array}{l} H^+ \\ fast \end{array}$$

$$(\phi CH=CHCO_2H)^+H \xrightarrow{e^-} \phi \overset{\bullet}{C}HCH_2CO_2H \xrightarrow{dimerizes} (\phi CHCH_2CO_2H)_2$$

$$\downarrow \begin{array}{l} 1.\ e^- \\ 2.\ H^+ \end{array}$$

$$\phi CH_2CH_2CO_2H$$

D. Onium Compounds

Electrolysis of certain quaternary ammonium nitrates in dimethyl-formamide or N,N-dimethylacetamide gave moderate yields (Table I) of coupled products (60). Yields were not optimized.

TABLE I

Electrolysis of $R_1R_2R_3R_4N^+$ in Anhydrous Media

$R_1R_2R_3R_4N^+$					
R_1	R_2	R_3	R_4	Coupled product	Yield
C_6H_5	$C_6H_5CH_2$	CH_3	CH_3	Bibenzyl	35.0
Fluorenyl	CH_3	CH_3	CH_3	Bifluorenyl	26.1
Benzyl	C_2H_5	C_2H_5	C_2H_5	Bibenzyl	31.9
p-MeO-$C_6H_4CH_2$	CH_3	CH_3	CH_3	4,4'-Dimethoxybibenzyl	15.0
Cinnamyl	CH_3	CH_3	C_6H_5	Bicinnamyl	6.0
α-Phenylethyl	CH_3	CH_3	CH_3	2,3-Diphenylbutane	30.0

The authors considered two mechanisms. Their experimental evidence

$$R_1R_2R_3R_4N^+ \xrightarrow{1e^-} R_1R_2R_3N + R_4\cdot\,;\ 2R_4\cdot \longrightarrow R_4R_4 \qquad (10)$$

$$R_1R_2R_3R_4N^+ \xrightarrow{2e^-} R_1R_2R_3N + R_4^-$$

$$R_4^- + R_1R_2R_3R_4N^+ \longrightarrow R_4R_4 + R_1R_2R_3N \qquad (11)$$

favored mechanism of eq. (10). The electrolysis of optically active d-α-phenylethyltrimethylammonium nitrate gave only racemic 2,3-diphenyl-butane. The mechanism of eq. (11) involves a nucleophilic displacement of the anion, R_4^-, on the ammonium ion which should yield optically active inverted product.

Horner et al. (61) found no coupled products as a result of electrolysis of quaternary phosphonium and arsonium salts in aqueous or alcoholic media in divided cells. Finkelstein (62), however, on electrolyzing benzyl-triphenylphosphonium nitrate in dimethylformamide obtained a 31% yield of bibenzyl.

E. Aromatic Nitro Compounds and Derivatives

Electroreduction of nitro compounds in alkaline medium is an excellent synthetic procedure for preparing azoxy-, azo-, and hydrazocompounds (3):

$$2ArNO_2 + 6F + 6H_2O \xrightarrow[\text{NaOH}]{} \underset{\underset{O}{\downarrow}}{ArN}{=}NAr + 6\,OH^- \qquad (12)$$

(32)

$$2ArNO_2 + 8F + 8H_2O \xrightarrow[\text{NaOH}]{} ArN{=}NAr + 8\,OH^- \qquad (13)$$

(33)

$$2ArNO_2 + 10F + 10H_2O \xrightarrow[\text{NaOH}]{} ArNHNHAr + 10\,OH^- \qquad (14)$$

(34)

The yields in all three cases are generally 70–95%. The azoxy compounds (4) usually precipitate out of solution. They have been prepared at nickel cathodes. Certain sterically hindered and o- or p-substituted nitro compounds give poor results. To achieve reduction to **33** it is necessary to keep the intermediate azoxy compounds hot and in solution by adding organic solvents or hydrotropic agents. This means also serves to promote reoxidation of over-reduced product (hydrazo-) by the nitro or nitroso precursor. Nickel or phosphor bronze cathodes have been used. To achieve **34**, lead, tin, zinc, monel metal and lead electroplated on iron have been used.

Illustrative of intramolecular reductive coupling of this type is the preparation of benzo-[c]-cinnoline from 2,2'-dinitrobiphenyl (63):

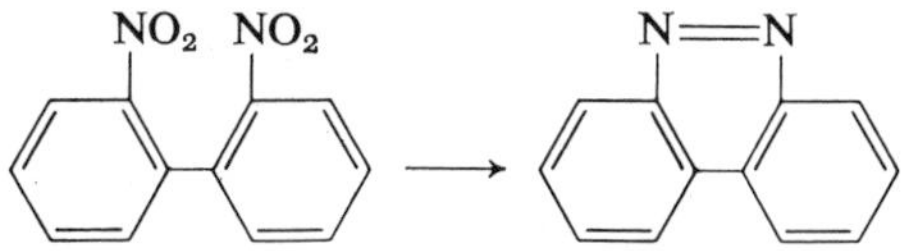

Udupa and co-workers have reported a series of studies involving the use of rotating cathodes to accelerate reductions. In the reduction of nitrobenzene to hydrazobenzene (rearranged to benzidine for isolation and determination) they found (64) that, if the lead-coated iron-disk-type cathode was rotated at 2000 rpm, an 88% yield of product could be obtained at high current densities (30 amp/dm^2). Results with a stationary cathode under these conditions were much poorer. The authors conclude that rotation favors the migration of nitro compound to and hydrazo compound away from the cathode and thus permits rapid reduction and little overreduction. A similar technique has been used in the reduction of o-nitroanisole (65).

The electrochemical reduction of aromatic nitro compounds is probably the most extensively studied electrochemical system. It is interesting that, in spite of the fact that the main features of the above EC reactions were elucidated by older classical techniques, new light has been cast on the mechanism by the developing area of ion-radical chemistry (66). The

$$ArNO_2 \xrightarrow{e^-} ArNO_2^{\cdot -} \xrightarrow{H^+} Ar\dot{N}O_2H \xrightarrow{e^-} Ar\bar{N}O_2H \xrightarrow[-H_2O]{H^+} ArNO$$

$$ArNO \xrightarrow{e^-} ArNO^- \xrightarrow{H^+} Ar\dot{N}OH \xrightarrow{e^-} Ar\bar{N}OH \xrightarrow{H^+} ArNHOH$$

$$ArNHOH \xrightarrow{e^-} Ar\dot{N}HOH \xrightarrow[-H_2O]{H^+} Ar\dot{N}H \xrightarrow{e^-} Ar\bar{N}H \xrightarrow{H^+} ArNH_2 \qquad (15)$$

reduction of nitrobenzene to aniline has been shown to involve a series of one-electron reductions followed by protonation (eq. (15)). The formation of coupled products occurs by the reaction of nitrosobenzene with phenylhydroxylamine to form azoxybenzene—only a C step is involved—which is subsequently reduced stepwise first to azobenzene and then to hydrazobenzene. The use of controlled potential electrolysis was imperative for the understanding of this reaction scheme.

$$ArNO_2 \xrightarrow{2e^-} ArNO \xrightarrow{2e^-} ArNHOH \xrightarrow{2e^-} ArNH_2$$

$$\downarrow ArNHOH$$

$$ArN{=}NAr \xrightarrow{2e^-} ArN{=}NAr \xrightarrow{2e^-} ArNHNHAr$$
$$\overset{|}{O}$$

The mixed reductive coupling of benzalaniline with acrylonitrile (eq. (16)) and of azobenzene with ethyl acrylate (eq. (17)) have been reported (7).

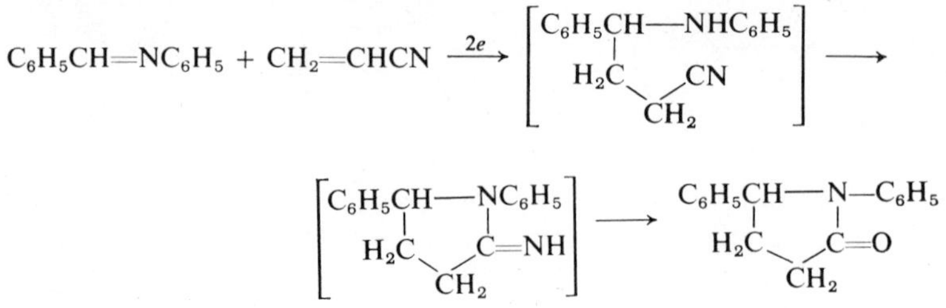

$$C_6H_5CH{=}NC_6H_5 + CH_2{=}CHCN \xrightarrow{2e^-} \begin{bmatrix} C_6H_5CH{-}NHC_6H_5 \\ H_2C \quad CN \\ CH_2 \end{bmatrix} \longrightarrow$$

$$\begin{bmatrix} C_6H_5CH{-}NC_6H_5 \\ H_2C \quad C{=}NH \\ CH_2 \end{bmatrix} \longrightarrow \begin{matrix} C_6H_5CH{-}N{-}C_6H_5 \\ H_2C \quad C{=}O \\ CH_2 \end{matrix} \quad (16)$$

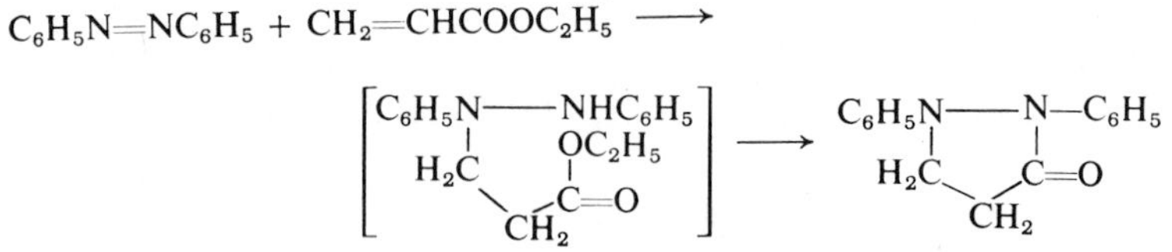

$$C_6H_5N{=}NC_6H_5 + CH_2{=}CHCOOC_2H_5 \longrightarrow$$

$$\begin{bmatrix} C_6H_5N{-}NHC_6H_5 \\ H_2C \quad OC_2H_5 \\ CH_2 \quad C{=}O \end{bmatrix} \longrightarrow \begin{matrix} C_6H_5N{-}N{-}C_6H_5 \\ H_2C \quad C{=}O \\ CH_2 \end{matrix} \quad (17)$$

Henning Lund (67) has developed a tremendous variety of syntheses of heterocyclic compounds based on intramolecular reductive coupling. The reducible group may be, e.g., nitro, nitroso, azo, azo-methine; an electrophilic center within the same molecule is attacked by the reduced intermediate to yield the products. The following are typical examples:

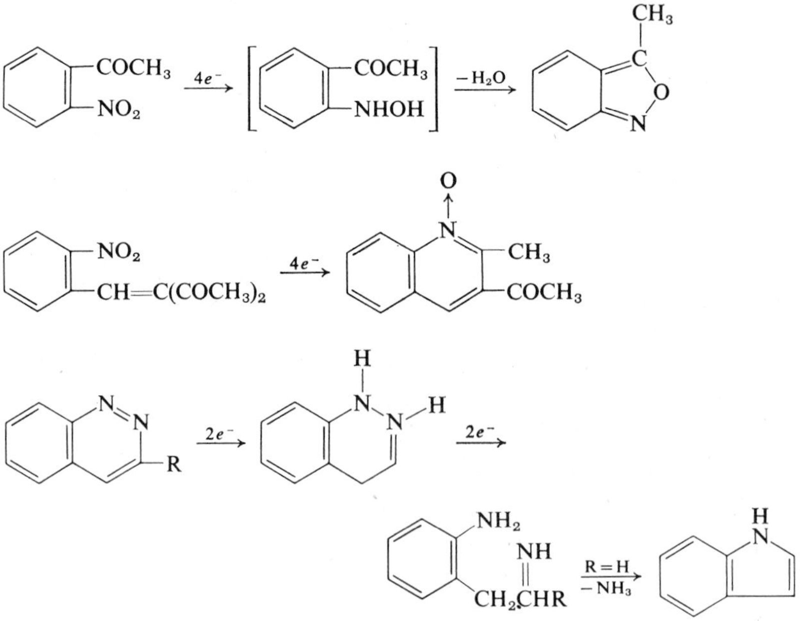

F. Olefins

Reduction of stilbene at mercury in a divided cell with tetrabutyl-ammonium iodide as supporting electrolyte yielded up to 30% of the reduced dimer 1,2,3,4-tetraphenylbutane when anhydrous dimethylformamide was used as a solvent but only the reduced monomer, i.e., 1,2-diphenylethane when the less "aprotic" solvent, acetonitrile, was used (68). Triphenylethylene and tetraphenylethylene gave no reduced dimeric products.

Reduction of a mixture of stilbene and an excess of ethyl iodide in acetonitrile yielded a small quantity of the mixed reductively-coupled product, 1,2-diphenylbutane. If carbon dioxide was bubbled through the

$$C_6H_5CH{=}CHC_6H_5 \xrightarrow[EtI]{2e^-} C_6H_5CH_2\underset{\underset{C_2H_5}{|}}{C}HC_6H_5$$

catholyte during stilbene reduction in dimethylformamide, a 92% yield of *meso*-diphenylsuccinic acid was obtained (69); when the more "protic"

$$C_6H_5CH{=}CHC_6H_5 + 2CO_2 + 2e^- \longrightarrow C_6H_5\underset{\underset{COO^-}{|}}{C}H{-}\!\!-\!\!\underset{\underset{COO^-}{|}}{C}H{-}C_6H_5$$

acetonitrile was used as a solvent (69a), the monocarboxylic acid was obtained.

When stilbene in anhydrous THF using sodium tetraphenylboron as supporting electrolyte was reduced to the anion radical stage, the current shut off and carbon dioxide then bubbled through the catholyte, *meso*-diphenylsuccinic acid and not 1,2,3,4-tetraphenyladipic acid was obtained (70). This occurred presumably because of electron transfer from a stilbene anion radical to the intermediate carboxylate anion radical:

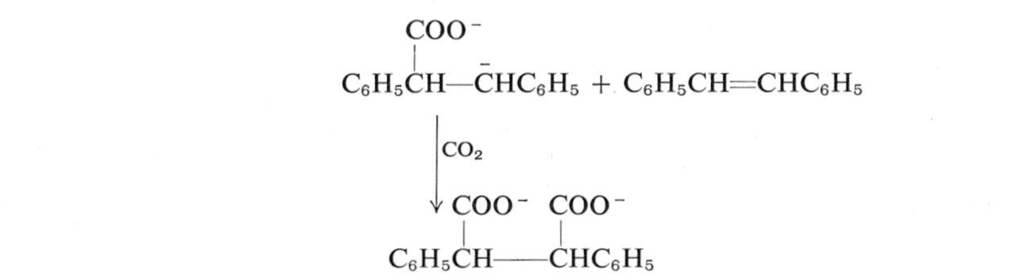

$$(C_6H_5CH{-}CHC_6H_5)^{\overline{\cdot}} + CO_2 \longrightarrow$$

$$C_6H_5\underset{\underset{COO^-}{|}}{C}H{-}\overset{\cdot}{C}HC_6H_5 \xrightarrow{\;\;\;/\!\!\!\!\to\;\;\;} C_6H_5\underset{\underset{C_6H_5}{|}}{\underset{\underset{COO^-}{|}}{C}}HCH{-}\!\!-\!\!\underset{\underset{C_6H_5}{|}}{C}H{-}\underset{\underset{COO^-}{|}}{C}HC_6H_5$$

$$\Big\downarrow (C_6H_5CH{-}CHC_6H_5)^{\overline{\cdot}}$$

$$C_6H_5\underset{\underset{COO^-}{|}}{C}H{-}\overset{-}{C}HC_6H_5 + C_6H_5CH{=}CHC_6H_5$$

$$\Big\downarrow CO_2$$

$$C_6H_5\underset{\underset{COO^-}{|}}{C}H{-}\!\!-\!\!\underset{\underset{COO^-}{|}}{C}HC_6H_5$$

As part of a larger program on reductive coupling reactions of activated olefins (71), the highly conjugated olefins 9-benzalfluorene (**35**), 2-phenyl-1,3-butadiene (**36**) and 6,6-diphenylfulvene (**37**) were electrolyzed in concentrated aqueous solutions of hydrotropic quaternary ammonium salts containing DMF as co-solvent. In spite of the presence of proton donors in the catholyte (contrast Wawzonek et al. work cited above) **35** and **36** formed their reductively-coupled products ("hydrodimers"). Due to steric effects **37** formed only reduced monomer. Olefins **35** and **37** readily formed cross-coupled products with acrylonitrile (**38** and **39**, respectively). Electrolysis of **35** and an excess of ethyl acrylate give a poor yield of **40**.

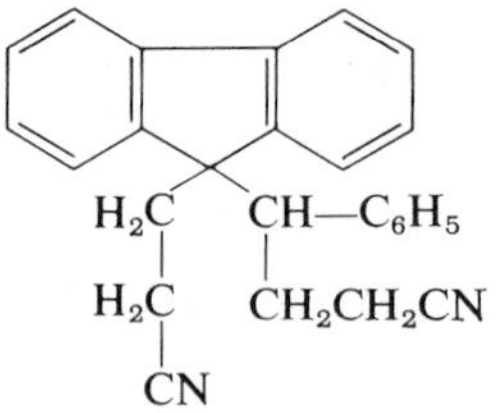

(**38**) 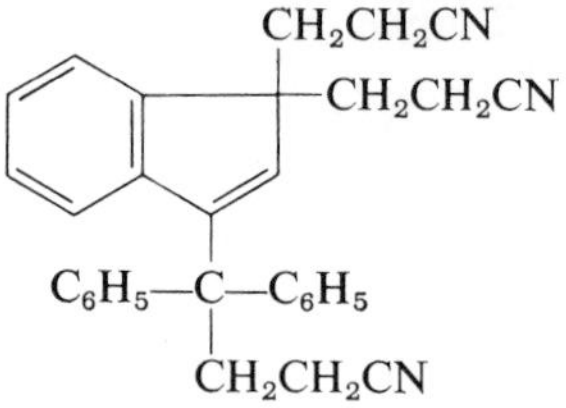   (**39**)

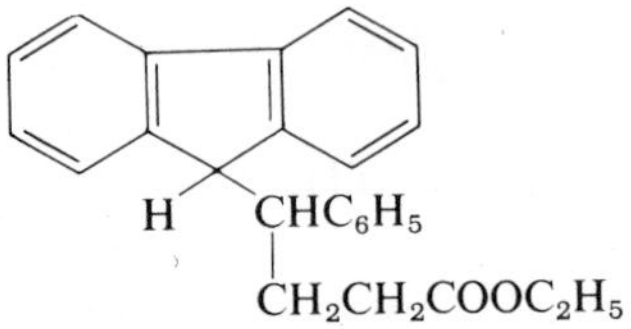

(**40**)

There is now a considerable literature (72) on the initiation of polymerization of vinyl aromatic compounds by anion radicals produced by electron transfer to the monomer from an alkali metal or from an alkali metal aromatic anion radical (e.g., sodium naphthalenide or sodium biphenylide). The initial dimeric species formed is a product of reductive coupling; conceptually it may arise by radical coupling of two anion radicals (eq. (19)) or by attack of an anion radical upon starting monomer (eq. (20)):

$$R-CH=CH_2 \xrightarrow{1e^-} (RCHCH_2)^- \qquad (18)$$
$$(\textbf{41}) \qquad\qquad (\textbf{42})$$

$$2(RCHCH_2)^{\cdot-} \longrightarrow \bar{R}CHCH_2CH_2\bar{C}HR \qquad (19)$$
$$(\textbf{43})$$

$$(RCHCH_2)^{\cdot-} + RCH=CH_2 \longrightarrow R\dot{C}HCH_2CH_2\bar{C}HR \qquad (20)$$
$$(\textbf{44})$$

$$R\dot{C}HCH_2CH_2\bar{C}HR \xrightarrow{1e^-} \bar{R}CHCH_2CH_2\bar{C}HR$$

43 can initiate anionic polymerization to produce "living" polymers. Whether **42** can initiate anionic and free-radical polymerization from the respective available sites is more controversial (73) than whether **44** can do so. It is also possible for **44** to dimerize by radical coupling to produce a tetrameric dianionic initiator (73).

The literature on studies of the electrochemical analogs of the above systems is still sparse (74–76). An electrochemical-chemical study of the reductive coupling of 1,2-diactivated olefins (see Section IV-G) indicated that eq. (20) rather than eq. (19) was the major route to dimeric products. This may point the way to determining the pathways to polymerization by direct experimental means rather than by inferences from kinetic data.

G. Activated Olefins

α,β-Unsaturated carbonyl compounds have been discussed in Section IV-B.

Since 1963 (71), there has appeared an extensive series of papers dealing with the electrolytic reductive coupling of activated olefins $\overset{\diagdown}{\underset{\diagup}{C}}{=}\overset{|}{C}{-}X$ (in which X is an electron-attracting group not reduced under the conditions of the electrolysis) under neutral or slightly alkaline aqueous conditions using high hydrogen overvoltage cathodes and quaternary ammonium supporting electrolytes. Optimum yields of coupled products are obtained at partial conversion; more exhaustive electrolyses led to increasing quantities of $\overset{\diagdown}{\underset{\diagup}{C}}H{-}\overset{|}{C}HX$.

The initial investigation concerned the "hydrodimerization" of acrylonitrile to adiponitrile (eq. (21), X = CN). The method was later extended to other derivatives of α,β-unsaturated acids including cyclic compounds (77), to derivatives of 1,3-butadiene, to Michael acceptors in

$$2CH_2{=}CHX \xrightarrow[2H_2O]{2e^-} X(CH_2)_4X + 2OH^- \tag{21}$$

general, except for nitroolefins (78), to mixed reductive couplings (eq. (22)), and to the intramolecular reaction (eq. (23)) which leads to cyclic products (79). The most recent study in this series has concerned the behavior of 1,2-diactivated olefins (19).

$$\overset{\diagdown}{\underset{\diagup}{C}}{=}\overset{|}{C}{-}X + \overset{\diagdown}{\underset{\diagup}{C}}{=}\overset{|}{C}{-}Y \longrightarrow X{-}\overset{|}{C}H{-}\overset{|}{C}{-}\overset{|}{C}{-}\overset{|}{C}H{-}X + X{-}\overset{|}{C}H{-}\overset{|}{C}{-}\overset{|}{C}{-}\overset{|}{C}HY$$

$$+ Y{-}\overset{|}{C}H{-}\overset{|}{C}{-}\overset{|}{C}{-}\overset{|}{C}H{-}Y \tag{22}$$

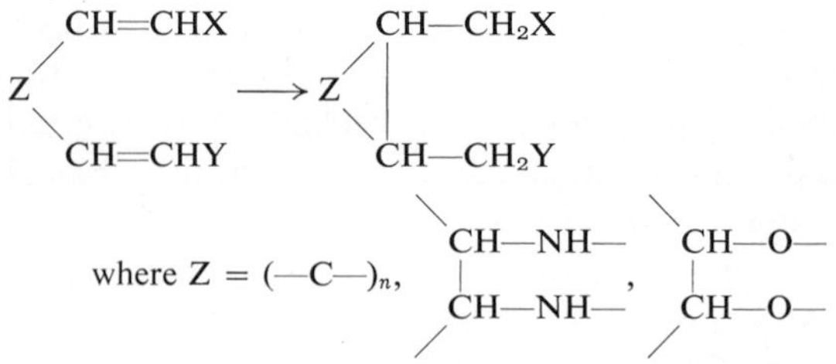

$$Z \underset{CH=CHY}{\overset{CH=CHX}{\diagup\!\!\!\diagdown}} \longrightarrow Z \underset{CH-CH_2Y}{\overset{CH-CH_2X}{\diagup\!\!\!\diagdown}} \tag{23}$$

$$\text{where } Z = (-C-)_n, \quad \underset{CH-NH-}{\overset{CH-NH-}{\diagup\!\!\!\diagdown}}, \quad \underset{CH-O-}{\overset{CH-O-}{\diagup\!\!\!\diagdown}}$$

The macroscale electrolyses of the monoactivated olefins (eq. (21)) were carried out usually at a potential slightly anodic to $E_{1/2}$. Representative hydrodimerizations which give good enough yields to be considered of preparative value are collected in Table II.

TABLE II

Hydrodimerizations of Some Activated Olefins

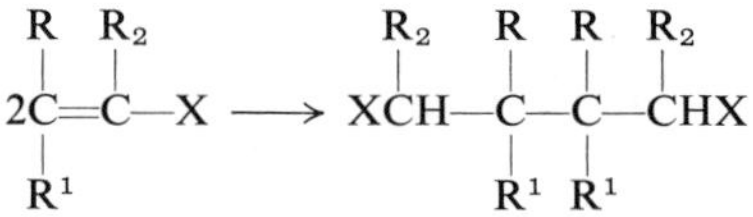

$$2\underset{R^1}{\overset{R}{\underset{|}{\overset{|}{C}}}}=\underset{}{\overset{R_2}{\underset{}{\overset{|}{C}}}}-X \longrightarrow X\underset{R^1}{\overset{R_2}{CH}}-\underset{R^1}{\overset{R}{\underset{|}{\overset{|}{C}}}}-\underset{}{\overset{R}{\underset{|}{\overset{|}{C}}}}-\overset{R_2}{CHX}$$

Olefin				$-E(V)$ vs.	
R	R^1	R^2	X	SCE	Ref.
H	H	H	CN	1.81–1.91	80
H	H	CH_3	CN	2.01–2.05	77
CH_3	CH_3	H	CN	—	57
H	H	H	$COOC_2H_5$	1.85	77
H	H	CH_3	$COOC_2H_5$	—	81
CH_3	CH_3	H	$COOC_2H_5$	2.10–2.20	57
CH_3	H	H	$CONEt_2$	2.03–2.12	77
H	H	H	2-pyr-	1.6	82
H	H	H	4-pyr-	1.5	82
9-Fluorylidene		H	C_6H_5	1.5–1.6	83
H	H	CH_2CH_2CN	CN	1.8–1.9	84
H	H	H	$PO(OEt)_2$	2.0	78
H	H	H	$PO(C_6H_5)_2$	2.29	78
C_6H_5	C_6H_5	H	CN	—	85
NCH_2CH_2	H	H	CN	—	86

Experimental facts which must be borne in mind in developing a mechanistic picture are:

1. The aliphatic monoactivated olefins show only one polarographic reduction wave under a variety of conditions. There has been only one report (87) of the observation of a feeble ESR signal for the intermediate anion radical of acrylonitrile.

2. The hydrodimerization is not inhibited by free radical traps.

3. The use of alkali metal rather than quaternary ammonium counter-ions leads to an increase in the formation of saturated monomer at the expense of hydrodimer (80,88). Addition of quaternary ammonium salts to electrolytes containing alkali metal salts improves the yield of hydrodimer. This phenomenon is observed also in amalgam reductions (89,90). The special role of quaternary ammonium ions has been extensively discussed (91–94).

4. Electrolysis at too low or too high a pH, particularly in the case of acrylonitrile, yields oligomers and polymers (95,96).

5. Electrolytes containing very low concentrations of proton donors lead to oligomers and polymers (96–98).

6. Hydrodimerization involves coupling of two molecules almost exclusively (99) through their respective β-positions.

The original hypotheses (71,80,100) concerning the mechanism of hydrodimerization, particularly of acrylonitrile, were advanced in order to rationalize the experimental facts and to provide a basis for predicting the scope of the reaction (71). They were not based on experiments designed as mechanistic probes. They recognized that hydrodimerization was not a radical coupling reaction and that a reduced intermediate (dianion **45**, anion **46**, or anion radical **47**) anionically attacked an electrically neutral acrylonitrile molecule to yield the coupled intermediate which led to final product. Entities like **45** and **46** are obviously not those expected to show

$$(CH_2CHCN)^- 2R_4N^+, \quad (CH_2CH_2CN)^- R_4N^+, \quad (CH_2CHCN)^{-} R_4N^+$$

$$\textbf{(45)} \qquad\qquad\qquad \textbf{(46)} \qquad\qquad\qquad \textbf{(47)}$$

as great stability in homogeneous solution chemistry as are alternate formulations; particular electric field and adsorption effects were invoked to support their transitory existence (87,91). The number of electrons taken up by the acrylonitrile molecule ranges from 0.1 under anhydrous polymerizing condition to two electrons in acid solution where propionitrile is the reaction product. Beck has done an extensive mechanistic study of the acrylonitrile system. From the dependence of Tafel curves on the concentration of acrylonitrile and water, he concluded that the rate-determining step in the reduction of acrylonitrile contained one molecule of acrylonitrile, one molecule of water, and one electron. The postulated β-anion

$$CH_2{=}CHCN + H_2O + e^- \longrightarrow [\cdot CH_2CH_2CN] \longrightarrow \textbf{46}$$

was rationalized on the basis of an unusual effect of the electric field (91). The reaction of **46** with an acrylonitrile rather than a water molecule has been explained on the basis of adsorption of hydrophobic quaternary ammonium ions on the electrode (91,93).

Molecular orbital treatment has persuaded Lazarov et al. (101) of the possible intermediacy of **45** in the coupling reaction. Figeys and Figeys (102), however, derived from their molecular orbital calculations the conclusion that **47** was the primary intermediate—which is not consistent with Beck's data—and that the subsequent steps of the hydrodimerization were:

$$47 + CH_2\!\!=\!\!CHCN \longrightarrow \cdot[NCCH]\!-\!CH_2CH_2\!-\![CHCN]^- \qquad (24)$$
$$(48)$$

$$48 + e^- \longrightarrow {}^-[NCCH\ ]\ CH_2CH_2\ [\ CHCN]^- \qquad (25)$$
$$(49)$$

$$49 + 2H_2O \longrightarrow \text{adiponitrile} + 2OH^- \qquad (26)$$

No allowances were made (if any should be made) for the fact that the electron transfers and initial coupling step must take place in the vicinity of the polarized cathode.

It is clear that additional mechanistically oriented experiments must be designed.

The work on 1,2-diactivated olefins (p. 218) led to conclusions which are of value in interpreting the mechanism of the reductive coupling reaction.

Mixed reductive couplings (eq. (22)) lead to three products when the reduction potentials of the two olefins are fairly close together and to only two when the reduction potentials are at least 0.2 V apart and the electrolysis is potentiostated at the more positive potential. Illustrative examples are given in Table III.

It is clear that the most easily reduced olefin provides the intermediate species for attack upon parent or second olefin. The distribution of products (hydrodimer vs. mixed coupled products) should depend on (*a*) relative concentration of olefins, (*b*) steric factors, and (*c*) relative susceptibility to attack within the electric field upon each olefin by the electrogenerated intermediate. More will be said about (*c*) in the discussion of 1,2-diactivated olefins (p. 218).

The electrohydrocyclization reaction (eq. (23)) has been studied most extensively when X = Y = COOR. It gives excellent yields of functionally substituted cyclopropanes, cyclopentanes, and cyclohexanes, moderate yields of cyclobutanes, poor yields of cycloheptanes, and no yields of larger ring compounds.

The mechanism of the electrolytic reductive cyclization of bis-activated olefins has been studied (eq. (27)). In each case, where cyclic products were obtained by macroelectrolysis, two polarographic waves were obtained. The $E_{1/2}$ value of the first wave was shifted in an anodic direction from the expected potential of the activated olefin, while the second wave was at

TABLE III

Some Mixed Reductive Couplings

Olefin I ($-E_{1/2}$ V)	Olefin II ($-E_{1/2}$ V)	Electrolysis at $-E$ (V vs. SCE)	Products	Ref.
$CH_2=CHCOOC_2H_5$ (−1.8)	$CH_2=CHCN$ (1.9)	1.83–1.85	Diethyl adipate + ADN + $CN(CH_2)_4COOEt$	103
$CH_2=\underset{NHCOCH_3}{C}-COOCH_3$ (−1.7)	$CH_2=CHCN$ (1.9)	1.75	$NC(CH_2)_3\underset{NHCOCH_3}{CH}-COOCH_3$	103
$CH_2=CHCH=CHCN$ (1.5)	$CH_2=CHCN$ (1.9)	<1.71	$NC(CH_2)_6CN$ (55%) + isomers [a]	82
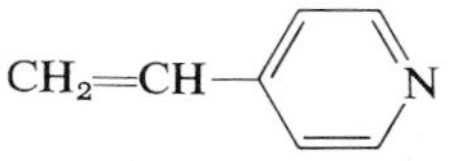 $CH_2=CH-$(pyridyl) N (1.5)	$CH_3COCH=CH_2$ (1.4)	1.4–1.5	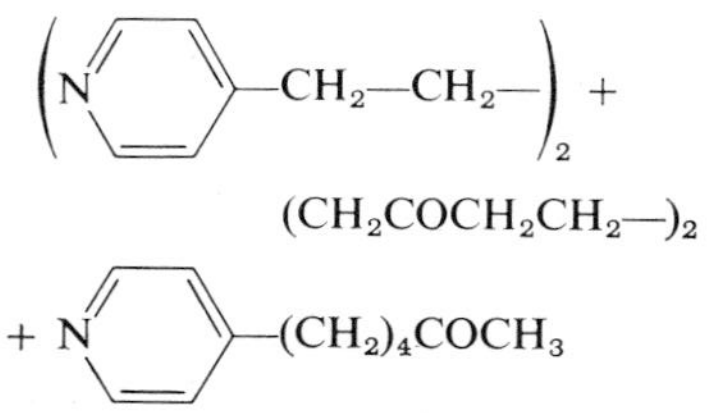 $\left(\text{N}\!\!-\!\!CH_2\!-\!CH_2-\right)_2 +$ $(CH_2COCH_2CH_2-)_2$ + $^+\text{N}-(CH_2)_4COCH_3$	82
$CH_2=\underset{CH_2CH_2CN}{C}-CN$ (1.8–1.9)	$CH_2=CHCN$ (1.9)	1.8–1.9	$NC(CH_2)_4CN + \left(\underset{CH_2CH_2CN}{NCCHCH_2-}\right)_2$ + $NCCH_2CH_2CH_2CHCN$	104
$\underset{CH-COOC_2H_5}{\overset{CH-COOC_2H_5}{\parallel}}$ (1.4)	$CH_2=CHCN$ (1.9)	1.3–1.4	$\left(\underset{CHCOOC_2H_5-}{CH_2COOC_2H_5}\right)_2$ + $\underset{CH_2CH_2CN}{\overset{CH_2CH_2CN}{CHCOOC_2H_5}}$ + no adiponitrile	103

[a] After hydrogenation of the coupled product.

$$\underset{\text{(CR}_2)_n}{\overset{\text{CH}=\text{CHX}}{\underset{\text{CH}=\text{CHX}}{\Bigg<}}} \xrightarrow{e^-} \underset{\text{(CR}_2)_n}{\overset{\text{CH}-\overset{(-)}{\text{C}}\text{HX}}{\underset{\text{CH}-\underset{\bullet}{\text{C}}\text{HX}}{\Bigg<}}} \xrightarrow[2\text{H}^+]{e^-} \underset{\text{(CR}_2)_n}{\overset{\text{CHCH}_2\text{X}}{\underset{\text{CHCH}_2\text{X}}{\Bigg<}}} \qquad (27)$$

nearly the expected potential. The relative heights of the two waves were found to be very similar to the relative yields of cyclic and linear products obtained in macroelectrolysis. It was concluded that the first polarographic wave was a concerted reductive cyclization, bond formation being responsible for the anodic shift observed. The second wave is then the normal reduction of an activated olefin to an anion radical. In order to substantiate this mechanistic proposal the macroelectrolysis of diethyl 2,6-octadiene-1,8-dioate was conducted at three different potentials along the polarographic wave. The yield of cyclic product decreased with increasing potential as predicted by the mechanistic proposal.

The diactivated olefins, $XCH=CHY$ (50) lend themselves particularly well to a mechanistic study in this area since the stability of the intermediate anion radicals allows them to be observed by electrochemical techniques (18,19). Cyclic voltammetry was used here to estimate the rate constants for the decomposition of these anion radicals and electron spin resonance spectroscopy was used to verify their intermediacy in the reactions. Representative olefins studied were: $C_6H_5COCH=CHCO_2C_2H_5$, $C_6H_5CH=CHCN$, $NCCH=CHCO_2C_2H_5$, $C_6H_5CH=CHCON(CH_3)_2$, $4\text{-PyCH}=CHCO_2C_2H_5$, $C_6H_5COCH=CHCN$. Two paths for the formation of dimeric products from the intermediate anion radicals are immediately suggested. The first path (eq. (28)) involves the radical dimerization of anion radicals, while the second path (eq. (29)) involves the anionic

$$\underset{(51)}{2XCHCHY^{\underline{\cdot}}} \longrightarrow \underset{(52)}{(XCHCHY)_2^=} \xrightarrow{2\text{H}^+} \underset{(53)}{(XCHCHY)_2H_2} \qquad (28)$$

$$e^-\Big\uparrow 2\text{H}^+$$

$$51 + 50 \longrightarrow \underset{(54)}{(XCHCHY)^-(XCHCHY)\cdot} \qquad (29)$$

attack of the anion radical on unreduced olefin. From the dependence of the observed rate constant for the disappearance of the anion radical on the concentration of olefin, the second appears to be the major path for dimer formation.

In this system, when $X \neq Y$, three dimeric products are possible: head-to-head, head-to-tail, and tail-to-tail dimers. The product distribution was rationalized on the basis of the anion-stabilizing ability of the groups in the acceptor part of the dimer and the relative radical stabilizing ability in the donor part (cf. **54**).

The reaction of the intermediate anion radical with water could not be eliminated since a proton source was required to prevent oligomerization of the diactivated olefins. This reaction gives a radical which is reduced to

$$(XCHCHY)^{\cdot-} + H_2O \longrightarrow (XCHCHY)\dot{H} \xrightarrow{e^-} (XCHCHY)H^-$$

an anion which can also react with the starting olefin to form dimer products. Little can be said about predicting the structure of the dimer arising by this route. Since the reaction could be subject to either kinetic or thermodynamic control, the protonation and reduction could yield either anion (α- to X or α- to Y). It was, however, demonstrated that the dimeric product distribution changed with changing concentration of water in the macroelectrolysis. The isomeric dimeric products obtained from the electrolysis of N,N-dimethyl-β-carbethoxyacrylamide were different when the water concentration was increased from 1 to $3M$.

Cross-coupled products between **50** and acrylonitrile were formed by the anionic attack of **51** upon acrylonitrile. It is interesting to note that mixed couplings—in which acrylonitrile must compete with **50** for the electrogenerated **51**—were obtained only when the reduction potentials of **50** and acrylonitrile were within 0.4 V. This observation was interpreted to be the direct result of an increase of reactivity of **50** toward nucleophilic attack due to its high polarization by the electric field. The increase in reactivity of the olefins studied parallels their polarographic reduction potential. While these observations and deductions were made mostly on diactivated olefins, they probably hold for monoactivated olefins as well.

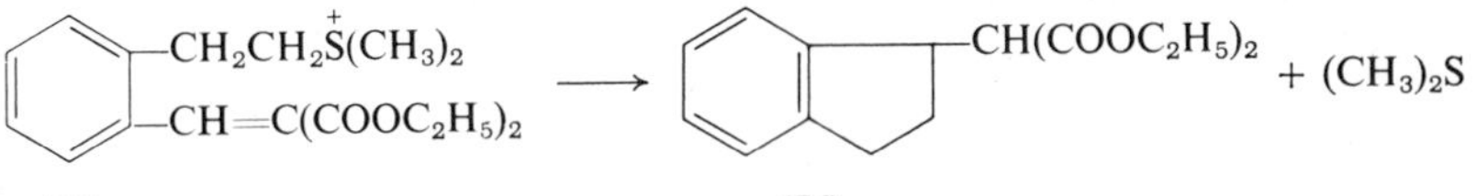

(55)   (56)

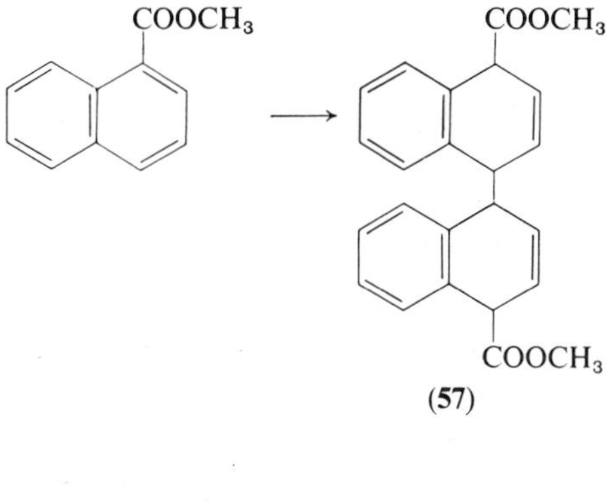

(57)

Two other types of reductive coupling, which have not been extensively investigated, are worth mentioning. The cinnamic ester derivative **55** on electrolysis forms **56** with expulsion of dimethyl sulfide (79). Electrolysis of methyl α-naphthoate in methanol–boric acid is reported to yield 20% of **57** in which coupling has occurred through the 4,4'-positions (105).

H. Organometallic and Organometalloid Compounds

The reductions of arsanilic acid in strong HCl at cathodes of high overpotential to **58** and of 3-nitro-4-hydroxyphenylarsenic acid to **59** (Salvarsan) are old examples of reductive couplings in this area.

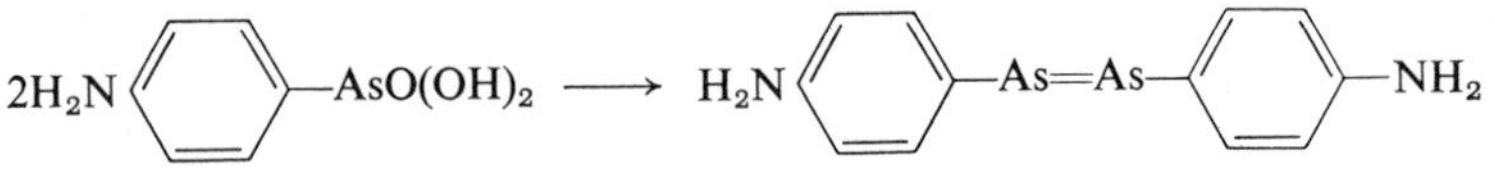

$$2H_2N\text{—C}_6H_4\text{—AsO(OH)}_2 \longrightarrow H_2N\text{—C}_6H_4\text{—As}{=}\text{As—C}_6H_4\text{—NH}_2$$

(58)

$$2 \underset{\substack{\text{AsO(OH)}_2}}{\overset{\substack{\text{OH}\\ \text{—NO}_2}}{\text{C}_6H_3}} \longrightarrow \text{HO—C}_6H_3(\text{NH}_2)\text{—As}{=}\text{As—C}_6H_3(\text{NH}_2)\text{—OH}$$

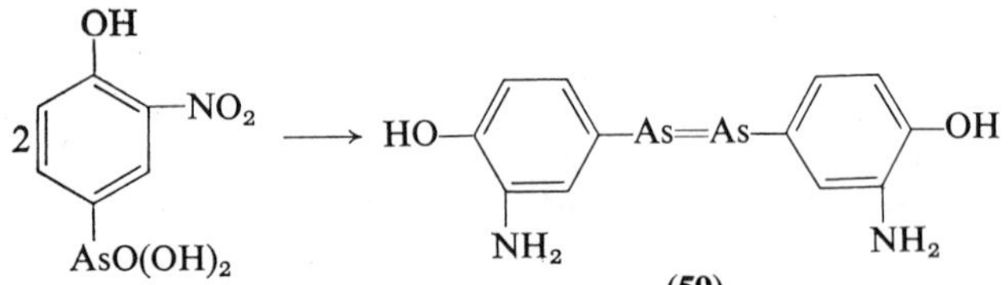

(59)

More recently Dessy and co-workers have reported on extensive study of organometallic electrochemistry. Applying the electrochemical techniques of polarography, controlled potential electrolysis, and triangular voltammetry to derivatives of Group IV-B elements, they have presented evidence (106) for the reductive coupling, e.g., of triphenyltin chloride to hexaphenylditin:

$$2Ph_3SnCl \xrightarrow{2e^-} Ph_3SnSnPh_3 + 2Cl^-$$

The data obtained by the authors do not permit them to distinguish between a route involving radical coupling and one involving nucleophilic attack of a tin anion on starting material:

$$Ph_3SnCl \xrightarrow{e^-} Ph_3Sn\cdot \times 2 \longrightarrow \text{product}$$

or

$$Ph_3SnCl \xrightarrow{2e^-} Ph_3Sn^- + Ph_3SnCl \longrightarrow \text{product}$$

I. 1-Halohydrocarbons

The bibliography on the electrochemical reduction of halogenated hydrocarbons is vast (e.g., 107,108). In keeping with the intended scope of this chapter this discussion will be limited to the proposals which have been made concerning the routes whereby 1-halohydrocarbons are converted, at least in part, to dimeric dehalogenated products at a cathode.

$$RX + 1e^- \longrightarrow (RX)^{\underline{\cdot}} \longrightarrow R\cdot + X^- \tag{28a}$$

$$2R\cdot \longrightarrow R:R \tag{29a}$$

$$RX + 2e^- \longrightarrow R^- + X^- \tag{30}$$

$$R^- + RX \longrightarrow R:R + X^- \tag{31}$$

$$RX + 1e^- \longrightarrow R\cdot + X^- \tag{32}$$

$$R\cdot \xrightarrow{e^-} R^- \tag{33}$$

$$RX + e^- \underset{\longleftarrow}{\overset{\longrightarrow}{\rule{0pt}{0pt}}} (RX)^- \tag{34}$$

$$2(RX)^- \longrightarrow R:R + 2X^- \tag{35}$$

$$RX \longrightarrow R^+ + X^- \tag{36}$$

$$R^+ + e^- \longrightarrow R\cdot \xrightarrow{e^-} R^-, \text{ etc.} \tag{37}$$

Equations (28a) and (29a) are presented as a *conjecture* by Feoktistov and Zhdanov to explain the phenomena observed in the polarographic reduction of β-iodopropionitrile, an alleged unique case of two successive one-electron reductions. In a later paper the same authors (109) report a polarographic study of both β-iodo and β-bromopropionitrile. The iodo compound is reduced in two one-electron stages (-1.2 and -1.5 V vs. SCE), and the intermediate radical has an opportunity to dimerize substantially (eq. (29a)) as well as to be reduced (eq. (33)). Even fast-sweep cyclic voltammetry failed to detect the intermediate radical. The bromo compound is reduced at a more negative potential (ca. -1.8 to -2.0 V) so that the reduction rate of the intermediate radical is overwhelmingly larger than its rate of dimerization and only a single two-electron wave appears. In each case there is an interesting dependence of the reduction potential upon the nature of the cation of the supporting electrolyte presumably due to the alteration of the structure of the double layer.

Marple et al. (110) present evidence from both polarography and from "bulk" electrolyses to indicate that bibenzyl is formed from benzyl bromide probably by steps (32) and (29a); they find no evidence for prior formation of $C_6H_5CH_2HgBr$ (cf. 122).

Klopman (111) obtained an unspecified yield of 4,4'-dinitrobenzyl by electrolysis of *p*-nitrobenzyl bromide in methanol. In a more general study Grimshaw et al. (112) reported a ρσ treatment of the half-wave potentials of a series of substituted benzyl bromides and also the results of macroscale electrolyses of some of them. Those with small negative or small positive σ values gave dibenzylmercuries in methanolic lithium bromide; those with large positive σ values (e.g., p-nitrobenzyl bromide) gave the reductively coupled products; those with intermediate σ values (3-bromo- and 3,4-dichlorobenzyl bromides) gave mixtures of the organometallic and the bibenzyl derivatives. In some cases the yields are good enough for preparative purposes. The ratio of the two types of product varied with change in supporting electrolyte and, with a given electrolyte, with the cathode voltage. The authors conclude that all products are formed via benzyl radicals and that it is not possible to reduce benzyl radicals to benzyl carbanions at mercury in aqueous or alcoholic solutions. However, the formation of a small yield of phenylacetic acid in the electrolysis of benzyl chloride in DMF containing tetrabutylammonium iodide while CO_2 was passed through the solution was interpreted as indicating the intermediacy of benzyl carbanions (113) under those conditions. Further information concerning the reducibility of benzyl radical at mercury (in a very different system) is given by Weedon and associates (114). They reduced benzyl methyl ethers substituted by electron-withdrawing groups. The solvent was methanol and the supporting electrolyte sodium acetate. The alkoxy group was replaced by hydrogen. The course of the reaction must be (a) a one-electron reduction followed by (or concerted with) displacement of methoxide followed by reduction to a benzyl carbanion or (b) a two-electron reduction-displacement.

The reduction of alkyl halides in DMF at a graphite cathode yielded dimeric products (115). Butane was obtained from ethyl iodide with 40% current efficiency and octane from butyl bromide with 2% current efficiency. Radical intermediates are postulated as well as the formation of a preliminary complex between DMF and the alkyl halide:

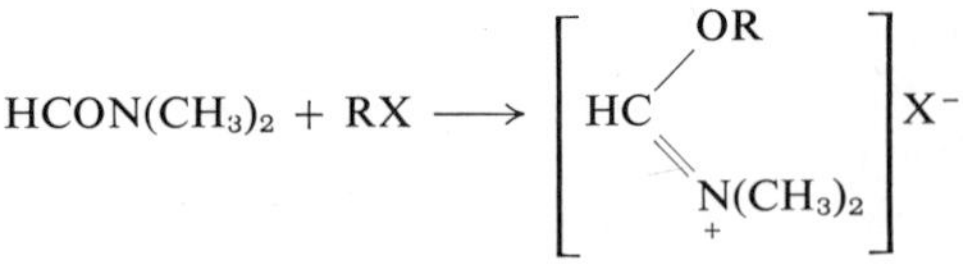

$$HCON(CH_3)_2 + RX \longrightarrow \left[HC \begin{array}{c} OR \\ \diagdown \\ N(CH_3)_2 \\ + \end{array} \right] X^-$$

(60)

Reduction of 60 does not now, of course, involve cleavage of the R—X bond.

Electroreduction of α,α'-dibromo-p-xylene 61 (116) yielded 5–10% of 2,2-paracyclophane 62 and largely poly-p-xylylene 63. Covitz (116) has proposed a two-electron reductive cleavage of one of the bromines and in a concerted reaction intramolecular nucleophilic displacement of the second bromine to yield the key intermediate, p-xylylene.

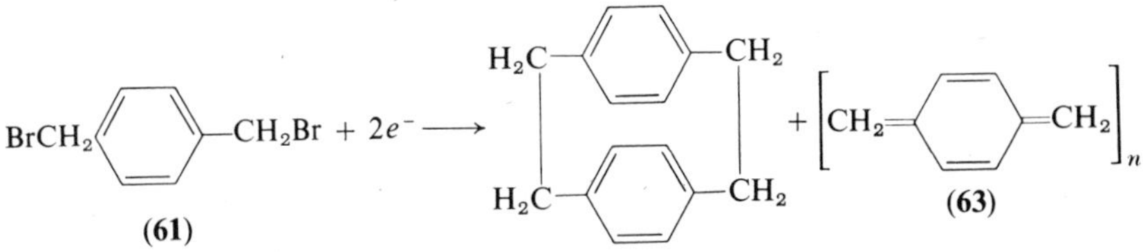

$$BrCH_2-\langle\text{ring}\rangle-CH_2Br + 2e^- \longrightarrow$$

(61)

$$\begin{array}{c} H_2C-\langle\text{ring}\rangle-CH_2 \\ | \\ H_2C-\langle\text{ring}\rangle-CH_2 \end{array} + \left[CH_2=\langle\text{ring}\rangle=CH_2 \right]_n$$

(62)   (63)

In related intramolecular reactions Rifi (117) has prepared cyclo-propane, cyclobutane, bicyclobutane, and spiropentane:

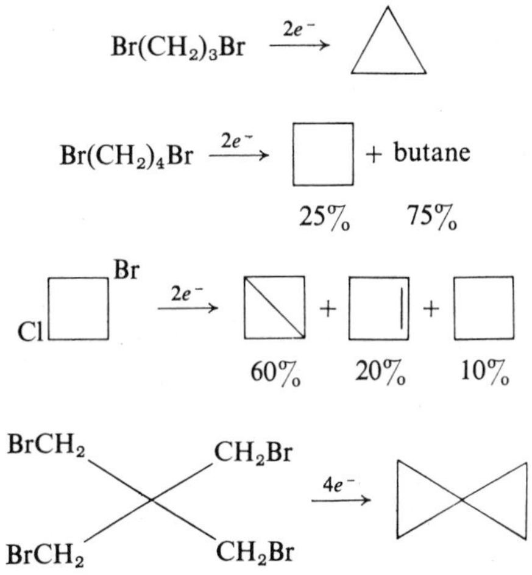

$$Br(CH_2)_3Br \xrightarrow{2e^-} \triangle$$

$$Br(CH_2)_4Br \xrightarrow{2e^-} \square + \text{butane}$$

25%   75%

$$\underset{Cl}{\overset{Br}{\square}} \xrightarrow{2e^-} \boxtimes + \square + \square$$

60%   20%   10%

$$\begin{array}{c} BrCH_2 \quad CH_2Br \\ \times \\ BrCH_2 \quad CH_2Br \end{array} \xrightarrow{4e^-} \bowtie$$

It is clear that the addition of one electron to the carbon halogen bond is sufficient to cause cleavage to a halide ion and a free radical (eq. (32)). The site of electron attack, particularly for hindered or bridgehead halogen compounds, is still a moot question (118–120). It has been postulated that the dimeric products occur from the reaction of two free radicals. There seems to be no evidence to support radical dimerization (eq. (29a)) rather than nucleophilic displacement (eq. (31)) as the route to dimeric products. Perhaps future work will define conditions under which each route may be used. Rifi has argued convincingly that small ring compounds are formed from 1,3-dibromopropane and 1,4-dibromobutane by eq. (31).

J. Pyridines and Quinolines

Pyridinium and quinolinium salts couple through an available 4-position to form bisdihydropyridines and dihydroquinolines respectively in moderate yield (3). When the 4-position is blocked, as in 4-methyl-quinolinium sulfate, coupling occurs in poor yield through the 2-position. Reduction of quinaldine in 10% alcoholic potassium hydroxide in a divided cell using a mercury cathode yielded an unusual reduced dimer to which structure **I** has been assigned (121).

(I)

References

1. N. L. Weinberg and H. R. Weinberg, *Chem. Rev.*, *68*, 449 (1968).
2. I. L. Knunyants and N. P Gambaryan, *Uspek. Khim.*, *23*, 781 (1954).
3. S. Swann, Jr., "Electrolytic Reactions," in *Technique of Organic Chemistry*, 2nd ed., Vol. 2, A. Weissberger, Ed., Interscience, New York, 1956, pp. 385–523; F. Fichter, *Organische Elektrochemie*, Theodor Steinkopff Verlag, Dresden and Leipzig, 1942.
4. M. J. Allen, *Organic Electrode Processes*, Reinhold Publ., New York, 1958.
5. F. D. Popp and H. P. Schultz, *Chem. Rev.*, *62*, 19 (1962).
6. M. Ya. Fioshin and A. P. Tomilov, *Khim. Prom.*, *1964*, (9), 9–17.
7. M. M. Baizer, J. D. Anderson, J. H. Wagenknecht, M. R. Ort, and J. P. Petrovich, *Electrochim. Acta*, *12*, 1377 (1967).
8. S. G. Mairanovskii, *Talanta*, *12*, 1299 (1965).
9. S. Wawzonek, *Talanta*, *12*, 1229 (1965).
10. P. J. Elving, *Talanta*, *12*, 1243 (1965).
11. P. Zuman, *Collection Czech. Chem. Commun.*, *25*, 3225 (1960).
12. J. Heyrovsky and J. Kuta, *Principles of Polarography*, Academic Press, New York, 1966; I. M. Kolthoff and J. J. Lingane, *Polarography*, 2nd ed., Interscience, New York, 1952; L. Meites, *Polarographic Techniques*, 2nd ed., Interscience, New York, 1965; P. Zuman, *Substituent Effects in Organic Polarography*, Plenum Press, New York, 1967; P. Zuman, *J. Polarog. Soc.*, *13*, 53 (1967).
13. F. C. Anson, Symposium, Division of Fuel Chemistry, American Chemical Society, Miami, Florida, April 9–14, 1967, Vol. 11, No. 1, p. 56; C. N. Reilley, *ibid.*, p. 43.
14. R. S. Nicholson and I. Shain, *Anal. Chem.*, *36*, 706 (1962).
15. R. S. Nicholson, *Anal. Chem.*, *38*, 1406 (1966).
16. R. S. Nicholson and I. Shain, *Anal. Chem.*, *37*, 178 (1965).
17. D. H. Geske and A. H. Maki, *J. Am. Chem. Soc.*, *82*, 2671 (1960).
18. J. M. Fritsch and J. P. Petrovich, unpublished results.
19. J. P. Petrovich, M. M. Baizer, and M. R. Ort, *J. Electrochem. Soc.*, *116* (1969).

20. J. M. Nelson and A. M. Collins, *J. Am. Chem. Soc.*, *46*, 2256 (1924).
21. S. Wawzonek and A. Gundersen, *J. Electrochem. Soc.*, *107*, 537 (1960).
22. J. H. Stocker and R. M. Jenevein, *J. Org. Chem.*, *33*, 294 (1968).
23. S. Swann, Jr., *Trans. Electrochem. Soc.*, *64*, 245 (1933).
24. M. J. Allen, *J. Org. Chem.*, *15*, 435 (1950).
25. M. J. Allen and A. H. Corwin, *J. Am. Chem. Soc.*, *72*, 114 (1952); N. J. Leonard, S. Swann, Jr., and G. Fuller, *J. Am. Chem. Soc.*, *75*, 5127 (1953).
26. M. J. Allen, *J. Am. Chem. Soc.*, *72*, 3797 (1950).
27. F. Escherlich and M. Moest, *Z. Elektrochem.*, *8*, 849 (1902).
28. R. M. Elofson, *J. Org. Chem.*, *25*, 305 (1960).
29. W. C. Albert and A. Lowy, *Trans. Electrochem. Soc.*, *75*, 367 (1939).
30. M. J. Allen, J. F. Fearn, and H. A. Levine, *J. Chem. Soc.*, *1952*, 2220.
31. H. A. Levine and M. J. Allen, *J. Chem. Soc.*, *1952*, 254.
32. K. M. Johnston and J. D. Stride, *Chem. Commun.*, *1966*, 325.
33. J. Jadot, *Bull. Soc. Chim. Belges*, *57*, 346 (1948); *Chem. Abstr.*, *43*, 5744f (1949).
34. P. F. Casals and J. Wiemann, *Bull. Soc. Chim. France*, *1967*, 3478.
35. D. H. Evans and E. C. Woodbury, *J. Org. Chem.*, *32*, 2158 (1967).
36. J. P. Stradins, *Electrochim. Acta*, *9*, 711 (1964).
37. J. Grimshaw and E. J. F. Rea, *J. Chem. Soc.*, *C1967*, 2628.
38. J. G. Faugere, R. Calas, and R. Lalande, *Compt. Rend.*, *262*, 124 (1966).
39. R. N. Gourley and J. Grimshaw, *J. Chem. Soc.*, Ser. C, *1968*, 2388.
40. B. J. Herold, *Tetrahedron Letters*, *1962*, 75.
41. P. Zuman, D. Barnes, and A. Ryvolová-Kejhorová, Faraday Society Meeting, Newcastle, April 1968; P. Zuman, *Collection Czech. Chem. Commun.*, *33*, 2548 (1968).
42. J. H. Stocker and R. M. Jenevein, *J. Org. Chem.*, *33*, 294 (1968).
43. K. Sugino and T. Nonaka, *Electrochim. Acta*, *13*, 613 (1968).
44. M. Nicolas and R. Pallaud, *Compt. Rend.*, *265C*, 1044 (1967).
45. T. Sekine, A. Yamura, and K. Sugino, *J. Electrochem. Soc.*, *112*, 439 (1965).
46. A. P. Tomilov and B. L. Klyuev, *Soviet Electrochem.*, *3*, 1042 (1967).
47. O. R. Brown and K. Lister, Faraday Society Meeting, Newcastle, April 1968.
48. V. G. Khomyakov and A. P. Tomilov, *Zh. Prikl. Khim.*, *36*, 378 (1963).
49. J. Wiemann, *Bull. Soc. Chim. France*, *1964*, 2545.
50. R. Pasternak, *Helv. Chim. Acta*, *31*, 48 (1948).
51. L. J. Sargent and U. Weiss, *J. Org. Chem.*, *25*, 987 (1960).
52. J. Wiemann and M. L. Bouguerra, *Ann. Chim.*, *3*, 215 (1967).
53. L. Holleck and D. Marquarding, *Naturwiss.*, *49*, 468 (1962).
54. J. Wiemann and M. L. Bouguerra, *Ann. Chim.*, *2*, 35 (1967).
55. A. Misono, T. Osa, and T. Yamagishi, *Kogyo Kagaku Zasshi*, *69*, 945 (1966); *Chem. Abstr.*, *66*, 75482e (1967).
56. M. M. Baizer and J. D. Anderson, *J. Org. Chem.*, *30*, 3138 (1965).
57. J. Wiemann and M. L. Bouguerra, *Compt. Rend.*, *C265*, 751 (1967).
58. C. L. Wilson and K. B. Wilson, *Trans. Electrochem. Soc.*, *84*, 153 (1943); E. P. Goodings and C. L. Wilson, *ibid.*, *88*, 77 (1945).
59. C. L. Wilson, *Trans. Electrochem. Soc.*, *75*, 353 (1939).
60. S. D. Ross, M. Finkelstein, and R. C. Petersen, *J. Am. Chem. Soc.*, *82*, 1582 (1960).
61. L. Horner, F. Röttger, and H. Fuchs, *Chem. Ber.*, *96*, 3141 (1963); L. Horner and A. Mentrup, *Ann. Chem.*, *646*, 65 (1961).

62. M. Finkelstein, *J. Org. Chem.*, *27*, 4076 (1962).
63. T. Wohlfahrt, *J. Prakt. Chem.*, (2) *65*, 295 (1902).
64. K. S. Udupa, G. S. Subramanan, and H. V. K. Udupa, *J. Electrochem. Soc.*, *108*, 373 (1961).
65. K. S. Udupa, G. S. Subramanan, and H. V. K. Udupa, *Bull. Chem. Soc. Japan* *35*, 1168 (1962).
66. G. A. Russell, *Science*, *161*, 423 (1968); E. T. Kaiser and L. Kevan, Eds., *Radical Ions*, Interscience, New York, 1968.
67. H. Lund, Preprints of a Symposium on The Synthetic and Mechanistic Aspects of Electro-organic Chemistry, U.S. Army Research Office, Durham, North Carolina, October 14–16, 1968, p. 197; H. Lund, *Acta Chem. Scand.*, *21*, 2525 (1967).
68. S. Wawzonek, E. W. Blaha, R. Berkey, and M. E. Runner, *J. Electrochem. Soc.*, *102*, 235 (1955); S. Wawzonek and J. W. Fan, *J. Am. Chem. Soc.*, *68*, 2541 (1946).
69. R. Dietz and M. Peover, *Faraday Soc. Discussions*, *45*, 154 (1968).
70. J. D. Anderson and M. M. Baizer, unpublished work.
71. M. M. Baizer, *Tetrahedron Letters*, *1963*, 973.
72. K. Higasi, H. Baba, and A. Rembaum, *Quantum Organic Chemistry*, Interscience, New York, 1965, pp. 304–332; A. Cserhezyl, J. Chaudhurl, E. Fanta, J. Jagur-Grodzinski, and M. Szwarc, *J. Am. Chem. Soc.*, *89*, 7129 (1967); S. C. Chadha, J. Jagur-Grodzinski, and M. Szwarc, *Trans. Faraday Soc.*, *1967*, 2994; Y. Iwakura, F. Toda, H. Katsuki, and H. Watanabe, *Polymer Letters*, *6*, 201 (1968).
73. M. Szwarc, *Makromol. Chem.*, *35*, 132 (1960).
74. B. L. Funt, "Electrolytically Controlled Polymerizations," in *Macromolecular Reviews*, Vol. 1, A. Peterlin, M. Goodman, S. Okamura, B. H. Zimm, and H. F. Mark, Eds., Interscience, New York, 1967, pp. 35–56.
75. D. Laurin and G. Parravano, *Polymer Letters*, *4*, 797 (1966).
76. J. D. Anderson, *J. Polymer Sci.*, *Ser. A-1*, *6* (11), 3185 (1968); N. Yamazaki, S. Nakahama, and S. Kambara, *J. Polymer Sci.*, *Ser. B*, *3*, 57 (1965).
77. M. M. Baizer and J. D. Anderson, *J. Electrochem. Soc.*, *111*, 223 (1964).
78. M. M. Baizer and J. D. Anderson, *J. Org. Chem.*, *30*, 3138 (1965).
79. J. D. Anderson, M. M. Baizer, and J. P. Petrovich, *J. Org. Chem.*, *31*, 3890 (1965); J. P. Petrovich, J. D. Anderson, and M. M. Baizer, *ibid.*, *31*, 3897 (1965).
80. M. M. Baizer, *J. Electrochem. Soc.*, *111*, 215 (1964).
81. M. L. Bouguerra and J. Wiemann, *Compt. Rend.*, *263C*, 782 (1966).
82. J. D. Anderson, M. M. Baizer, and E. J. Prill, *J. Org. Chem.*, *30*, 1645 (1965).
83. M. M. Baizer and J. D. Anderson, *J. Org. Chem.*, *30*, 1348 (1965).
84. M. M. Baizer and J. D. Anderson, *J. Org. Chem.*, *30*, 1351 (1965).
85. I. G. Sevast'yanova and A. P. Tomilov, *Soviet Electrochem.*, *3*, 494 (1967).
86. Yu. D. Smirnov, S. K. Smirnov, and A. P. Tomilov, *J. Org. Chem. U.S.S.R.*, *4*, 208 (1968).
87. F. Beck, *Chem. Ing. Technik*, *37*, 607 (1965).
88. F. Beck, *Ber. Bunsengesellschaft*, *72*, 379 (1968).
89. F. Matsuda, *Tetrahedron Letters*, *1966*, 6193.
90. Y. Arad, M. Levy, I. R. Miller, and D. Vofsi, *J. Electrochem. Soc.*, *114*, 899 (1967).
91. I. Gillet, *Bull. Soc. Chim. France*, *1968*, 2919.

92. A. P. Tomilov, E. V. Kryukova, V. A. Klimov, and I. N. Brogo, *Soviet Electrochem.*, *3*, 1352 (1967).
93. L. G. Feoktistov, A. P. Tomilov, and I. G. Sevast'yanova, *Soviet Electrochem.*, *1*, 1165 (1965).
94. M. M. Baizer and J. P. Petrovich, *J. Electrochem. Soc.*, *114*, 1023 (1967).
95. W. Kern and H. Quast, *Makromolekulare Chemie*, *10*, 202 (1953).
96. A. Zilkha and B. A. Feit, *J. Appl. Polymer Sci.*, *5*(15), 251 (1961); M. M. Baizer and J. D. Anderson, *J. Org. Chem.*, *30*, 1351 (1965).
97. J. W. Breitenbach and Ch. Srna, *Pure Appl. Chem.*, *4*, 245 (1962).
98. B. L. Funt and F. D. Williams, *J. Polymer Sci.*, *A2*, 865 (1964).
99. G. C. Jones et al., *Tetrahedron Letters*, *1967*, 615.
100. S. I. Zhdanov and I. G. Feoktistov, *Izvest. Akad. Nauk S.S.S.R., Otdel Khim. Nauk*, *1963*, 53.
101. S. Lazarov, A. Trifonov, and T. Vitanov, *Z. Physik. Chem. (Leipzig)*, *3–4*, 221 (1964).
102. M. Figeys and H. P. Figeys, *Tetrahedron*, *24*, 1097 (1968).
103. M. M. Baizer, *J. Org. Chem.*, *29*, 1670 (1964).
104. M. M. Baizer and J. D. Anderson, *J. Org. Chem.*, *30*, 1357 (1965).
105. G. T. Mondodoev et al., *Zh. Organ. Khim.*, *1*, 2008 (1965); *Chem. Abstr.*, *64*, 9651b (1966).
106. R. E. Dessy, W. Kitching, and T. Chivers, *J. Am. Chem. Soc.*, *88*, 453 (1966).
107. P. J. Elving and B. Pullman, *Advances in Chemical Physics*, Vol. 3, I. Prigogine, Ed., Interscience, New York, 1961, p. 1.
108. A. Streitwieser, Jr., and C. Perrin, *J. Am. Chem. Soc.*, *86*, 4938 (1964).
109. L. G. Feoktistov and S. I. Zhdanov, *Bull. Acad. Sci., U.S.S.R.*, *1962*, 2036; *Electrochim. Acta*, *10*, 657 (1965).
110. L. W. Marple, L. E. I. Hummelstedt, and L. B. Rogers, *J. Electrochem. Soc.*, *107*, 437 (1960).
111. G. Klopman, *Helv. Chim. Acta*, *44*, 1908 (1961).
112. J. Grimshaw and J. S. Ramsey, *J. Chem. Soc., Ser. B*, *1968*, 60.
113. S. Wawzonek, R. C. Duty, and J. H. Wagenknecht, *J. Electrochem. Soc.*, *111*, 74 (1964).
114. R. F. Garwood, N-U-Din, and B. C. L. Weedon, *Chem. Commun.*, *1968*, 923.
115. L. V. Kaabak, A. P. Tomilov, S. L. Varshavskii, and M. I. Kabachnik, *J. Org. Chem. U.S.S.R.*, *3*, 1 (1967).
116. F. H. Covitz, *J. Am. Chem. Soc.*, *89*, 5403 (1967).
117. M. R. Rifi, *J. Am. Chem. Soc.*, *89*, 4442 (1967).
118. J. W. Sease, P. Chang, and J. L. Groth, *J. Am. Chem. Soc.*, *86*, 3154 (1964).
119. F. L. Lambert, A. H. Albert, and J. P. Hardy, *J. Am. Chem. Soc.*, *86*, 3155 (1964).
120. P. Zuman, ref. 12, p. 322.
121. I. W. Elliott, E. S. McCaskill, M. S. Robertson, and C. H. Kirksey, *Tetrahedron Letters*, *1962*, 291.
122. A. Kirrmann and M. Kleine-Peter, *Bull. Soc. Chim. France*, *1957*, 894.

The Application of Radiation Chemistry to Mechanistic Studies in Organic Chemistry*

By Eleanor J. Fendler

Department of Chemistry, University of Pittsburgh, Pittsburgh, Pennsylvania

AND

Janos H. Fendler

Radiation Research Laboratories, Mellon Institute, Carnegie-Mellon University, Pittsburgh, Pennsylvania

CONTENTS

* The preparation of this review was supported, in part, by the U.S. Atomic Energy Commission.

I. INTRODUCTION

In recent years radiation chemists have obtained a considerable understanding of the nature and reactivities of the primary chemical species
produced by the irradiation of water, alcohols, hydrocarbons, and other
organic compounds. Sensitive new analytical techniques have allowed

quantitative determinations of the products formed in the reactions of these species with added solutes; and absolute rate constants have been determined for these reactions by pulse radiolytic techniques. The realization that the hydrated electron is, in fact, the simplest nucleophile and that the hydroxyl radical, the primary oxidizing species generated in water radiolysis, has electrophilic properties has unveiled new possibilities for mechanistic studies of the reactions of these species with organic solutes. Increasing numbers of such studies are being conducted in several laboratories *by radiation chemists*. The aim of the present review is to familiarize *organic chemists* with the vast potentialities of radiation chemistry as a tool for the study of organic reaction mechanisms, the availability of moderately priced equipment, and the advantages of such an interdisciplinary approach.

Since the usual media for publishing original work relevant to the present review are journals oriented toward physical chemistry and chemical physics such as *The Journal of Physical Chemistry*, *The Journal of Chemical Physics*, the physical section of *The Journal of the American Chemical Society*, *Transactions of the Faraday Society*, and *Radiation Research*, this work is only randomly, if ever, scanned by the busy physical-organic chemist. In addition, the rapid progress of radiation chemistry is evidenced by a spectacular increase in the quantity of published papers, new techniques, and novel applications. In 1950 the number of published papers in radiation chemistry did not exceed 50, whereas the corresponding number for 1968 is certainly in the thousands. To aid in guiding the novice through this maze of publications, recent advances in radiation chemistry are summarized regularly. Such reviews appear in the *Annual Review of Physical Chemistry* (1), the *Annual Review of Nuclear Science* (2), *Actions Chimique et Biologiques des Radiations* (3), *Current Topics in Radiation Chemistry* (4), *Advances in Radiation Biology* (5), *Advances in Radiation Chemistry* (6), *Progress in Reaction Kinetics* (7), and *Radiation Research Reviews* (8). The proceedings of several international conferences pertinent to radiation chemistry have been published recently (9–17).

Since the fundamental processes in radiation chemistry are treated with varying degrees of sophistication in several recent books and articles (18–26), only a brief outline of the experimental techniques will be presented here in Section II. The scope of this review is limited to processes in the liquid phase and to studies from which mechanistic conclusions can be drawn. Hence, the preparative aspects of radiation chemistry are not included. Examples illustrating the different approaches are emphasized and the underlying principles are discussed in each case. However, no attempt has been made to cover the literature *exhaustively*.

II. BRIEF OUTLINE OF RADIATION CHEMICAL TECHNIQUES

For the sake of simplicity, steady state and pulse radiolysis will be discussed separately. The terms refer, somewhat arbitrarily, to the duration of radiation with respect to the lifetime of the transient produced and to the sources and intensities of radiation. Steady-state irradiation, the time-honored technique, generally implies the irradiation of a sample by X- or γ-rays of moderate intensity or by a constant electron beam for times not less than 30 sec followed by an analysis of the products. Pulse radiolysis, on the other hand, utilizes high-intensity radiation of very short duration (10^{-5} sec and less), and its primary aim is, in many cases, the determination of the rate of formation or decomposition of a transient whose half-life is in the order of micro to milliseconds. In addition, intermittent radiation (27–31), produced by a rotating sector or similar device which interrupts a steady beam of γ-rays or electrons at a constant rate such that the usual exposure time is 10^{-3} sec or longer, is used to study the kinetics of longer lived species. In most cases, however, the combination of pulse or intermittent radiolysis with steady-state radiolysis is necessary in order to obtain mechanistic insight into radiation-induced reactions—the former being employed primarily for rate-constant determination and the latter at low dose rates for quantitative product analysis.

A. Steady-State Irradiation

The basic principles and techniques of steady-state irradiation are presented in the following discussion, parts 1–6. Additional information and details can be found in standard texts (18–26).

1. Sample Preparation

Since radiation-induced reactions involve free radicals, it is vitally important to use substrates and solvents of the highest purity. In many instances, special techniques are employed to achieve the purity required for reliable and reproducible results in radiation studies.

Water is purified by triple distillation in an all-glass system. One technique involves the distillation of ordinary distilled water from acidic dichromate, then from alkaline permanganate, and finally without added reagents through a column into a quartz receiver. In each distillation the vapor passes through a long, well-insulated column packed with glass helices and then passes into the next pot or the receiver through a glass tube heated to ca. 500°. A gentle stream of oxygen is passed through the entire system to oxidize any organic impurities (22). Alcohols are usually refluxed for several hours with 2,4-dinitrophenylhydrazine and a catalytic amount of sulfuric acid and distilled through a fractionating column under a stream of dry nitrogen. Dissolved oxygen and carbon dioxide which are good electron scavengers are removed by repeated freeze-pump-thaw cycles on a high vacuum line (32,33).

Organic and inorganic solutes also must be purified with unusual care. Depending on their nature, standard organic and inorganic techniques, such as fractional distillation, repeated recrystallization, and preparative gas or column chromatography, are available for purification (34).

In addition, special techniques are required for the cleaning and drying of glassware. The standard procedure includes washing the glassware first with chromic acid cleaning solution and then with a large quantity of triply distilled water and baking it in an annealing oven at ca. 500°. Often in the case of aqueous radiation studies, the irradiation vessel is filled with triply distilled water, preirradiated to remove surface contamination, emptied, and filled with the solution to be studied (22).

2. Radiation Sources

The most frequently used source for steady-state irradiation is the cobalt-60 γ-ray source. Its half-life of 5.27 years, the availability of suitably designed sources which give fields of uniform intensity, and moderate cost have resulted in its popularity. In order to provide adequate shielding, the design of the cobalt-60 source is either of the cave or the well type. The former is a large, shielded cave or room into which the cobalt source is raised from a well. Monitoring and safety arrangements are mandatory with this type of radiation facility to avoid inadvertent exposure to radiation. The advantage of this source arrangement is that large samples can be irradiated and that different radiation intensities, dose rates, can be obtained by placing the samples at different distances from the source. In the well-type source the cobalt-60 is shielded by a large block of lead into which the sample can be mechanically lowered. Occasionally, other artificial radionucleides such as cesium-137 are used as γ-ray radiation sources. The different types of commercially available γ-ray sources are listed by Spinks and Woods (20).

X-ray machines and Van de Graaff or other types of accelerators can also be used as steady-state radiation sources (19–22).

3. Units and Dosimetry

For all quantitative radiation studies, a knowledge of the amount of energy *absorbed by the irradiated sample* is required. As a consequence of the recommendations of the International Commission of Radiological Units and Measurements (35), the following terms and units are used most frequently. The *absorbed dose* equals the amount of energy imparted to the matter by the ionizing particles per unit mass of irradiated material. It is usually given in units of rads or eV/g. One *rad* is equivalent to 100 ergs/g or 10^{-5} joules/g and equals 2.4×10^{-6} cal/g and 6.24×10^{13} ρ eV/cm^3 (ρ = density in g/cm^3). The *absorbed dose rate* equals the absorbed dose per unit time, e.g., rad/min or rad/hr. The yield of products relative to the amount of radiation is expressed by the *G-value*. $G(X)$ and $G(-X)$ refer to the number of molecules of product X formed or decomposed, respectively, on irradiation per 100 eV of absorbed energy. For solution studies, substitution of the appropriate units into the above definition gives the following useful expression*:

$$G(X) = \frac{(X \text{ in moles liter}^{-1})(9.65 \times 10^8)}{(\text{absorbed dose in rads}) \, \rho}$$

Various physical and chemical methods are available for determining the dose, and hence the dose rate, of a given radiation source. In fact, several reference books have been published on the subject of dosimetry (36,37). The most frequently used chemical dosimeter is the Fricke solution, which is a $0.8N$ H_2SO_4 solution in triply distilled water containing $0.0014M$ $FeSO_4$ and $0.001M$ NaCl. The radiation chemistry of this solution which involves the oxidation of Fe^{2+} to Fe^{3+} is well understood (36,37), and the relationship $G(Fe^{3+}) = 15.5$ for γ-rays allows a simple determination of the absorbed dose in aqueous solutions from any source by measuring the amount of Fe^{3+} formed spectroscopically at 302–305 nm ($\varepsilon = 2193 \pm 6$ M^{-1} cm^{-1} at 25.0°) (38). Since the interaction of cobalt-60 γ-rays with organic compounds is almost exclusively by the Compton process, the absorbed dose rate in an organic liquid is related to that in water by:

$$D_{\text{org}} = D_{H_2O} \times \frac{(Z/M)_{\text{org}} \, \rho_{\text{org}}}{(Z/M)_{H_2O}}$$

where D_{org} and D_{H_2O} are the absorbed dose rates in eV/cc in the organic liquid and in water, respectively, Z is the total number of electrons in the molecule, and M is the molecular weight. With proper calibration of the source, absorbed doses can be readily determined to an accuracy of a few per cent or better.

4. Determination of Free Radical Intermediates and Stable Products

The inherent problems in product analysis in radiation studies are sensitivity and specificity. The concentration of the products to be analyzed

---

* For a derivation consult p. 108 of ref. 20.

is in many cases less than $10^{-5}M$. In addition, it is occasionally necessary to separate these low-concentration products from the reactants.

In order to elucidate the mechanism of a radiation-induced reaction, it is necessary to understand the numerous reaction paths involving free-radical intermediates through which the products are formed. Depending on the time scale of the radical reactions and on that required for their detection, the yield and nature of the free-radical intermediates can be determined by physical and chemical methods. Pulse radiolysis (39) (see Section II-B) and *in situ* ESR spectroscopy of radiolytically generated transients (40) are the standard physical techniques. The chemical methods involve the judicious selection of suitable radical scavengers which form relatively stable products with the radicals more rapidly than they are able to undergo radical–radical or other reactions (41,42). Extensive use has been made of iodine as a radical scavenger since the reaction

$$R\cdot + I_2 \longrightarrow RI + I\cdot$$

is rapid and the alkyl iodide can easily be determined by gas–liquid partition chromatography or by radio–gas–liquid partition chromatography using iodine-131 (41,42). Since the reaction of hydrogen iodide with radicals is also very fast, tritium iodide has been used successfully for alkyl radical yield determinations in irradiated hydrocarbons (43,44). However, care must be exercised to limit the concentration of the radical scavengers I_2 and HI to less than $10^{-3}M$ to avoid complications due to electron capture by the scavenger (41).

The free-ion yields (i.e., the yield of those charged species which have diffused into the bulk of the liquid during radiolysis rather than those which have reacted in the spurs (45–51)) in irradiated liquid hydrocarbons have been successfully measured using low concentrations (ca. 10^{-4} to $10^{-6}M$) of C-14 methyl bromide as the scavenger and analyzing the product, C-14 methane, by radio–gas–liquid partition chromatography (52).

Similarly sulfur hexafluoride (53), tetranitromethane (54), and nitrous oxide (55–57) have been used as electron scavengers to obtain the yield of hydrated and solvated electrons in water and other protic solvents by measuring the concentration of the fluoride ion, the nitroform, or the nitrogen produced.

When the mechanisms of radiation-induced reactions are understood either by the determination of the free-radical yields and reactivities or by the combination of such data, it is desirable to obtain the yields of the stable products. Both physical and chemical analytical techniques, such as ultraviolet-visible, fluorescence, and infrared spectrophotometry, thin-layer and gas–liquid partition chromatography, mass spectroscopy, and ion activity measurements with specific ion activity electrodes, have been useful to radiation chemists for this purpose. Regardless of the analytical technique employed, the importance of its specificity and a knowledge of the chemistry involved cannot be overemphasized. The early literature of radiation chemistry is plagued in many cases by questionable data. For example, the yields of phenol and biphenyl in irradiated dilute aqueous benzene solutions were determined by measuring the absorbance of the irradiated solution at a particular wavelength; however, the fact that dimeric phenols, possible products in the radiolysis, might also absorb in the same region was completely ignored (58,59). Similarly, numerous other erroneous conclusions based on inaccurate analyses can be found in the literature; the discovery of these is left as a challenge to the reader and researcher.

5. Importance of Small Conversions and Knowledge of the Chemistry of the System

Physical organic chemists will appreciate the advantage of sacrificing high yields,* high percentages of conversion, in order to simplify and elucidate the mechanisms of radiation-induced reactions. Indeed it is true that the greater the percentage of radiolytic conversion of reactants to products the more difficult it is to disentangle the reaction mechanism and the less purposeful is the experiment. Meaningful mechanistic studies of radiation-induced reactions involve less than 1% conversion of the radical scavengers to products. The primary reason for this is to avoid secondary or back reactions of the radiolytically produced intermediates and products. In practice reliable G-values are obtained from the initial slopes of yield-dose plots at low conversion and at as small doses as possible (preferably less than 10^4 rads). It was demonstrated recently for the reaction of the hydrated electron, e_{aq}^-, with scavengers that taking incorrect slopes of yield–dose plots can result in a 15% error in $G_{e_{aq}^-}$ (60,61). The advantages of working with relatively simple systems whose chemistry is fully understood is self-evident. Pitfalls in the determination of primary radical yields in irradiated aqueous solutions at different pH values and at different scavenger concentrations have been discussed recently (60,61).

---

* The term yield to the organic chemist implies the percentage of product as a function of the quantity of the reactant while the radiation chemist uses it to refer to the G-value, i.e., in radiation chemistry yield implies the quantity of product as a function of the amount of energy absorbed by the system.

6. Literature Compilations of G-Values

The known radiolytic yields of organic and inorganic compounds under different conditions have been tabulated by Haïssinsky and Magat for the period from 1905 to 1961 (62). It is advisable, however, to consult the original work for proper evaluation of the conditions and analytical techniques employed. In addition to the usual literature sources, *Nuclear Science Abstracts* can profitably be scrutinized for more recent data.

B. Pulse Radiolysis

Due to the necessity for conciseness, only a short outline of this extremely powerful and rapidly developing technique can be given here. A book (63) and several recent reviews (39,64–66) should be consulted for additional details.

1. Outline of the Underlying Principles

To a certain extent pulse radiolysis may be considered to be the radiation chemical analog of flash photolysis (67). The basic arrangement of the most frequently used system is illustrated schematically in Figure 1. The sample is introduced into an absorption cell and with a suitably

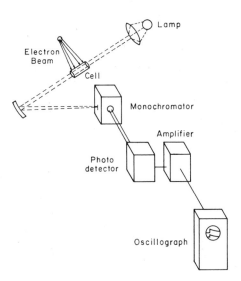

Fig. 1. Basic arrangement of the apparatus for pulse radiolysis using the kinetic spectrophotometric method.

arranged light source, monochromator, and detector, usually connected to an oscilloscope through an amplification system, the light transmittance at a selected wavelength is measured. Chemical reactions occurring in the sample during and after a short pulse of high-energy radiation cause a change in transmittance at a predetermined wavelength which can be observed and photographed at the oscilloscope as a function of time. Conductivity (68–70), polarography (71), fluorescence (72), and *in situ* ESR (40,73,74) have also been employed as detection techniques in pulse radiolytic studies. The hydrated electron generated by the pulse radiolysis of alkaline $1M$ solutions of methanol in water has been detected and studied by ESR with the use of a flow system (75). The observed spectra agreed well with those of the electron trapped in alkaline ice (76). However, due to its versatility spectrophotometry is the most commonly used technique at the present time.

2. *Available Equipment—its Use, Cost, and Limitations*

The equipment used for pulse radiolytic studies can be described and discussed most conveniently by considering the radiation sources and the auxiliary detection equipment separately. The major requirement for the radiation source is that it can produce pulses whose duration is negligible in comparison with the half-lives of the species investigated and whose intensity is great enough to produce detectable changes in the concentration of these species, their intermediates, or their products. Additionally it is desirable to have a short rise and fall time of the pulse and to produce uniformly absorbed radiation. The pulsed Van de Graaff generators, X-ray machines, and microwave linear accelerators meet these requirements. Generally they produce pulses of 0.3 μsec to several milliseconds duration with a 0.2–0.4 μsec rise and fall time and currents of approximately 1–200 mA. The price of linear accelerators or pulsed Van de Graaff generators inevitably limits their use. However the commercial availability of relatively inexpensive electron generators, such as the Febetron* whose cost is similar to that of a moderately priced mass or NMR spectrometer and which is relatively simple to operate and maintain, certainly places pulse radiolytic equipment within the reach of college or university organic laboratories.

The purpose of the auxiliary detection system is to provide a means of measuring the concentration changes induced by pulse radiation. Of the

* Field Emission Corporation, McMinnville, Ore., 97128, produces several models of the Febetron, one capable of generating 2 MeV and another 600 KeV, which cost $49,500 and $14,900, respectively.

systems, spectrophotometric detection is the most commonly used since it can achieve a high degree of sensitivity while keeping the signal-to-noise ratio low. The use of spectrophotometric detection systems and the technical difficulties involved in photoelectric recording of data in pulse radiolytic studies have been discussed by Keene (65,77). Several recent innovations have extended the scope of spectrophotometric detection. For example, with suitable electronic circuitry it is possible to detect optical density changes of one part in 10^5–10^6 (78), with repetitive pulse devices a rapid scan of the spectrum over a wide wavelength region is possible, and using a suitable monochromator changes in the concentrations of transient species can be followed simultaneously at a number of wavelengths.

Radiation chemists have been generating pulses of nanosecond (10^{-9} sec) duration* to study electron and proton transfer reactions which are too fast to measure by conventional methods. Investigations which utilize this technique are subsequently discussed.

3. Treatment of Pulse Radiolysis Data

Figure 2 illustrates a typical oscillographic trace and the resulting second-order plot. Analysis of the data from pulse radiolytic studies is carried out by the standard techniques of kinetic analysis (79,80). In most

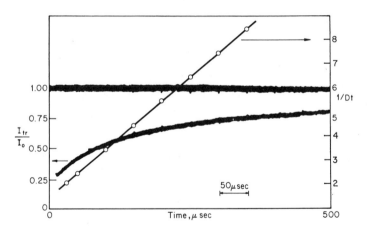

Fig. 2. Disappearance of the $(HO)C_6H_6\cdot$ radical as a function of time; test of the second-order rate law. The upper trace defines 100% transmission. The ordinate on the right represents the reciprocal of the absorbance at time t. Taken from ref. 82.

* The generation of pulses of ca. 20 psec (20×10^{-12} sec) has been reported recently (78a).

cases the concentration changes can be expressed by first, pseudo first, or second-order rate equations. The advantages of a first or pseudo first-order kinetic situation are that a knowledge of the concentration, or the extinction coefficient, of the species involved is not necessary and that a uniform distribution of the absorbed dose is not required for a determination of the rate constant of the reaction. Most of the second-order rate constants for the reactions of the hydrated electron, e_{aq}^-, with added solutes have been determined under pseudo first-order conditions. For pseudo first-order reactions the concentration of the reactant is, of course, involved and it is important to test the constancy of the second-order rate constant over a sufficiently large range of its concentration. For truly second-order kinetics and for a complete understanding of the spectrum of the transient in the presence of other species, the extinction coefficient must be known or be determined. If the transient species reacts quantitatively with known stoichiometry to form a stable product whose extinction coefficient is known or can be determined, the extinction coefficient of the transient can be calculated. This technique has been used to determine the extinction coefficient of the solvated electron in aliphatic alcohols (81) by allowing the solvated electron to react with biphenyl to form biphenylide ion whose extinction coefficient is known. Alternatively, if the radiation chemical yield of the precursor of a transient is known, the concentration and hence the extinction coefficient of the transient can be obtained from a knowledge of the absorbed dose. The determination of the extinction coefficient of the hydroxycyclohexadienyl radical is a typical example of the use of this method (82).

To simplify and facilitate the calculation of rate constants from oscillographic traces, especially in those cases where the kinetics are second or higher order or those instances where they are more complex and of mixed order, curve-fitting techniques or other forms of computer analysis have been employed (83,84).

4. Bibliographies of Absolute Rate Constant Data

Anbar and Neta have compiled the absolute rate constants for the reactions of the hydrated electron, hydroxyl radical, and hydrogen atom with a large number of organic and inorganic solutes (85). This compilation includes the literature data up to 1966 and is currently being revised (86). Due to the extremely rapid progress of radiation chemistry and the uncertainties involved in some of the early work it is advisable both to consult the original papers and to search the literature for more recent rate-constant data.

III. RADIATION-INDUCED REACTIONS OF ORGANIC COMPOUNDS IN DILUTE AQUEOUS SOLUTIONS

Just as a considerable portion of the kinetic investigations in physical organic chemistry involve water as the solvent, the radiation chemistry of water and of aqueous solutions has been studied more extensively than that of any other liquid or solvent, and hence it is better understood. The omnipresence of water and the search for an understanding of complex biological systems, of which water is the primary component, provided much of the impetus for these studies. In recent years the data in aqueous radiation chemistry concerning free-radical reactions has accumulated very rapidly. It is the primary purpose of this section to examine the types of linear free energy and spectral correlations which can be obtained from this data, the similarities and differences between radiation and chemically induced free-radical reactions, and the existence and reactivities of radical transients in oxidation and reduction processes which have only been postulated in studies of the chemically induced reactions.

A. Fundamentals of Aqueous Radiation Chemistry

The vast amount of original work in this important aspect of radiation chemistry can be bewildering to the proselyte. To a certain extent, a general appreciation of the whole field of aqueous radiation is a necessary prerequisite for a profitable use of this accumulated research. The subject of aqueous radiation chemistry prior to the development of pulse radiolysis and to the general acceptance of the importance of reactions of the hydrated electron has been treated comprehensively (22,23). A recent and useful review by Anbar (19) can serve as a starting point for an understanding of the current status of aqueous radiation chemistry. Hence, only the fundamental concepts and essential techniques of aqueous radiation chemistry are summarized here.

1. Yields of Primary Species in Irradiated Water

The primary chemical result of the irradiation of water after the deposition of energy is the formation of the following species:

$$H_2O \xrightarrow{\quad\rightsquigarrow\quad} e_{aq}^- + \cdot H + \cdot OH + H_2 + H_2O_2 + H_3^+O \qquad (1)$$

With gamma rays the primary products given in reaction (1) are formed inhomogeneously in small, widely spaced clusters in what has been known as spurs within a time scale of 10^{-10} to 10^{-8} sec (87). The events which take place from 10^{-15} to 10^{-8} sec after the energy deposition have been treated theoretically (23,45,87), but clearly they are outside the scope of the present discussion. After the spur is formed, these entities diffuse into

TABLE I

Selected Rate Constants for the Primary Species in Water[a]

Reaction	Equation number used in this review	Rate constant,[b] M^{-1} sec^{-1}	pH	Ref.
$e_{aq}^- + H_3O^+ \rightarrow \cdot H + H_2O$	2	$(2.07 \pm 0.08) \times 10^{10}$	2.1–4.3	90
		$(2.36 \pm 0.24) \times 10^{10}$	4.0–4.6	91
		$(2.26 \pm 0.21) \times 10^{10}$	4.1–4.7	92
$e_{aq}^- + e_{aq}^- \rightarrow H_2 + 2OH^-$	3	$(0.9 \pm 0.15) \times 10^{10}$ [c]	10.9	93
		$(1.1 \pm 0.15) \times 10^{10}$ [c]	13.3	94
$e_{aq}^- + H_2O_2 \rightarrow \cdot OH + OH^-$	4	$(1.23 \pm 0.14) \times 10^{10}$	7	91,93,95
$e_{aq}^- + \cdot H \rightarrow H_2 + OH^-$	5	$(2.5 \pm 0.6) \times 10^{10}$	10.5	91,93,95
$e_{aq}^- + \cdot OH \rightarrow OH^-$	6	$(3.0 \pm 0.7) \times 10^{10}$	10.5	91,93,95
$e_{aq}^- + \cdot O^- \rightarrow 2OH^-$	7	$(2.2 \pm 0.6) \times 10^{10}$	13	94
$e_{aq}^- + H_2O \rightarrow \cdot H + OH^-$	8	16.0 ± 1.0	8.4	96
$e_{aq}^- + O_2 \rightarrow \cdot O_2^-$	9	$(1.88 \pm 0.2) \times 10^{10}$	7	79,91,95,97
$e_{aq}^- + N_2O \rightarrow N_2 + \cdot OH + OH^-$	10	$(8.67 \pm 0.6) \times 10^9$	7	90,93
$e_{aq}^- + SF_6 \rightarrow SF_6^-$	11	$(1.65 \pm 0.10) \times 10^{10}$	7	53
$e_{aq}^- + CO_2 \rightarrow CO_2^-$	12	$(7.7 \pm 1.1) \times 10^9$	—	93,95
$\cdot H + \cdot H \rightarrow H_2$	13	1.5×10^{10}	0.1–1.0	98
		1.3×10^{10}	0.4–3.0	99
		1.0×10^{10}	2	100
		$(1.5 \pm 0.1) \times 10^{10}$	3	101
$\cdot H + OH^- \rightarrow e_{aq}^- + H_2O$	14	1.8×10^7	11–13	94
$\cdot H + \cdot OH \rightarrow H_2O$	15	$(0.7–3.2) \times 10^{10}$	3	99
		7×10^9	7	102
$\cdot H + O_2 \rightarrow \cdot HO_2$	16	2.6×10^{10}	0.4–3.0	99
		1.9×10^{10}	2	100
$\cdot H + H_2O_2 \rightarrow \cdot OH + H_2O$	17	$(9.0 \pm 1) \times 10^7$	2.1	100

		Value	Temp.	Ref.
18	$\cdot H + H_3O^+ \rightarrow H_2^+ + H_2O$	2.6×10^3 [d]	3.5–11	103
19	$\cdot H + N_2O \rightarrow N_2 + \cdot OH$	$\sim 1.2 \times 10^4$ [d]	3.5–11	103
20	$\cdot OH + \cdot OH \rightarrow H_2O_2$	5×10^9	7	102
21	$\cdot OH + OH^- \rightarrow \cdot O^- + H_2O$	3.6×10^8 [e]	—	103
22	$\cdot OH + H_2O_2 \rightarrow H_2O + \cdot HO_2$	4.5×10^7	7	104
23	$\cdot OH + H_2 \rightarrow \cdot H + H_2O$	$(6.0 \pm 2.0) \times 10^7$	7	102
		4.5×10^7 [e]	7	104
24	$\cdot OH + CH_3OH \rightarrow \cdot CH_2OH + H_2O$	4.8×10^8	7	105
25	$H_3O^+ + OH^- \rightarrow 2H_2O$	1.43×10^{11} [f]	7	106,107

[a] Determined at room temperature.

[b] Determined by pulse radiolysis unless stated otherwise.

[c] Rate constant, k, defined by $d(X)/dt = k(X)^2$, where $X = e_{aq}^-$ (eq. (3)), $X = \cdot H$ (eq. (13)), and $X = \cdot OH$ (eq. (20)).

[d] Determined photochemically.

[e] Determined by competition kinetics.

[f] Determined by T-jump technique.

the bulk of the solvent or competitively react with each other and with the solvent to form H_2 and H_2O_2 or to reform water (see Table I).

In pure, degassed, triply distilled water, no net change is observed with the exception of the formation of small amounts of H_2 and H_2O_2 since the radicals and molecular products formed according to eq. (1) undergo secondary reactions to reform water. Therefore, in order to obtain the yields of the species on the right-hand side of eq. (1), the addition of scavengers which form stable, easily analyzable products is necessary. In the last 15 years, a great deal of research effort has been expended in order to accurately determine the primary radical yields in irradiated aqueous solutions using different scavengers (22,60,88,89). The primary yields* generally accepted today are: $G_{e_{aq}^-} = 2.8 \pm 0.1$, $G_H = 0.6 \pm 0.1$, $G_{OH} = 2.4 \pm 0.3$, $G_{H_2} = 0.45$, and $G_{H_2O_2} = 0.71$. These values do not strictly follow the material balance equation (22):

$$G_{e_{aq}^-} + G_H + 2G_{H_2} = G_{OH} + 2G_{H_2O_2}$$

due to the uncertainties in the G_{OH} value and to our choice of $G_{e_{aq}^-} = 2.8$ (53,54,61) rather than $G_{e_{aq}^-} = 2.3$.† However, an extremely accurate and careful determination of the initial yields of ferrous sulfate and hydrogen peroxide with a Cary Model 16 spectrophotometer in irradiated aqueous solutions of ferrous sulfate and sodium formate at doses as low as 30–150 rads led to $G_{e_{aq}^-} = 2.76 \pm 0.07$ and $G_{OH} = 2.74 \pm 0.08$ (61). It is always desirable to obtain a material balance between the primary yields and the yields of the products formed. For example, if the radiation-induced hydroxylation of benzene in dilute, air-saturated solution exclusively produces phenol, then its expected yield is 2.4/100 eV.

2. Reactions of Primary Species and the Use of Solutes which Alter their Proportions

Table I contains the rate constants for the most important reactions of the primary species. For a tabulation of the rate constants for other reactions, ref. 85 and the original sources should be consulted.

It is apparent from Table I that changes in the pH will profoundly affect the chemistry of the system. At low pH values, e_{aq}^- is converted into

---

* The notation G_x is used to designate the primary radical yield and distinguishes it from the observed yield, $G(X)$. For example, $G(H_2)$ is rarely equal to G_{H_2} since the hydrogen is formed not only by primary recombination of e_{aq}^-, $\cdot H$, and $\cdot OH$ but also via reactions of these species with the solutes or with other species formed during radiolysis. In an alternative notation, the yield of the radical and molecular products is denoted by lower case g's, e.g., $g(e_{aq}^-)$, $g(OH)$, and $g(H_2O_2)$ (23).

† For a detailed discussion of the determination of the primary radical and molecular yields in aqueous solution, see ref. 89.

$\cdot$H (eq. (2)), and at high pH values $\cdot$H is converted into e_{aq}^- (eq. (14)). Indeed, a knowledge and appreciation of the rate constants for the primary chemical species with each other (Table I) and with dissolved solutes (85) are commonly and profitably employed to simplify aqueous, radiation-induced reactions. For example, when it is desirable to study exclusively the reaction of e_{aq}^- with solutes, advantage can be taken of eq. (23) by saturating the triply distilled water with H_2 and, concurrently, of eq. (14) by making the solution alkaline. Alternatively, the system can be simplified to contain primarily e_{aq}^- by adding methanol (eq. (24)) and adjusting the pH to ca. 10 (eq. (14)). In order to investigate the reactions of $\cdot$H with solutes, the triply distilled water is made acidic (eq. (2)) and is saturated with hydrogen (eq. (23)). Hydroxyl radical reactions are studied in the presence of nitrous oxide since the reaction given in eq. (10) not only eliminates e_{aq}^- but doubles the amount of hydroxyl radical in the system. This reaction is, therefore, very conveniently employed in radiation-induced hydroxylation studies. It is evident from the preceding discussion that reasonable care must be exercised in order to design meaningful experiments in aqueous radiation chemistry.

B. Linear Free Energy Correlations

Linear free-energy correlations between structures and reactivities have been used successfully for over 30 years in organic chemistry for predicting rates and equilibria and for unifying a number of seemingly independent parameters into a logical framework (108–113). Substituent effects in some free-radical reactions have been correlated by means of $\sigma\rho$ relationships in cases where the radical species possessed some degree of polar character (112), but most of this work has been carried out in nonaqueous solvent systems. However, by using pulse radiolytic techniques, absolute rate constants for the reactions of e_{aq}^-, $\cdot$OH, $\cdot$H, or other species with dissolved solutes can be determined in water and in other solvents without the complications inherent in competition kinetics. From these absolute rate constants useful and important linear free-energy correlations can be obtained. Although such correlations have been found in recent investigations and are discussed here, this subject is, as yet, far from being completely exploited.

1. Linear Free-Energy Correlations for the Reactions of e_{aq}^- with Organic Compounds

The hydrated electron, e_{aq}^-, is perhaps the most unique chemical species available to organic chemists for mechanistic studies. It can be considered to be the most elementary nucleophile, consisting of a unit negative charge which is chemically unbound to any other atoms or molecules but which is highly dispersed over a number of solvent molecules (114). The hydrated electron can also be viewed as the simplest reducing agent known. Its slow reaction with water (eq. (8)) and broad absorption maximum centered at 600 nm with a high extinction coefficient has allowed its spectroscopic detection (115–117) and the measurement of its reaction rates with solutes (85,118,119). The reaction of the hydrated electron is essentially an ideal charge-transfer process involving localization of the negative charge on an acceptor molecule. The probability of this process depends on the free energy change required in the system which, in turn, is governed by the electron density and polarizability of the acceptor. The information obtained from the reactivities of e_{aq}^- with organic compounds, therefore, has far-reaching significance especially with regard to the chemical properties of the reactants and to reactivity theories in general and allows testing of the validity of these theories up to the limit of diffusion-controlled reactions. The reactivities of the hydrated electron with aromatic compounds have been reviewed recently (118,119).

Anbar and Hart obtained a good $\sigma\rho$ correlation for the reaction of the hydrated electron with substituted aromatic compounds of the type C_6H_5X

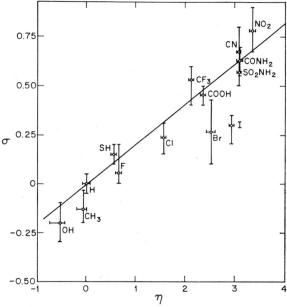

Fig. 3. Plot of η as a function of σ. $\eta = \log(k_{C_6H_5X + e_{aq}^-}/k_{C_6H_6 + e_{aq}^-})$. Taken from ref. 120.

(Fig. 3) (120). The second-order rate constants for these reactions cover a wide range of values, from 4×10^6 M^{-1} sec^{-1} for $e_{aq}^- + C_6H_5OH$ to 3×10^{10} M^{-1} sec^{-1} for $e_{aq}^- + C_6H_5NO_2$. The slope of the line in Figure 3, ρ, is 4.8 (120), a value which is within the range of ρ values generally obtained for other aromatic substitutions (121). A potential energy diagram for the interaction of e_{aq}^- with a substituted aromatic compound (Fig. 4) is similar to those which describe electrophilic (122) and nucleophilic (123) substitution at an aromatic carbon atom. Additionally, the formation of aromatic carbanions of finite stability, which has been demonstrated by pulse radiolysis (124) and by chemical means (125), resembles the formation of Meisenheimer complexes in nucleophilic aromatic substitution (123). However, the reactions of the hydrated electron with aromatic compounds are also similar to electrophilic aromatic substitutions in some respects. The preferred interaction of e_{aq}^- with aromatic compounds occurs with the π-orbital system of the aromatic ring, and better $\sigma\rho$ correlations are obtained by using σ values derived from electrophilic substitutions than by using σ-para values from nucleophilic substitutions. Iodo and bromobenzene are somewhat exceptional in that they react faster with e_{aq}^- than would be expected from the effects of these substituents on the electron density of the ring, i.e., from their σ values. This deviation can be rationalized by assuming that there is some contribution to the overall rate from attack of the hydrated electron on the halogen substituent.

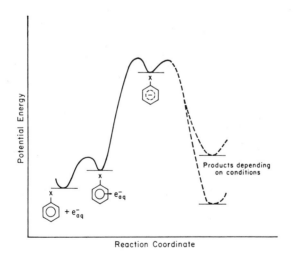

Fig. 4. Potential energy diagram for the interaction of the hydrated electron with aromatic compounds.

The similarities between the interaction of the electron with aromatic systems and other aromatic substitution reactions have been investigated further by studies of possible $\sigma\rho$ correlations for the reaction of e_{aq}^- with p-substituted benzoic acids, m-substituted toluenes, and p-substituted phenols (120). Again, excellent correlations were obtained between η and σ. However, in contrast to aromatic substitutions such as halogenation and nitration, the ρ values for the reaction of e_{aq}^- with these compounds varied considerably. This nonadditivity of η values is explicable in terms of the different factors which govern the reactivity of the electron with mono and with disubstituted benzenes. The overall π-electron distribution is the important factor in the case of the former while charge localization at one ring position may predominate for the latter. This explanation is substantiated by a comparison of the rate constants for the reactions of e_{aq}^- with disubstituted benzenes, C_6H_4XY. Generally,

$$k_{(o-XC_6H_4Y + e_{aq}^-)} < k_{(m-XC_6H_4Y + e_{aq}^-)} = k_{(p-XC_6H_4Y + e_{aq}^-)};$$

however, when X is O^- and Y is COO^-, F, or CN, the reactivity order becomes

$$k_{(o-XC_6H_4Y + e_{aq}^-)} > k_{(m-XC_6H_4Y + e_{aq}^-)} > k_{(p-XC_6H_4Y + e_{aq}^-)}.$$

Quantum mechanical calculations should throw considerable light on the intricacies of the behavior of the electron with mono and disubstituted aromatic compounds, but such calculations are still awaiting the interested theoreticians.

Although the addition of an electron to the aromatic ring is well documented (118–120), evidence concerning the fate of this transient has only become available recently (126). 1,4-Cyclohexadiene, cyclohexene, cyclohexane, and dimers have been identified by gas–liquid partition chromatography in irradiated aqueous hydrogen saturated (100 atm, eq. (23)) solutions of benzene at pH 13 (eq. (14)) (126). Based on the identity of the products in H_2O and D_2O and the relative product yields as a function of dose, a mechanism for the reduction of benzene by the hydrated electron was postulated (126) involving electron addition followed by isomerization and protonation (eq. (26)) and subsequent reaction of the hydrocyclohexadienyl radicals with e_{aq}^- (eq. (27)) and disproportionation (eq. (28)).

1,3-Cyclohexadiene could also be formed in reactions analogous to (27) and (28); however, since the reactivity of this isomer with e_{aq}^- is several orders of magnitude greater than that of 1,4-cyclohexadiene and benzene (126), the 1,3-cyclohexadiene produced would be rapidly reduced to cyclohexene or transformed into dimer. Depending on the conditions 1,4-cyclohexadiene ultimately forms cyclohexane or dimers. Electron-withdrawing

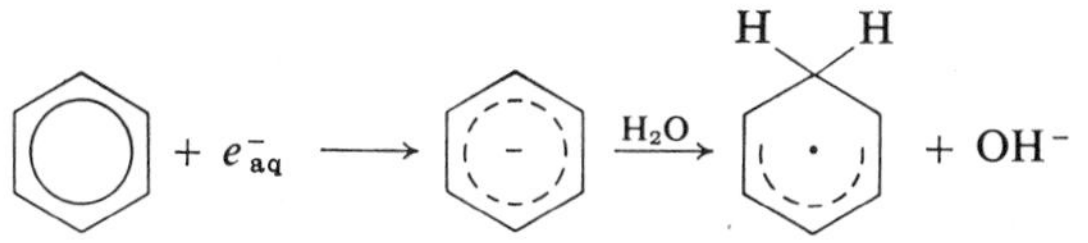

$$\text{(benzene)} + e_{aq}^- \longrightarrow \text{(adduct)} \xrightarrow{H_2O} \text{(cyclohexadienyl)} + OH^- \qquad (26)$$

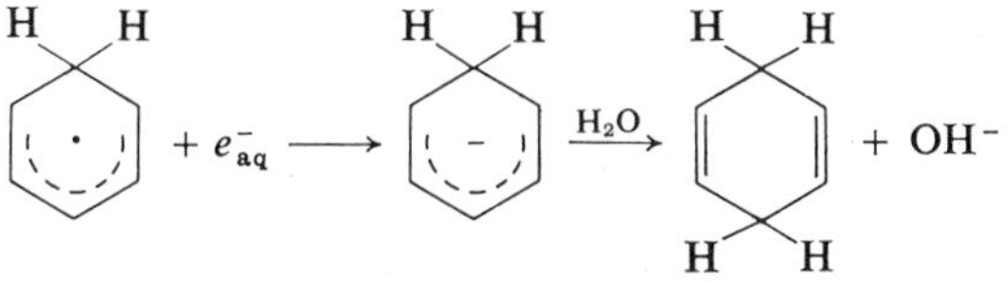

$$\text{(cyclohexadienyl radical)} + e_{aq}^- \longrightarrow \text{(anion)} \xrightarrow{H_2O} \text{(cyclohexadiene)} + OH^- \qquad (27)$$

$$2\ \text{(cyclohexadienyl)} \longrightarrow \text{(benzene)} + \text{(cyclohexadiene)} \qquad (28)$$

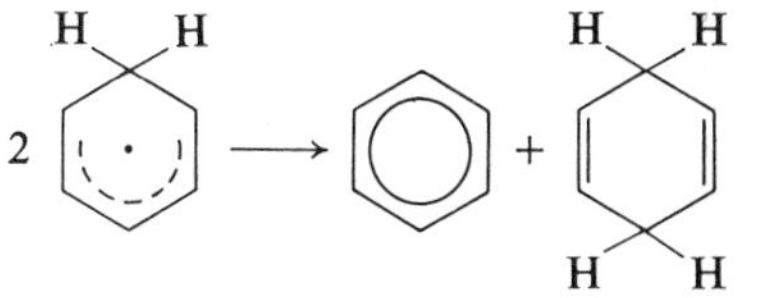

substituents would be expected to increase the lifetime of the electron adduct by stabilizing its negative charge; in fact, carbanions of finite stability have been reported in the cases of trinitrophenol (127), polycyclic aromatic compounds (128,129), and benzoic, trimesic (127), and phthalic (97) acids. Hence, the kinetics and mechanisms of these short-lived aromatic transients provide excellent research opportunities for the physical organic chemist.

A satisfactory correlation with Taft's σ^* constants (110) has been obtained for the relative rates of the reactions of e_{aq}^- with substituted halo-aliphatic compounds using butyl chloride as the standard (Fig. 5) (130). The reactivity of RX toward e_{aq}^- increased in the order $F \ll Cl < Br < I$, and no detectable solvent effect was observed when the reaction of e_{aq}^- with $C_6H_5CH_2Cl$ was carried out in water, methanol, and ethanol (130). These results are only compatible with a mechanism in which the formation of $C_6H_5CH_2Cl^-$ is rate determining and in which the charge on the anion resides mainly on the halogen atom. Therefore, a parallel between the reaction of e_{aq}^- with alkyl halides and S_N1 or S_N2 reactions at a saturated carbon atom (131) cannot be drawn easily. However, the mechanisms of electron addition to alkyl halides and to substituted aromatic compounds are similar in that attack in both cases occurs at the position of lowest electron density. Since a halogen atom bound to carbon can act as an electrophilic center only when the attacking agent possesses extremely strong reducing properties, the hydrated electron in its reaction with alkyl halides is not only the most reactive nucleophile but is also the most elementary reducing agent. Thus, in principle, the reactivities of alkyl

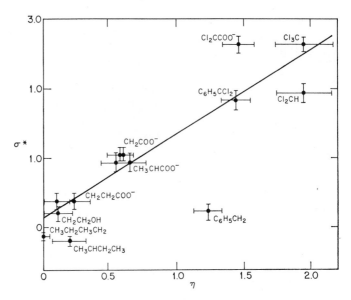

Fig. 5. Plot of η as a function of σ^*. $\eta = \log (k_{RX + e_{aq}^-}/k_{Cl(CH_2)_3CH_3 + e_{aq}^-})$. Taken from ref. 130.

halides with e_{aq}^- can be used as a measure of the relative reduction potentials of these compounds. Again, the higher reactivity of $C_6H_5CH_2Cl$ with e_{aq}^- than is expected from its σ^* value has been rationalized by postulating an additional interaction of the hydrated electron with the aromatic ring (130).

The reactivity of the hydrated electron toward carbonyl compounds, R_1COR_2, was found to be dependent to a considerable extent on the nature of the substituents R_1 and R_2 (84). While a decrease in the electron-withdrawing power of R_1 and/or R_2 resulted in a *decrease* in the reactivity of e_{aq}^- with aldehydes and ketones, it enhanced the rate constants, $k_{(e_{aq}^- + R_1COR_2)}$, in the case of amides and esters. Indeed, a good linear correlation was obtained between $\log k_{(e_{aq}^- + R_1COR_2)}$ and the sum of Taft's substituent constants $\sigma_{R_1}^* + \sigma_{R_2}^*$ for aldehydes and ketones and for amides and esters with slopes of -0.74 and $+1.2$, respectively (84). These results were rationalized in terms of the predominance of inductive effects in the case of aldehydes and ketones but of mesomeric effects for amides and esters (84). The site of electron attack was postulated to be the carbonyl oxygen atom for the former compounds and the nitrogen or alkoxy oxygen atom for the latter ones. The available rate constants for the reaction of e_{aq}^- with substituted carboxylic acids suggest, however, that both mesomeric

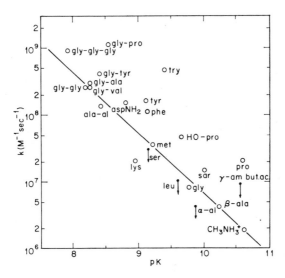

Fig. 6. Plot of rate constants for the reaction of the protonated forms of amino acids and peptides versus the pK of the amino group. Glycine = gly; proline = pro; tyrosine = tyr; tryptophane = try; alanine = ala; valine = val; asparagine = asp; phenylalanine = phe; methionine = met; serine = ser; lysine = lys; leucine = leu; and γ-amino buteric acid = γ-am but ac. Taken from ref. 132.

and inductive effects contribute to the overall electron affinities of these compounds (84), but dissection of the electron affinities into contributing inductive, mesomeric, steric, and other effects may not be valid and considerable caution must be exercised in the physical interpretation of reactivity differences.

For the reaction of e_{aq}^- with amino acids and peptides, a correlation has been obtained between the rate constants and the dissociation constant of the amino group (Fig. 6) which suggests that for these amino acids the reactivity of the side chain is important (132).

The activation energy for the reaction of e_{aq}^- with a large number of organic compounds has been determined both by competition kinetics (118,133) and by pulse radiolysis (118,134,135). In the great majority of cases the activation energy, E, was found to be 3.5 ± 0.5 kcal mole^{-1} (118, 133–135),* but the entropies of activation, $\Delta S^{\ddagger}$, were found to range from 0 to −25 eu (133). The mechanistic significance of these activation parameters has been discussed recently (118).

_____
* For a different view, see ref. 135a.

2. Linear Free-Energy Correlations for the Reactions of the Hydrogen Atom with Organic Compounds

Hydrogen atom, $\cdot$H, is formed in the radiolysis of water (eq. (1)). In fact, until the discovery of the hydrated electron, it was considered to be the sole reducing species formed in eq. (1) (22). The experimental difficulties involved in simplifying the system to contain exclusively hydrogen atom and the absence of a well-defined absorption spectrum of $\cdot$H such as that of e_{aq}^- prevented the rapid accumulation of rate constants for the reactions of $\cdot$H with organic compounds. A suitable matrix for the hydrogen atom is obtained by working in acidic media (eq. (2)) containing alcohol (eq. (24)) or hydrogen (eq. (23)). The use of alcohol to scavenge the hydroxyl radical has the disadvantage that allowance must be made for the reaction of the alcohol with the hydrogen atom $(k_{(\cdot H + CH_3OH)} = 1.6 \times 10^6 \ M^{-1}$ sec$^{-1})$ (100) and that it is difficult to assess the reactivity the $\cdot$CH$_2$OH formed in reaction (24) with the dissolved organic compounds. The problem of the appreciable rate of the reaction of the hydrogen atom with the alcohol is circumvented, however, by the use of CD$_3$OD instead of CH$_3$OH as the hydroxyl radical scavenger since the rate constant for $\cdot$OH + CD$_3$OH $\rightarrow$ $\cdot$CD$_2$OH + H$_2$O is $4.1 \times 10^8 \ M^{-1}$ sec^{-1} (136) but that for $\cdot$H + CD$_3$OH $\rightarrow$ $\cdot$CD$_2$OH + HD is only $8 \times 10^4 \ M^{-1}$ sec^{-1} (137). Since molecular hydrogen is not an effective hydroxyl radical scavenger (eq. (23)) the high concentrations and, consequently, pressures (ca. 30–50 atm) necessary render this scavenger cumbersome.

The observed rate constants for the reactions of $\cdot$H with a number of haloaliphatic compounds have been separated into those for hydrogen abstraction (RX + $\cdot$H $\rightarrow$ H$_2$ + R$'$X) and dehalogenation (RX + $\cdot$H $\rightarrow$ HX + $\cdot$R) by competitive kinetics (138). The rate constants for the hydrogen atom abstraction reaction correlated better with those for the reaction of $\cdot$OH with the same compounds than with those for the analogous reaction of e_{aq}^-. The electron density of the site of attack and the stability of the haloaliphatic radical formed determine the rates of hydrogen atom abstraction by both hydrogen atoms and hydroxyl radicals. Using isopropanol as a competitor in the presence of acetone as the electron scavenger, the order of reactivity of $\cdot$H with substituted aromatic compounds, C$_6$H$_5$X, was found to be C$_6$H$_5$OH > C$_6$H$_5$NH$_2$ > C$_6$H$_5$NO$_2$ > C$_6$H$_5$COCH$_3$ > C$_6$H$_5$COO$^-$ > C$_6$H$_5$CH$_2$COO$^-$ > C$_6$H$_5$NHCOCH$_3$ > C$_6$H$_5$CN > C$_6$H$_5$SO$_3^-$ (139) (Table II). A linear free-energy correlation, with considerable scatter, was obtained between Hammett's σ values and log $k_{(\cdot H + C_6H_5X)}$ with a slope -0.7 (139). From the qualitative agreement between the reactivity order in the reaction of $\cdot$H with C$_6$H$_5$X and the

TABLE II

Rate Constants for the Reactions of Hydrogen Atoms with Aromatic Compounds

Compound	$k, M^{-1} sec^{-1}$ [a]	$k, M^{-1} sec^{-1}$ [b]
C_6H_6	—	$(1.00 \pm 0.1)10^9$ [c]
C_6H_5OH	4.20×10^9	$(1.8 \pm 0.3)10^9$ [d]
C_6H_5COOH	8.70×10^8	$(1.00 \pm 0.15)10^9$
C_6H_5CN	4.53×10^8	$(6.80 \pm 1.0)10^8$
$C_6H_5NO_2$	1.66×10^9	$(1.04 \pm 0.15)10^9$
	—	5.6×10^9 [e]
$C_6H_5SO_3H$	4.30×10^8	$(8.2 \pm 1.3)10^8$
$C_6H_5CH_3$	—	$(1.8 \pm 0.2)10^9$ [c]
$C_6H_5CH_2COOH$	7.70×10^8	$(1.01 \pm 0.15)10^9$
$C_6H_5COCH_3$	1.26×10^9	$(1.10 \pm 0.17)10^9$
$C_6H_5NH_2$	1.76×10^9	—
$C_6H_5NHCOCH_3$	6.70×10^8	—
$p\text{-}ClC_6H_4COOH$	—	$(1.13 \pm 0.2)10^9$
$p\text{-}NO_2C_6H_4COOH$	—	$(9.8 \pm 1.5)10^8$
$p\text{-}HOC_6H_4COOH$	—	$(1.45 \pm 0.2)10^9$

[a] Obtained by competition kinetics, ref. 139.
[b] Obtained by pulse radiolysis, ref. 136, unless specified otherwise.
[c] Ref. 140.
[d] Ref. 141.
[e] Ref. 142.

order of reactivity of these compounds toward electrophilic aromatic substitution, the electrophilic nature of the hydrogen atom was inferred (139).

Evidence has been presented recently for spectra of the hydrogen atom with λ_{max} at 200 nm and $\varepsilon = 900 \pm 30$ (101) obtained in pulse-irradiated water containing $10^{-3}M$ $HClO_4$ (eq. (2)) and 35 atm of H_2 (eq. (23)). The decay of the absorption in the absence of other solutes followed second-order kinetics from which $2k_{13} = (1.5 \pm 0.1)10^{10}$ was calculated; this value was found to be in good agreement with those previously reported (98,99). Unfortunately, the wavelength of the absorption maximum of the hydrogen atom coupled with the small extinction coefficient does not permit easy pulse-radiolytic investigations of the rate of decrease in its absorption in the presence of many organic solutes, especially aromatic ones. Successful pulse-radiolytic studies of the reactions of $\cdot H$ with dissolved aromatic compounds have, however, been accomplished by observing the rate of formation of substituted hydrocyclohexadienyl radicals (136,140). In these carefully designed experiments, the hydrated electron was converted to $\cdot H$ by H_3O^+ (eq. (22)) and the hydroxyl radical was scavenged by CD_3OD (eq. (24)) (136). An analog computer was used to correct the experimental rate constants for the competing hydrogen atom recombination reaction (eq. (13)) at low organic solute concentrations (136). In Table II, the available rate constants for the reaction of the hydrogen atom with aromatic compounds are given. The differences between the values reflect the relatively large errors in the rate constants obtained by competition kinetics and render the validity of the Hammett correlation somewhat tenuous. Careful collection of more data is necessary in order to obtain meaningful linear free-energy correlations for this reaction. By following the formation of the substituted hydrocyclohexadienyl radical pulse radiolytically in H_2O and D_2O, an isotope effect of $k_{(\cdot H + C_6H_5X)}/k_{(\cdot D + C_6H_5X)} = 1.32 \pm 0.05$ was found for the reaction of $\cdot H$ with C_6H_5COOH and $C_6H_5SO_3H$ (136). From the lack of an observed isotope effect in the reaction of $\cdot OH$ and $\cdot OD$ with these compounds in H_2O and D_2O, respectively, it was inferred that solvent deuterium isotope effects are unlikely, and hence the observed isotope effect for the hydrogen atom reaction was interpreted in terms of a rate-determining electrophilic attack of the hydrogen atom on the aromatic nucleus. However, considerable caution must be exercised in attributing undue mechanistic significance to such small isotope effects.

The reactivity of the hydrogen atom with inorganic, organic, and biochemical solutes has been studied by generating $\cdot H$ from a stream of hydrogen in an electrodeless discharge system (143–151). This approach has the advantage that the hydrogen atom is the only species present without the necessity of scavenging the other species produced by the radiolysis of water, but it eliminates the possibility of absolute rate-constant measurements for rapid reactions.

3. Linear Free-Energy Correlations for the Reactions of the Hydroxyl Radical with Organic Compounds

The hydroxyl radical has been demonstrated to be the primary oxidizing species generated in the radiolysis of water (10,22). It possesses some electrophilic character (152,153), a property which has been inferred from its relative reactivities and from the isomeric ratios of its hydroxylation products with substituted aromatic compounds. The broad absorption of the hydroxyl radical and its low extinction coefficient (102), however, do not permit direct pulse-radiolytic observation of the rate of decrease in its absorption in the presence of organic solutes. Nevertheless, the relative rate constants for the reaction of $\cdot OH$ with a number of organic solutes are known from competition studies (105).

Taking advantage of the rapid reaction of p-nitrosodimethylaniline with $\cdot OH$ (154), Anbar and co-workers (155) studied the relative reactivity of the hydroxyl radical with a series of substituted benzenes, C_6H_5X, and benzoate ions, $C_6H_4XCOO^-$. Good $\sigma\rho$ correlations were obtained between σ_v or σ_m and η, where

$$\eta = \log \frac{k_{(\cdot OH + C_6H_5X \text{ or } C_6H_4XCOO^-)}}{k_{(\cdot OH + C_6H_6 \text{ or } C_6H_5COO^-)}}$$

and gave a ρ value of $+0.4$ for each of the two $\sigma\eta$ plots. Because the validity of the p-nitrosodimethylaniline competition method has been questioned (156), the absolute rate constants for these reactions have also been determined using pulse radiolysis by following the rate of appearance of the substituted hydroxycyclohexadienyl radicals (157). From these rate constants, a good Hammett $\sigma\rho$ plot was obtained (Fig. 7), and its ρ value of $+0.5$ agreed well with that determined by the competition techniques (155). Unlike the case of the reactions of e_{aq}^- with substituted aromatic compounds, the η values for these aromatic hydroxylation reactions were found to be additive, i.e., the same ρ value was obtained for the reaction of $\cdot OH$ with both C_6H_5X and $C_6H_4XCOO^-$. The implication of this difference is that the reaction of the hydroxyl radical with substituted aromatic compounds resembles electrophilic aromatic substitution to a greater extent than the corresponding reaction of e_{aq}^- resembles either electrophilic or nucleophilic substitution. In addition, these results imply that the rate-determining step is the addition of $\cdot OH$ to the aromatic ring

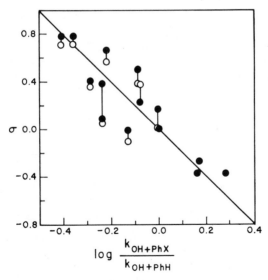

Fig. 7. Plot of log $k_{OH + PhX}/k_{OH + PhH}$ as a function of σ. Taken from ref. 157.

to form the hydroxycyclohexadienyl radical. Pulse radiolytic observation of hydroxycyclohexadienyl radical formation and the lack of deuterium isotope effects in the reactions of benzene (82) and nitrobenzene (158) with ·OH substantiate this assumption. However, the ρ values obtained for radiation-induced aromatic hydroxylations are somewhat smaller than those for other electrophilic aromatic substitutions. These lower ρ values indicate the relatively low sensitivity of radiation-induced hydroxylations of aromatic compounds to changes in substituents.

Attempts to correlate the rates of hydroxyl radical reactions for a large number of aliphatic compounds with substituent constants were less successful (159). The rate constants for the reactions of ·OH with aliphatic carboxylate ions increased with increasing chain length, but trimethyl-acetate and β,β,β-trimethylpropionate ions reacted faster and cyclic carboxylate ions reacted slower than expected. Charge transfer from the carboxylate group, resulting in decarboxylation, and steric effects were invoked to rationalize these discrepancies (159).

The reactivities of the hydroxyl radical with substituted aliphatic carboxylate ions, alcohols, polyhydric alcohols, ethers, and methanes were calculated from the sums of the partial reactivities of the C—H bonds at the different positions in the molecule. The specific rate constants calculated by this method agreed reasonably well with the observed ones (159). The reactivity of the hydrogen atom with aliphatic compounds generally parallels that found for the hydroxyl radical reactions (159). The rate-determining step for these reactions is hydrogen abstraction and the driving force is the resonance stabilization of the resulting free radical. The mechanisms for the reactions of e_{aq}^- and hydroxyl radicals (or hydrogen atoms) with aliphatic compounds are, therefore, different in that the rate-determining step for the former involves an electron transfer and the latter an atom-transfer process.

From the preceding discussion it is evident that although the hydroxyl radical is uncharged (160,161), because of its high electron affinity, it possesses considerable electrophilic properties. However, further work on the reactivity of the hydroxyl radical with both aliphatic and aromatic compounds is necessary to clarify and substantiate the details of the effect of this contention.

4. Linear Free-Energy Correlations for Radiation-Induced Radical Reactions

In an attempt to understand the reactivity of the intermediates generated by the addition of primary radiolytic species to substituted aromatic compounds, the reactivities of hydroxycyclohexadienyl radicals have been examined (162). By saturating triply distilled water with N_2O (eq. (10)), over 99% of the hydrated electron is converted into hydroxyl radical which then reacts (eq. (29)) to form a substituted hydrocyclohexadienyl radical:

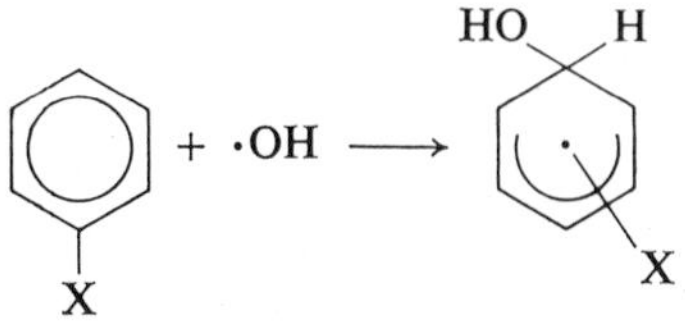

$$\text{(structure)} + \cdot\text{OH} \longrightarrow \text{(structure)} \qquad (29)$$

Taking advantage of the absorption of the cyclohexadienyl radical, its rate of reaction with oxygen (eq. (30)) and its rate of disappearance in the absence of scavengers (eq. (31)) were measured by pulse-radiolytic techniques (162).

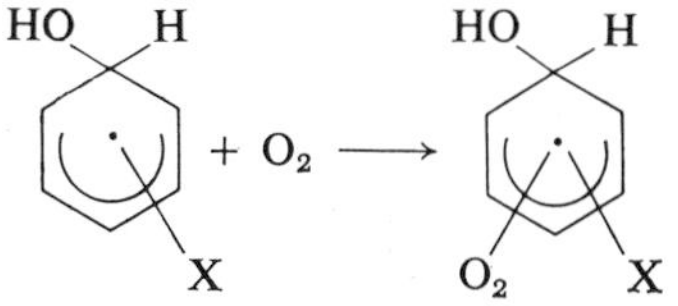

$$\text{(structure)} + O_2 \longrightarrow \text{(structure)} \qquad (30)$$

$$2 \, \text{(structure)} \longrightarrow \text{Products} \qquad (31)$$

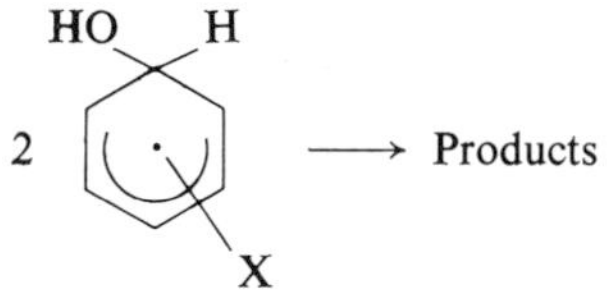

Table III summarizes the experimental results.

The rate constants in Table III increase as the electron-donating power of the substituent increases. This trend is in agreement with the general tendency of electron-withdrawing groups to hinder and electron-donating groups to facilitate radical reactions (163,164). Plotting Hammett's σ-para and σ-meta substituent constants for electron-withdrawing groups and only σ-meta constants for electron-donating groups against the logarithm of the relative rate constants, $\log (k_{(\cdot \text{HOC}_6\text{H}_5\text{X} + \text{O}_2)}/k_{(\cdot \text{HOC}_6\text{H}_6 + \text{O}_2)})$ or $\log (k_{(2 \cdot \text{HOC}_6\text{H}_5\text{X})}/k_{(2 \cdot \text{HOC}_6\text{H}_6)})$, linear free-energy plots were obtained with slopes of -1.0 and -0.75, respectively (162). The reactivities of $\cdot \text{HOC}_6\text{H}_6$, $\cdot \text{HOC}_6\text{H}_5\text{CH}_3$, and $\cdot \text{HOC}_6\text{H}_5\text{C}_2\text{H}_5$ with oxygen (reaction (30)) were much greater and the relative rate constants for reaction (31) were smaller than would have been expected from these correlations. Since the activation energies for the reactions of these radicals (Table III) do not deviate significantly from the correlation plot of activation energies versus Hammett's substituent constants, the reactivity differences for these radicals were attributed to entropy effects originating from different degrees of hydrogen bonding (162).

TABLE III

Rate Constants for the Reactions of Substituted Hydroxycyclohexadienyl
Radicals with Oxygen and with Themselves[a]

Substituent X	$\cdot HOC_6H_5X + O_2$ $10^{-6}k_{30}$, $M^{-1} sec^{-1}$	$\cdot HOC_6H_5X + \cdot HOC_6H_5X$ $10^{-8}k_{31}$, $M^{-1} sec^{-1}$	Activation energy for reaction (31), kcal mole^{-1}
NO$_2$	2.5	5.0	6.0
CN	2.9	3.9	4.7
COC$_6$H$_5$	3.7	3.7	4.7
COCH$_3$	3.8	4.2	4.3
SO$_2$NH$_2$	4.2	3.7	5.4
COOH	5.3	4.0	4.3
I	7.0	4.0	5.0
Cl	8.0	4.9	3.5
Br	9.6	4.2	3.9
OH	4.9	11.0	6.7
OCH$_3$	16.4	5.7	5.8
NH$_2$	25.4	16.0	3.9
N(CH$_3$)$_2$	23.2	7.3	4.6
H	400	4.6	3.5
CH$_3$	390	3.9	3.0
C$_2$H$_5$	555	4.2	3.6

[a] Ref. 162.

The formation of nitrobenzene radical anion, $C_6H_5\overset{\cdot}{N}O_2{}^-$, by an electron-transfer process from an α-alcohol radical (eq. (32)) has been investigated by pulse radiolysis (165).

$$R_1R_2\overset{\cdot}{C}OH + C_6H_5NO_2 \longrightarrow C_6H_5\overset{\cdot}{N}O_2{}^-, + H^+ + R_1R_2CO \qquad (32)$$

The good correlation obtained between Taft's σ^* value and the rate constants for a number of α-alcohol radicals indicates the importance of inductive effects in this reaction. Similarly, the dissociation constants for α-alcohol radicals (eq. (33)) correlated well with the electron withdrawing power of the substituents R_1 and R_2 as measured by their σ^* constants (166).

$$R_1R_2\overset{\cdot}{C}{-}OH \rightleftarrows R_1R_2\overset{\cdot}{C}{-}O^- + H^+ \qquad (33)$$

The dissociation constants for a number of radical protonation equilibria were found to correlate well with ΔH_R (167), which is the difference between the sums of all the occupied energy levels in the ground state for the protonated and basic forms of the radical calculated by HMO treatment (163).

5. Spectral Correlations

With the use of pulse-radiolytic techniques, the absorption spectra of numerous transient aliphatic and aromatic radicals have become available. Two recent reviews summarize the accumulated data (168,169). The effects of solvents (170,171) and temperature (170,172) on the absorption maximum of the hydrated electron spectrum have been investigated. A correlation between the spectral maximum of the solvated electron and that of iodide ion has been obtained in different solvents (171). In addition, the spectral maximum of the solvated electron in different solvents has been correlated with the dielectric constant of the medium (170). However, such a correlation is not generally true since it breaks down for water, ammonia, alcohols, and dioxane–water mixtures (173) indicating the importance of the local molecular environment and preferential solvation. Nevertheless, these correlations can be used to predict the absorption maximum of the solvated electron.

Extensive correlations have been obtained between the absorption maxima of substituted and unsubstituted aromatic compounds and those of their hydroxyl, hydro, and electron adducts (Table IV) (174,175). The bathochromic shifts of the aromatic compounds due to substituents relative

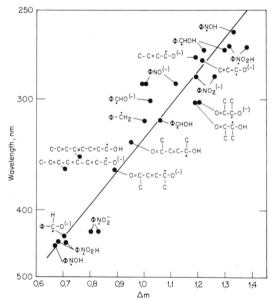

Fig. 8. Plot of absorption maxima of radicals against their HMO transition energies, Δm. Taken from ref. 167.

TABLE IV

Absorption Maxima and Relative Shifts of the Primary Band of C_6H_5X, HOC_6H_5X, $\cdot C_6H_6X$, and $\cdot C_6H_5X^-$ [a]

Aromatic compound	C_6H_5X λ,nm	A [b]	HOC_6H_5X λ,nm	B [c]	B/A	$\cdot C_6H_6X$ λ,nm	C [d]	C/A	$\cdot C_6H_5X^-$ λ,nm	D [e]	D/A
Nitrobenzene	268.5	0.232	410[f]	0.238	1.02	410[f]	0.238	1.02	285[f]	0.017	0.091
Benzophenone	252.0	0.192	380[f]	0.176	0.92	—	—	—	—	—	—
Benzaldehyde	249.5	0.186	380	0.175	0.94	380	0.175	0.94	295	0.017	0.091
Acetophenone	245.5	0.174	370	0.154	0.88	370	0.154	0.89	305	0.055	0.310
Benzoic acid	230.0	0.116	345	0.094	0.81	345	0.094	0.82	—	—	—
			350[f]	0.106	0.91						
Aniline	230.0	0.115	350[f]	0.106	0.92	—	—	—	—	—	—
Benzonitrile	224.0	0.098	340	0.082	0.84	340	0.082	0.89	310	0.063	0.645
Benzoate ion	224.0	0.098	328	0.047	0.48	—	—	—	313	0.072	0.740
			332[f]	0.054	0.55						
Benzenesulfonamide	217.5	0.064	330[f]	0.052	0.81	—	—	—	—	—	—
Anisole	217.0	0.062	330[f]	0.051	0.82	—	—	—	—	—	—
Benzenesulfonic acid	215.0	0.053	320[f]	0.0219	0.41	—	—	—	—	—	—
Bromobenzene	211.0	0.036	325	0.037	1.05	320	0.0323	0.90	—	—	—
Phenol	210.5	0.029	320	0.025	0.86	—	—	—	—	—	—
Chlorobenzene	209.5	0.028	322[a,f]	0.031	1.11	318	0.0265	0.95	—	—	—
Iodobenzene	207.0	0.017	318[f]	0.0157	0.92	—	—	—	—	—	—
Toluene	206.5	0.017	317	0.0125	0.71	315	0.0155	0.92	—	—	—
Benzene	203.5	—	313	—	—	310	—	—	290	—	—

[a] Ref. 174, unless specified otherwise.

[b] $A = \dfrac{\nu_{C_6H_6} - \nu_{C_6H_5X}}{\nu_{C_6H_6}}$.

[c] $B = \dfrac{\nu_{HOC_6H_6} - \nu_{HOC_6H_5X}}{\nu_{HOC_6H_6}}$.

[d] $C = \dfrac{\nu_{\cdot C_6H_7} - \nu_{\cdot C_6H_6X}}{\nu_{\cdot C_6H_7}}$.

[e] $D = \dfrac{\nu_{C_6H_6^{\cdot-}} - \nu_{C_6H_5X^{\cdot-}}}{\nu_{C_6H_6^{\cdot-}}}$.

[f] Ref. 175.

to the frequency of the primary band of benzene, A in Table IV, correlated reasonably well with those of the substituted radicals relative to the cyclo-hexadienyl radical adduct frequencies, B, C, and D in Table IV. Better correlations were obtained with the compounds and radical adducts con-taining uncharged substituents than with those containing charged ones such as COO^-. These results indicate that the structures of the hydroxy, hydro, and electron adducts of these aromatic compounds are similar with various substituents and that the relative energy differences of the adducts as a function of the substituents are proportional. A good correlation has been obtained between the absorption maxima of pulse-radiolytically generated transient radicals and their HMO transition energies (Fig. 8) (167). The slope, β_0, of the plot in Figure 8 was found to be 71 kcal mole^{-1} (167), a value somewhat higher than that previously reported (163). Obviously, spectral correlations are a powerful technique in that they allow reasonable predictions of the spectra of unknown transients; in fact, the predicted absorption maxima have been verified for a number of substituted hydroxycyclohexadienyl radicals (157).

6. Applicability of Linear Free-Energy Relationships to Radiation-Induced Reactions

In constructing and using linear free-energy relationships, it must be remembered that the reactivity indexes, σ, σ^*, and similar constants, are based on a standard reaction; for example, σ values are based on the dis-sociation constants of substituted benzoic acids. Therefore, the application of a reactivity constant based on one model to an entirely different system may not be meaningful or valid. It is important to realize that linear free-energy correlations are empirical relationships and that their validity, necessarily based on a large body of supporting experimental data, may or may not withstand rigorous mathematical treatment. Only when accurate and detailed electron-distribution data become available (109), will it be possible to develop sound theories of the effects of structure on reactivity and sensitivity for radical reactions. So far, the paucity of data on linear free-energy correlations for radical reactions has prevented tests of their validity and generality. The fact that good linear free-energy correlations have been obtained for the radiation-induced processes out-lined in this section indicates, nevertheless, that this treatment of the data is useful and meaningful at least for those reactions in which polar effects predominate. Assuming that an unbiased statistical treatment has been used to determine the straight line in the linear free energy plot, significant deviations from it may suggest a variety of important factors influencing the reaction in the case of the anomalous substituent—for example, the incursion of a concurrent reaction pathway or of a different reaction mechanism. Indeed, in the reactions of the hydrated electron with substituted benzenes (120) and with haloaliphatic compounds (130), the deviation of bromo and iodobenzene in the former case and of benzyl chloride in the latter case suggested an additional reaction pathway. In making predictions based on linear free-energy correlations, the assumption should not be made that all compounds in a series will necessarily give good agreement with a linear free-energy relationship which is based on a few members of that series.

Using pulse-radiolytic techniques, it is certainly possible to select suitable model reactions involving the primary chemical species, especially e_{aq}^-, $\cdot OH$, or $\cdot H$, or their transients from which substituent constants can be derived for radiation-induced and for other radical reactions. Such substituent constants derived from reactions involving radical species could be used more profitably for linear free-energy relationships of radical reactions than those such as σ, σ^*, or their variants which are based on ionic equilibria and reactions.

C. Mechanisms of Radiation-Induced Oxidation and Reduction Processes

There are numerous oxidation and reduction reactions in organic chemistry whose mechanisms can be studied profitably by the use of radiation chemistry, especially pulse radiolysis. It has been shown that the hydrated electron, the simplest nucleophile, possesses considerable reducing properties and that the hydroxyl radical acts as an electrophile and as an oxidizing agent. Hence, the kinetics and mechanisms of many organic oxidation and reduction reactions can be investigated by generating these primary oxidizing and reducing species radiolytically from water in the presence of organic solutes. This application of radiation chemistry to organic mechanistic problems is only in its infancy, but the tremendous power of this tool is clearly indicated by the achievements discussed in this section.

1. Oxidation–Reduction Processes in the Nitrobenzene–Nitrosobenzene–Phenylhydroxylamine System

The oxidation–reduction processes involving nitrobenzene (142,165), nitrosobenzene (69), and phenylhydroxylamine (176) have been studied in detail. Chart I summarizes the observed reactions, and reference is made to the equation numbers in this chart throughout the ensuing discussion.

The rate constant for the reaction of e_{aq}^- with nitrobenzene was determined to be $k_{34} = 3.0 \times 10^{10}\ M^{-1}\ sec^{-1}$ (165). In alkaline solution the

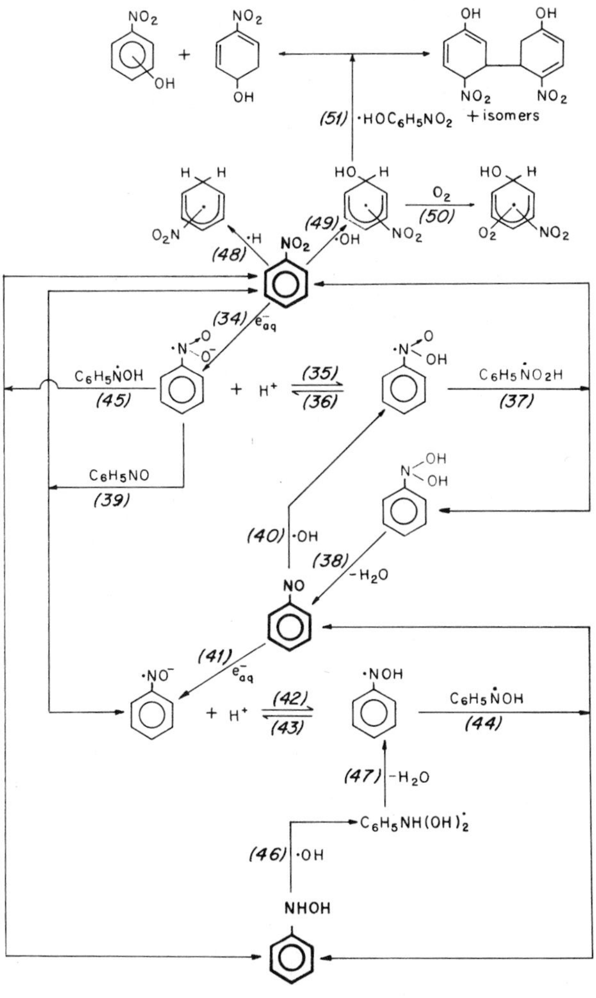

Chart I

product of this reaction, $C_6H_5\overset{\cdot}{N}O_2{}^-$, has a lifetime of several seconds and an absorption maximum at 285 nm. The ultraviolet and ESR spectra of $C_6H_5\overset{\cdot}{N}O_2{}^-$ are in accord with the assigned structure in which the electron resides on the nitro group. On the other hand, the addition of a hydrogen atom (eq. (48)) or a hydroxyl radical (eq. (49)) results in the formation of nitro-substituted hydro or hydroxycyclohexadienyl radicals.

The nitrobenzene radical anion, $C_6H_5\overset{\cdot}{N}O_2{}^-$, can also be formed by an electron transfer from an isopropanol radical (eq. (32)). Nitrous oxide saturated (eq. (10)) dilute nitrobenzene solutions containing ca. $10^{-1}M$ isopropanol (analog of eq. (24)) allowed a study of this system. The rate constant for the decomposition of $C_6H_5\overset{\cdot}{N}O_2{}^-$ was found to be pH dependent (165). This behavior suggested that protonation of the radical anion to form a species absorbing strongly at 275 nm, $C_6H_5\overset{\cdot}{N}O_2H$ (eq. (35)), was occurring. Since the protonated and unprotonated forms of the nitrobenzene radical anion absorb differently at 300 nm, observation of the spectra after the pulse at that wavelength as a function of pH allowed the measurement of the dissociation constant (Fig. 9). The pK, log k_{35}/k_{36}, was found to be 3.2. In alkaline solution the anionic form, $C_6H_5\overset{\cdot}{N}O_2{}^-$, is the predominant species, but in acid solution two molecules of the protonated radical, $C_6H_5\overset{\cdot}{N}O_2H$, disproportionate to form nitrobenzene and $C_6H_5N(OH)_2$ (eq. (37)). At pH 2 the rate constant for this reaction, k_{37}, was determined to be $6.8 \times 10^8\ M^{-1}\ sec^{-1}$. The absorption spectrum of the products of reaction (37) agreed with that of nitrosobenzene. Therefore, nitrosobenzene is formed from $C_6H_5N(OH)_2$ by the elimination of a water molecule (eq. (38)). This reaction was found to be acid catalyzed and the

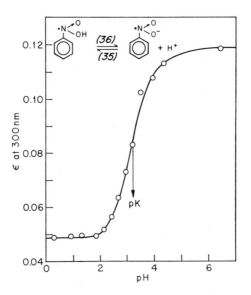

Fig. 9. Absorbance of the nitrobenzene radical anion at 300 nm vs. pH. Taken from ref. 165.

rate constant calculated from the appearance of nitrosobenzene, k_{38}, is $4.6 \times 10^3 \ M^{-1/2}\text{sec}^{-1} \cdot [\text{H}^+]^{1/2} + 10^2 \ \text{sec}^{-1}$.*

In a neutral nitrous oxide saturated solution (eq. (10)) of ca. $10^{-4}M$ nitrobenzene containing $10^{-1}M$ isopropanol (analog of eq. (24)), the reaction of only the nitrobenzene radical anion, $C_6H_5\overset{\cdot}{N}O_2{}^-$, with added nitrosobenzene, 1.3–$7.3 \times 10^{-5}M$ (eq. (39)), could be observed at 285 nm; k_{39} was found to be $4.1 \times 10^7 \ M^{-1} \ \text{sec}^{-1}$.

Using the thiocyanate competition method (105) the rate constant for the reaction of hydroxyl radical with nitrosobenzene (eq. (40)) in a nitrous oxide-saturated solution (eq. (10)) was measured; $k_{40} = 1.1 \times 10^{10}$ $M^{-1} \ \text{sec}^{-1}$. In neutral solution the product of this reaction immediately dissociates forming the nitrobenzene radical anion, $C_6H_5\overset{\cdot}{N}O_2{}^-$ (eq. (36)). This fact was established both by observing a spectrum at the end of the reaction sequence, eqs. (40) and (41), identical to that of $C_6H_5\overset{\cdot}{N}O_2{}^-$ and by obtaining identical rate constants for k_{36} from spectroscopic and conductivity measurements of the changes in concentration of the reactant and the product, respectively (69). The nitrobenzene radical anion produced in the presence of nitrosobenzene will react, of course, by eq. (39) as mentioned previously. The G-value of 6.0, determined by gas–liquid partition chromatography, for the nitrobenzene formed in this reaction supports this mechanism, i.e., eq. (40) $\rightarrow$ (36) $\rightarrow$ (39) (69).

The rate constant for the reaction of the hydrated electron with nitrosobenzene, eq. (41), was determined by following the rate of disappearance of the hydrated electron at its absorption maximum, 720 nm, in the presence of $10^{-1}M$ methanol as the $\cdot$OH scavenger (eq. (24)); k_{41} was found to be $4.3 \times 10^{10} \ M^{-1} \ \text{sec}^{-1}$ (69). As in the case of nitrobenzene, the electron adds to the substituent forming, in this case, the nitrosobenzene radical anion, $C_6H_5\overset{\cdot}{N}O^-$. This species, like its nitrobenzene analog, can be protonated, and the pK for the dissociation of the protonated radical $C_6H_5\overset{\cdot}{N}OH$, $\log k_{42}/k_{43}$, was measured and was calculated to be 11.7.

In a $10^{-4}M$ nitrosobenzene solution in the absence of scavengers, reaction (41) followed by (42) and reaction (40) followed by (36) led to the formation of $C_6H_5\overset{\cdot}{N}OH$ and $C_6H_5\overset{\cdot}{N}O_2{}^-$. The rate constant in neutral

---

* Reactions analogous to eqs. (34), (35), (36), (37), and (38) have been observed recently by pulse radiolysis for a number of substituted nitrobenzenes ($C_6H_4XNO_2$; X=OH, CHO, CH$_3$, CN). Good linear free-energy correlations between the observed rate constant's for each of these processes and Hammett's σ values were obtained (176a).

solution for the reaction of these two radicals (eq. (45)) forming nitro-benzene and phenylhydroxylamine was determined to be $k_{45} = 2.1 \times 10^8$ $M^{-1} \sec^{-1}$.

The rate constant for the reaction of the hydroxyl radical with phenyl-hydroxylamine (eq. (46)) was measured in a nitrous oxide-saturated solution (eq. (10)) using pulse radiolysis (176) both by following the appearance of the transient adduct $C_6H_5NH(OH)_2^{\cdot}$ at 290 nm and by the thiocyanate competition method (105). Using the former technique $k_{46} = 1.5 \times 10^{10} M^{-1} \sec^{-1}$ and with the latter $k_{46} = 1.2 \times 10^{10} M^{-1}$ $\sec^{-1}$. From the available experimental data the site of $\cdot$OH attack cannot be specified; however, the adduct $C_6H_5NH(OH)_2^{\cdot}$ forms $C_6H_5\overset{\cdot}{N}OH$ by the elimination of a molecule of water (eq. (47)). This reaction is both acid and base catalyzed, and at pH 7, $k_{47} = 1.8 \times 10^5 \sec^{-1}$. Subsequently two molecules of $C_6H_5\overset{\cdot}{N}OH$ disproportionate (eq. (44)) to give nitroso-benzene and phenylhydroxylamine. This reaction is also pH dependent, and the largest rate constant, $k_{44} = 5 \times 10^7 M^{-1} \sec^{-1}$, was measured at pH 11.7.

The addition of hydroxyl radical to nitrobenzene in a nitrous oxide-saturated solution (eq. (10)) resulted in the formation of a nitro-substituted hydroxycyclohexadienyl radical whose absorption maximum is at 410 nm (eq. (49)) (142). A similar spectrum was obtained from the irradiation of degassed solutions at low pH, where e_{aq}^- is converted to $\cdot$H (eq. (2)), implying that the hydrogen atom also adds to the aromatic ring forming a nitro-substituted hydrocyclohexadienyl radical. By following the increase in absorbance at 410 nm under different conditions, k_{48} and k_{49} were calculated to be $5.6 \times 10^9 M^{-1} \sec^{-1}$ and $4.7 \times 10^9 M^{-1} \sec^{-1}$, re-spectively. The decrease in absorbance at 410 nm gave clean second-order kinetics from which $k_{51} = 6.0 \times 10^8 M^{-1} \sec^{-1}$ was calculated. Since the observed G-value for the formation of the nitro-substituted hydroxycyclohexadienyl radicals (eq. (49)) is 6 but that for the formation of the isomeric nitrophenols in reaction (51) is less than 1 (158), the reaction of this radical (eq. (51)) involves not only disproportionation but also dimerization and possibly other reactions. Indeed, the absorption spectrum after the irradiation indicated the presence of additional products which could also exist in protonated or ionic forms. The most probable reaction would result in the formation of isomeric hydroxy and nitro-substituted cyclohexadiene dimers; however, these compounds are unstable.

In oxygen-saturated solution the rate constant for the addition of oxygen to the nitro-substituted hydroxycyclohexadienyl radicals forming peroxy radicals (eq. (50)) was found to be $(2.5 \pm 0.6) \times 10^6 M^{-1} \sec^{-1}$.

A linear free-energy correlation for the reaction of substituted hydroxy-cyclohexadienyl radicals with oxygen (see Section III-B-3) was obtained (162) and indicated that electron-withdrawing substituents decrease the rate of this reaction. However, the subsequent fate of the peroxy radicals formed in this reaction is not known, but it is very probable that they are unstable.

2. Radiation-Induced Oxidation and Reduction of Phenol, Nitrophenol, Hydroquinone, Quinone, and Related Systems

A detailed pulse-radiolytic investigation of the reactions of e_{aq}^-, $\cdot H$, and $\cdot OH$ with phenol has been reported recently (141). The rate constant, k_{52}, for the reaction

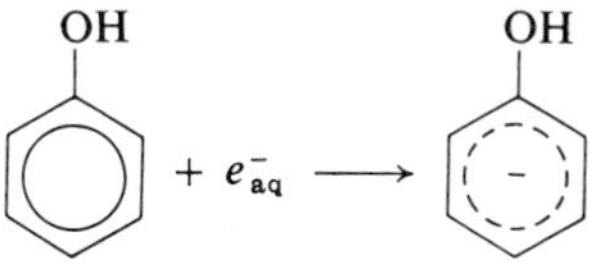

$$(52)$$

was determined to be 1.8×10^7 M^{-1} $\sec^{-1}$ at pH 6.3–6.8 by measuring the rate of disappearance of the absorption at 700 nm due to the hydrated electron. Similarly at pH 2 (eq. (2)), the rate of increase in the absorbance at 330 nm gave the rate constant for the reaction

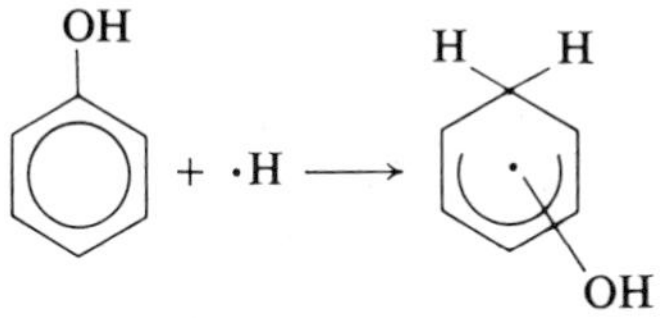

$$(53)$$

$k_{53} = 1.8 \times 10^9$ M^{-1} $\sec^{-1}$.

The absorption increase at 330 nm in a $10^{-3}M$ phenol solution saturated with nitrous oxide (eq. (10)) was attributed to the formation of the dihydroxycyclohexadienyl radical:

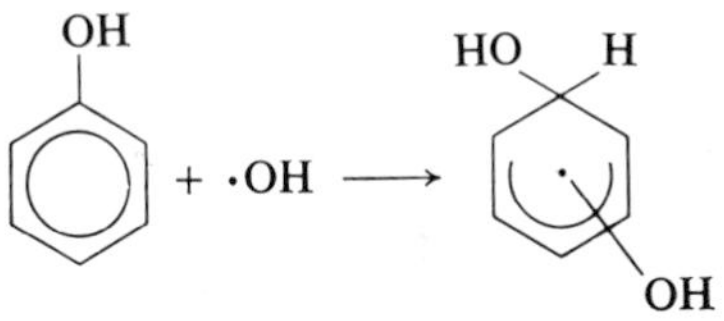

$$(54)$$

At pH 7.4–7.7 the rate constant, k_{54}, was determined to be 1.4×10^{10} M^{-1} $\sec^{-1}$. Subsequently, a new transient appeared at 400 nm at the expense of the absorption at 330 nm. Since the phenoxyl radical, $C_6H_5\dot{O}$, has an absorption maximum at 400 nm (177) and since this radical has been identified in mixtures of aqueous titanous ion and hydrogen peroxide containing phenol (178), the elimination of a molecule of water from the dihydroxycyclohexadienyl radical is well substantiated:

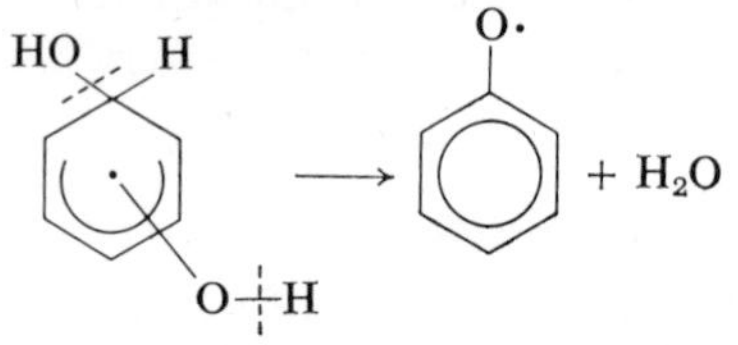

$$+ H_2O \qquad (55a)$$

Depending on the point of attack of the hydroxyl radical, the formation of isomeric 1,1-, 1,2-, 1,3-, and 1,4-dihydroxycyclohexadienyl radicals can be envisaged. However, water elimination (eq. (55a)) probably proceeds at the same rate for all these isomers since the rate of decomposition of the dihydroxycyclohexadienyl radicals followed at 330 nm equalled in all cases the rate of phenoxyl radical appearance followed at 400 nm (140). These results are compatible with a mechanism in which the driving force for water elimination (eq. (55b)) is the gain in resonance energy from the rehybridization which results in the formation of a stable aromatic structure.

In neutral solution the rate constant for reaction (55) was determined to be 5×10^3 sec^{-1}. Acids and bases as well as HPO_4^{2-}, H^+, and OH^- were found to catalyze the water elimination. Therefore, it is probable that this process occurs by separate eliminations of OH^- and H^+:

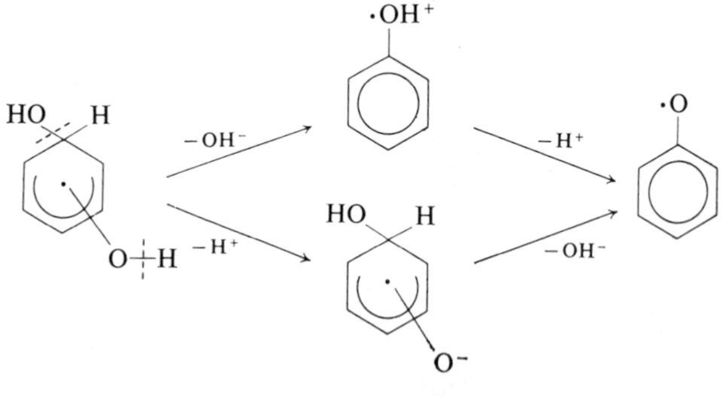

$$(55b)$$

This type of elimination may well represent a general case for the reactions of substituted hydroxycyclohexadienyl radicals:

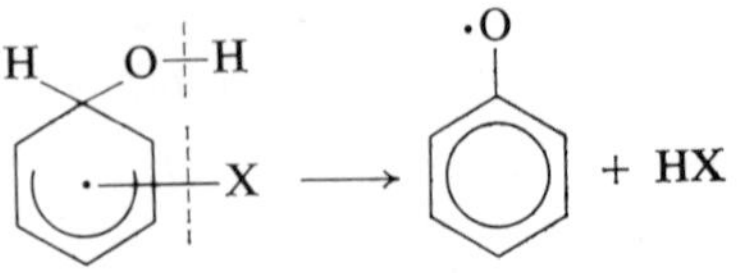

$$+ HX \qquad (56)$$

The presence of nitrate (158), chloride (179), and fluoride (180) ion in irradiated solutions of nitrobenzene, chlorobenzene, and fluorobenzene supports this postulation.

Since pulse radiolytic data on p-nitrophenol has become available recently (181), interesting comparisons can be made between the reactions of e_{aq}^-, $\cdot$H, and $\cdot$OH with nitrobenzene (166), phenol (141), and p-nitrophenol (181) and between the reactivities of their transients. Chart II summarizes the radiation-induced reactions of p-nitrophenol in aqueous solutions. The hydrated electron reacts with p-nitrophenol (eq. (57)) in a

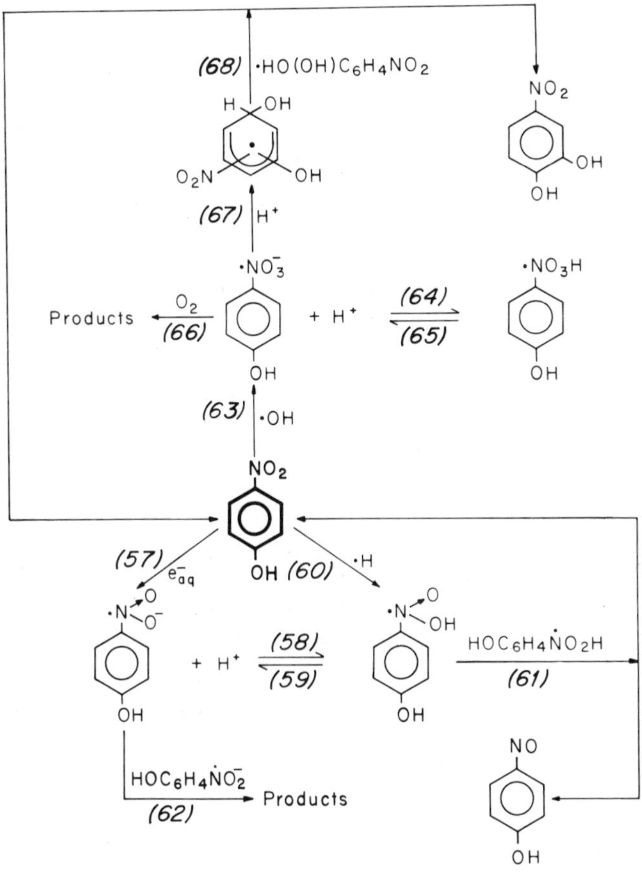

Chart II

manner analogous to its reaction with nitrobenzene (eq. (34)).* In both cases the radical anion of the nitro group is formed. The rate constants for reaction (57) were identical, within the stated experimental error, whether they were determined by following the disappearance of e_{aq}^- at 700 nm or whether they were obtained from the increase in the absorbance at 290 nm due to the formation of $HOC_6H_4\dot{N}O_2{}^-$. The determined value, $k_{57} = (3.5 \pm 0.6) \times 10^{10} \ M^{-1} \sec^{-1}$, indicates that e_{aq}^- reacts at the same rate with p-nitrophenol as with nitrobenzene ($k_{34} = 3.0 \times 10^{10} \ M^{-1} \sec^{-1}$). An electron can also be transferred to p-nitrophenol from an isopropanol radical ion. Only this species exists in a nitrous oxide-saturated solution (eq. (10)) of p-nitrophenol containing isopropanol (analog of eq. (24)). In the isopropanol–nitrous oxide system at low pH values, the protonation equilibrium (eqs. (58) and (59)) was observed. The pK for this radical ($\log k_{58}/k_{59}$) is 4.1, which is somewhat higher than that obtained for nitrobenzene (165). In alkaline solution $HOC_6H_4\dot{N}O_2{}^-$ is relatively stable; in an electrochemically generated solution of $HOC_6H_4\dot{N}O_2{}^-$, k_{62} was found to be 20 $M^{-1} \sec^{-1}$ (182). However, at pH 2.5 the radical anion, like the analogous nitrobenzene radical anion, is protonated and readily disproportionates forming p-nitrophenol and p-nitrosophenol; $k_{61} = 6 \times 10^9 \ M^{-1} \sec^{-1}$.

The paths of the reactions of hydrogen atoms and hydroxyl radicals with nitrobenzene and with p-nitrophenol are, however, considerably different. Substituted cyclohexadienyl radicals are formed in the case of the former (eqs. (48) and (49)) while the aromatic structure remains essentially intact in the latter case (eqs. (60) and (63)). The lack of absorption maxima characteristic of the cyclohexadienyl structure in the pulse-irradiated, nitrous oxide-saturated solutions of p-nitrophenol (174),† and the observed first-, rather than second-, order kinetics for the decay of the hydroxyl radical adduct corroborated these structural assignments (181). The rate constant for the reaction of hydroxyl radical with p-nitrophenol (eq. (63)) was determined by following the appearance of the transient at 290 nm; $k_{63} = (3.8 \pm 0.6) \times 10^9 \ M^{-1} \sec^{-1}$, a value only slightly lower than that obtained for the corresponding reaction of nitrobenzene (eq. (49)).

In acidic media the transient anion radical $HOC_6H_4\dot{N}O_3{}^-$ is protonated, and the pK for the dissociation of the protonated species, $\log k_{64}/k_{65}$ has

---

* For a detailed discussion of the processes involved in the pulse radiolytic reduction of the isomeric nitrophenols, consult ref. 181a.

† However, absorption maxima at 400 nm ($\epsilon = 1200$ at pH 5.1 and $\epsilon = 7600$ at pH 9.0 and 12.0) and at 310 nm ($\epsilon = 9200$ at pH 9.0 and 12.0) have been observed recently (182a).

been determined to be 5.3 $\pm$ 0.2 (181). Computer analysis of the data for the decay of $HOC_6H_4\dot{N}O_3{}^-$ (eq. (67)) and product analyses (183) were compatible with an intramolecular rearrangement to a dihydroxynitrocyclohexadienyl radical which in turn disproportionates forming p-nitrophenol and 3,4-dihydroxynitrobenzene (eqs. (67) and (68)). Oxygen reacts with the dihydroxynitrocyclohexadienyl radical (eq. (65)) presumably in a manner similar to its reaction with nitrobenzene (165) (eq. (50)). Since the pK of p-nitrophenol is 7.15, the extent to which reactions (57) and (63) are due to the reaction of the p-nitrophenolate ion rather than to p-nitrophenol is not known. However, this uncertainty concerning the ionic form of the reactive species does not invalidate the proposed reaction mechanisms. The differences between the reactivities of nitrobenzene and p-nitrophenol may originate from differences in their electron and charge distributions. Studies of other related systems will undoubtedly throw more light on this problem.

The oxidation and reduction reactions of quinone and hydroquinone involving radical intermediates have been investigated for some time (184–199). The role of the perhydroxy radical, $\cdot HO_2$, in the radiation-induced reactions of substituted quinones and hydroquinones has been extensively investigated (197–199). The G-values for the disappearance of the hydroquinones and those for the formation of the quinones have been determined spectrophotometrically together with $G(-O_2)$ and $G(H_2O_2)$ in air-saturated $0.8N$ sulfuric acid solutions of substituted hydroquinones (197,198). In air-saturated acidic aqueous solutions, the hydrated electron is converted to the perhydroxy radical (via eqs. (2) and (16)); this species, the hydroxyl radical, and hydrogen peroxide constitute the oxidizing agents present in irradiated air-saturated acidic aqueous solution. The experimental results indicated that the hydroxyl radical reacted quantitatively with these compounds but that the reactivity of the perhydroxy radical depended on the oxidation–reduction potential of the substituted hydroquinone (199). Subsequent to the formation of the oxidizing species, $\cdot HO_2$, $\cdot OH$, and H_2O_2 (eqs. (1), (2), and (16)), reactions (69)–(75) can be envisaged.

Four different mechanisms have been postulated to account for the radiation-induced oxidation of these substituted quinones and hydroquinones (199). In mechanism I, the reaction sequence is eqs. (1), (2), (16), and (69) followed by either eq. (70) or (71) and terminated by eq. (72). Using this reaction scheme good agreement was obtained between the calculated and observed yields for the radiation-induced oxidation of 2,5-dimethylhydroquinone and 1,2,4-trihydroxybenzene (197). In mechanism II, the perhydroxy radical oxidizes the semiquinone radical (eqs. (1),

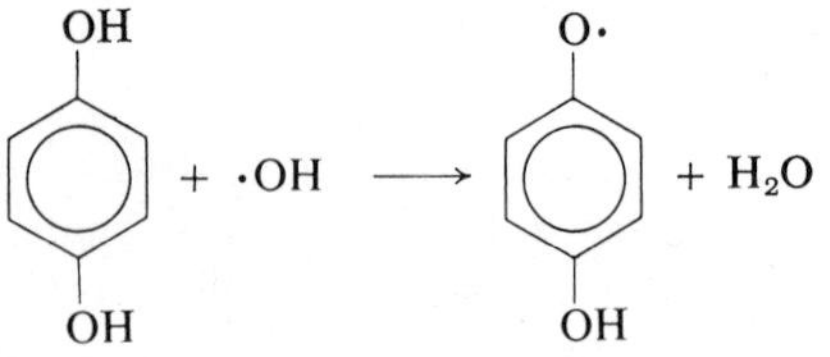

$$\cdot HO_2 + \cdot HO_2 \longrightarrow O_2 + H_2O_2 \qquad (73)$$

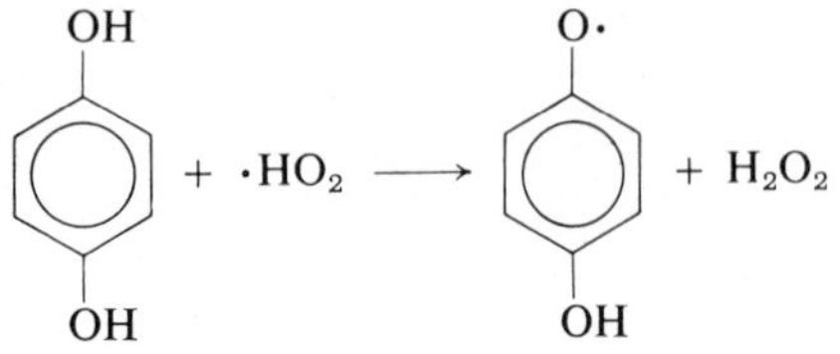

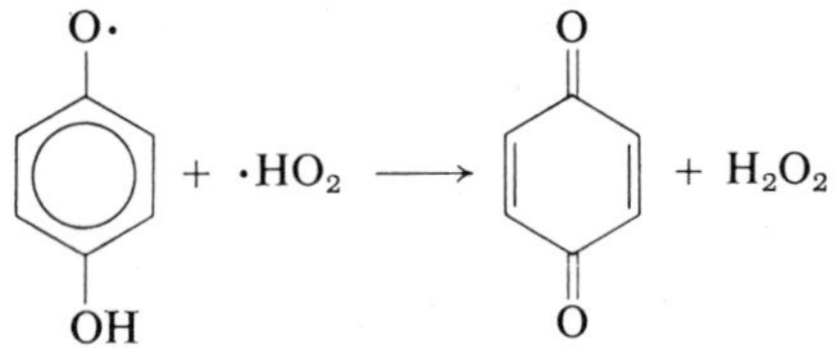

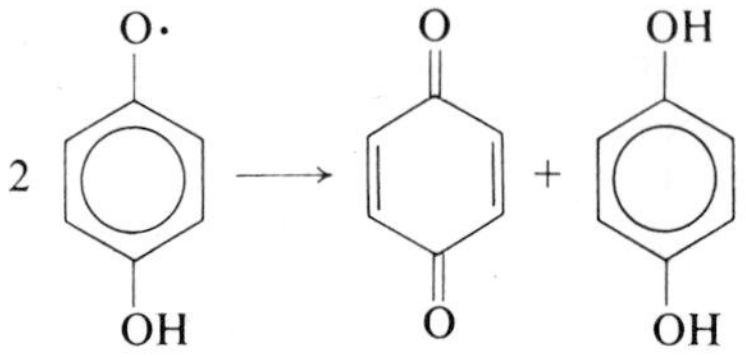

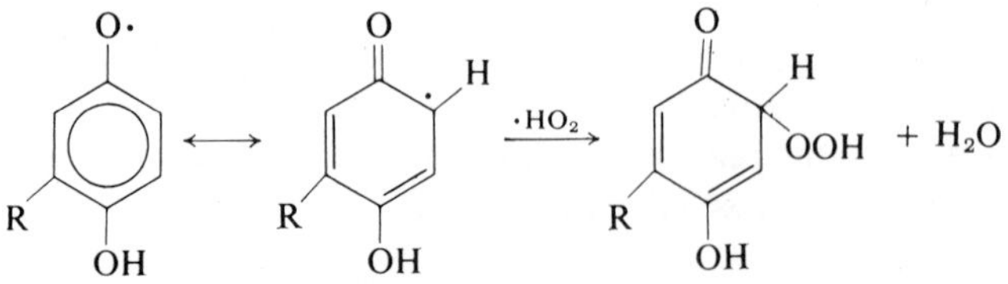

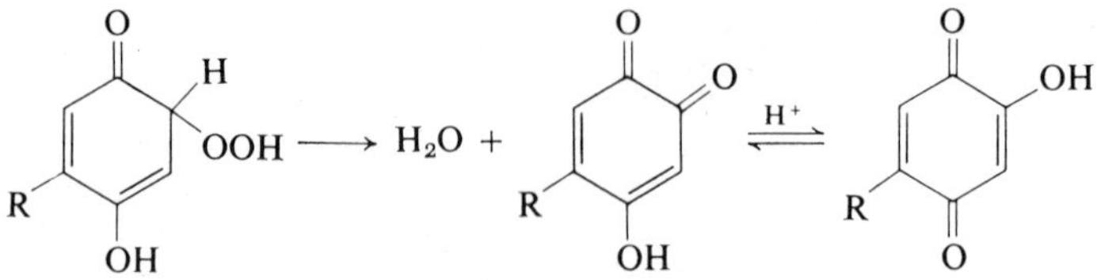

(2), (16), (69), and (71)) but is not a strong enough oxidizing agent to react with the hydroquinone, i.e., reaction (70) does not take place. The radiation-induced oxidation of 4-t-butylcatechol is an example of this case since the observed yields are compatible with the material balance equations derived from this mechanism (198). In mechanism III, the perhydroxy radical is unable to oxidize either the hydroquinone or the semiquinone radical, i.e., reactions (70) and (71) do not occur, and hence the steps subsequent to eqs. (1), (2), (16), and (69) are disproportionation of the perhydroxy radical and of the semiquinone radical (eqs. (72) and (73)). The reaction of 2,5-dichlorohydroquinone is an example of this mechanism (199). In mechanism IV the perhydroxy radical acts as a hydroxylating agent and the reaction sequence is eqs. (1), (2), (16), (69), (70), (72), (74), and (75). Since hydroquinone and its mono-substituted derivatives readily react with the perhydroxy radical, the extent of hydroxylation depends on the rates of these competing processes and the ratio of hydroquinone to semiquinone radical. The observed yields for hydroquinone, chlorohydroquinone, toluhydroquinone, and 1,2,4-trihydroxybenzene were found to be consistent with this mechanism (199). Additional research of this type would allow the accumulation of sufficient data to test the validity of a relationship between the oxidation–reduction potential of substituted hydroquinones and their radiolytic yields or their reactivity with the perhydroxy radical. Although the transient absorption of pulse radiolytically generated $\cdot HO_2$ and its anion, O_2^{-}, have been reported (200), no absolute rate constants are available for its reactions with dissolved organic compounds.

Pulse radiolytic studies of the interaction of $\cdot OH$ with hydroquinone and quinone in aqueous solution has, however, helped to elucidate the mechanisms of these oxidation–reduction processes (201). Chart III summarizes the reaction paths which have been studied. Since e_{aq}^- reacts considerably faster with quinone than with hydroquinone ($k_{76} = 1.25 \times 10^9 \ M^{-1} \sec^{-1}$ and $k_{83} < 10^7 \ M^{-1} \sec^{-1}$) (202), and since the reaction of the hydroxyl radical with quinone is an order of magnitude slower than with hydroquinone ($k_{84} = 1.2 \times 10^9 \ M^{-1} \sec^{-1}$ and $k_{80} = 1.2 \times 10^{10} \ M^{-1} \sec^{-1}$) (201), in an aqueous mixture of $10^{-3} M$ quinone and $10^{-2} M$ hydroquinone the single transient absorbing species 1 μsec after the pulse is the semiquinone radical, $HOC_6H_4\dot{O}$ (201). The absorption maximum for this species was found at 415 nm; this observation and assignment was confirmed by obtaining an identical spectrum from a pulse-irradiated solution of hydroquinone saturated with nitrous oxide (201) (eq. (10)). The semiquinone anion radical undergoes rapid protonation, and the pK for this process (log k_{77}/k_{78}) was found to be 4.0. Depending

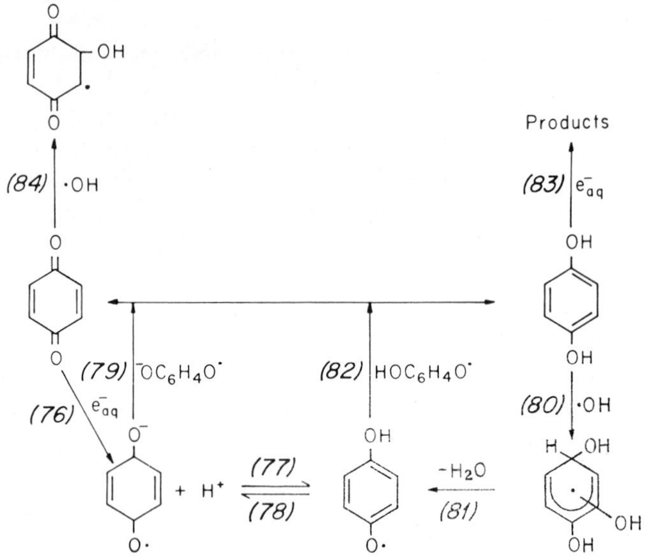

Chart III

on the pH of the solution, either two molecules of semiquinone radical (eq. (82)) or of its anion (eq. (79)) disproportionate to give quinone and hydroquinone ($k_{79} = 1.7 \times 10^8$ and $k_{82} = 1.09 \times 10^9 \ M^{-1} \sec^{-1}$). The hydroxyl radical adds to both quinone and hydroquinone; however, the latter reaction is followed by the elimination of a water molecule (eq. (81)) forming the semiquinone radical. Reaction (81) is both acid and base catalyzed and resembles that of the water elimination in the case of the dihydroxycyclohexadienyl radical formed by the addition of hydroxyl radical to phenol (eq. (55)) (141).

Using the titanous ion–hydrogen peroxide system, the oxidation of substituted phenylacetic acids was found to give different products depending on the pH of the solution (203). Using ESR techniques, it was found that the initial step of the reaction is the formation of the hydroxycyclohexadienyl radical (eq. (85)). The substituted hydroxycyclohexadienyl radical undergoes acid-catalyzed bond fission with the loss of H^+, CO_2, and OH^- forming the resonance-stabilized benzyl radical (eq. (86)). This elimination process is similar to that observed in the pulse radiolysis of phenol (141) and hydroquinone (201). The initial formation of hydroxycyclohexadienyl radicals in the reactions of benzoate (204) and salicylate

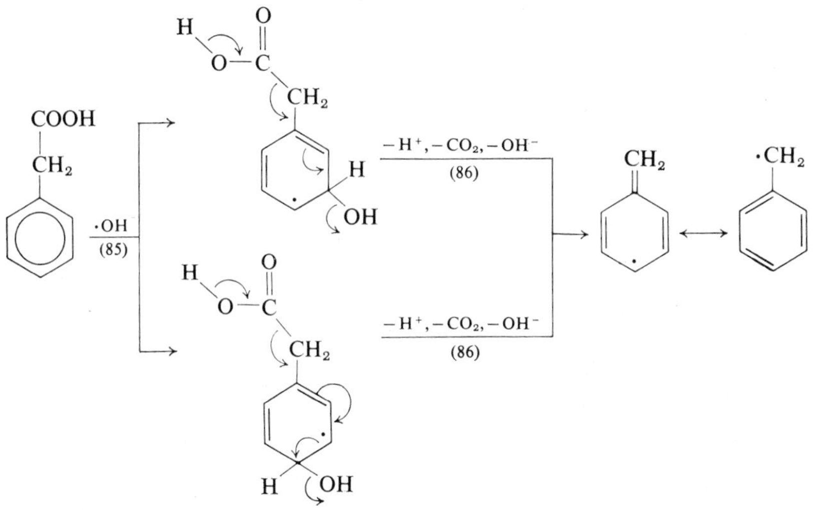

(205) ions with hydroxyl radical was verified in pulse radiolytic studies. However, details of the subsequent reactions have not been elucidated.

3. Reactions of the Hydrated Electron and Hydroxyl Radical with Nitromethane

The tautomeric and protic equilibria of the different forms of nitromethane (206,207) provide an interesting system for radiolytic oxidation–reduction studies. The hydrated electron reacts readily with both the nitro and the anionic form of nitromethane to form the nitromethane anion radical (eqs. (87) and (88) in Chart IV). By following the pseudo first-order decay of the hydrated electron at 720 nm, the rate constants $k_{87} = 2.1 \times 10^{10} \ M^{-1} \sec^{-1}$ and $k_{88} = 6.6 \times 10^9 \ M^{-1} \sec^{-1}$ were obtained and indicate that the undissociated nitromethane is somewhat more reactive than the anion (208). The structure of the product of these reactions, the nitromethane anion radical, was substantiated by the ESR spectra of *in situ* irradiated solutions of aqueous nitromethane (Fig. 10) (209). The absorption maximum of the nitromethane anion radical at 270 nm allows the observation of the subsequent reactions of this species. In acid solution this radical anion undergoes protonation (eq. (89)), and the pK (log k_{89}/k_{90}) of 4.4 for this equilibrium indicates that the nitromethane radical is a moderately strong acid. Both the neutral and anionic forms of the nitromethane radical readily transfer an electron to tetranitromethane (eqs. (91) and (92)). In this case, the dissociated form reacts more rapidly; $k_{91} = 1.2 \times 10^9 \ M^{-1} \sec^{-1}$ and $k_{92} = 1.1 \times 10^8 \ M^{-1} \sec^{-1}$ (208). However,

$$pK = 6.9$$

$$H_3C-N \underset{O}{\overset{O}{<}} \;\rightleftharpoons\; H_2C=N \underset{O}{\overset{OH}{<}}$$

$$pK = 10.2 \qquad\qquad pK = 3.2$$

$$(87)\,e_{aq}^- \qquad H^+ + H_2C = N \underset{O}{\overset{O^-}{<}} \xrightarrow[(93)]{\cdot OH} HO-CH_2-\dot{N} \underset{O}{\overset{O^-}{<}}$$

$$(88)\; e_{aq}^-, H_2O \qquad\qquad (94)$$

$$H^+ + H_3C - \dot{N} \underset{O}{\overset{O^-}{<}} \;\overset{(89)}{\underset{(90)}{\rightleftharpoons}}\; H_3C - \dot{N} \underset{O}{\overset{OH}{<}}$$

$$(91) \qquad\qquad (92)$$

$$C(NO_2)_4 \qquad\qquad C(NO_2)_4$$

$$? \qquad ? \; HOCH_2\dot{N}O_2^-$$

$$NO_2^- + HO\dot{C}H_2 \qquad HOCH_2NO_2 + CH_2NO_2^- + \bar{O}H$$

$$H_3CNO_2 + C(NO_2)_3^{\cdot -} + NO_2$$

Chart IV

both hydrogen atoms and hydroxyl radicals react very slowly with nitromethane $(k_{(CH_3NO_2 + \cdot OH)} < 5 \times 10^6 \; M^{-1} sec^{-1})$ (208) which suggests that hydrogen atom abstraction cannot occur readily with this species. The hydroxyl radical, however, readily *adds* to the C=N bond of the nitromethane anion (210) (eq. (93)). The rate constant for this reaction was measured by following the formation of the transient anion radical at 280 nm; $k_{93} = (8.5 \pm 1.5) \times 10^9 \; M^{-1} sec^{-1}$ (153). The structure of the product of this reaction, the hydroxynitromethane anion radical, was determined from its ESR spectrum (209) (Fig. 10). The transient hydroxynitromethane anion radical decays with mixed first- and second-order kinetics (eq. (94)). The precise nature of the first-order process is obscure, but it is probable that the anion radical decomposes forming the methanol radical, $\cdot CH_2OH$, and NO_2^-. By the addition of $NaClO_4$, the second-order rate constant was found to be linearly dependent on a function of the ionic strength of the solution. This dependence was interpreted to indicate that two species of like charge participate in the second-order decay process and that the hydroxynitromethane anion radical possesses a unit negative charge (210). The nature of the products formed in the second-order reaction is not known, but it is probable that disproportionation occurs leading to $HOCH_2NO_2$, $CH_2NO_2^-$, and $\bar{O}H$. Dimerization followed by elimination is also a possibility (210).

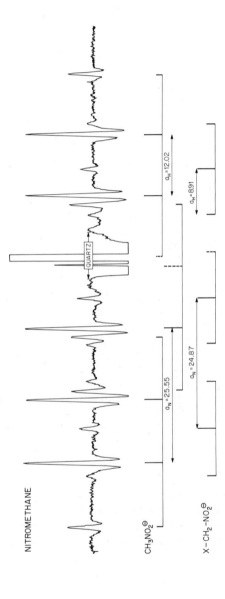

Fig. 10. Second-derivative ESR spectrum (normal double-modulation method) of $5 \times 10^{-3} M$ nitromethane solution, at pH 11.9, during irradiation. The spectra of $CH_3\dot{N}O_2^-$ and $XCH_2\dot{N}O_2^-$ ($X = HOCH_2$) are indicated. The magnetic field increases to the right. Taken from ref. 209.

4. Scope for Further Research

The radiation-induced oxidation and reduction of nitrobenzene, nitrosobenzene, phenylhydroxylamine, phenol, p-nitrophenol, hydroquinone, quinone, benzoate and salicylate ions, and nitromethane indicate the power of radiation chemistry in the elucidation of the mechanisms of these reactions. It is evident that rate and equilibrium constants which are unobtainable by other means can be measured with the use of pulse radiolysis. Obviously there are numerous other oxidation and reduction processes in organic chemistry which can be the subject of profitable research with the use of radiation chemistry; for example, the reactions of aliphatic and aromatic aldehydes and the oxidation and reduction of substituted anilines and nitriles. The primary requirement for undertaking such research is some knowledge of the chemistry of the system. In addition, if pulse radiolysis with spectrophotometric detection is contemplated, it is desirable to produce transients whose spectra are known, can be determined from independent sources such as flash photolysis studies, or can be predicted from spectral linear free-energy correlations. The use and combination of ESR,* polarography, and other techniques will undoubtedly lead to significant new areas of research.

D. Radiation-Induced Substitution of Aromatic Compounds in Dilute Aqueous Solutions

The effect of radiation on aromatic compounds in dilute aqueous solution has been studied for over 30 years (22). A dilute aqueous solution of benzene was suggested as a dosimeter (211,212), and research on the effects of X- and γ-rays on substituted aromatic compounds was initiated soon after World War II by Weiss and his school (22,213). However, the early work was inevitably hampered by the lack of analytical techniques sensitive enough to allow the optimal small percentages of radiolytic conversion and by the lack of basic knowledge on the radiation chemistry of water and aqueous systems. The development of the theory of aqueous radiation chemistry, the acceptance of the concept of the hydrated electron, the determination of valid G-values for the primary species, and the availability of pulse radiolysis for absolute rate constant determinations provided the necessary stimuli and techniques for vigorous work and bestowed respectability on quantitative mechanistic studies of radiation-induced reactions.

---

* The ESR spectra of a number of transient aliphatic and aromatic radicals have been observed in the *in situ* irradiation of the parent organic compounds in *dilute aqueous solutions* (210a).

1. Comparison of Radiation, Chemically, and Biochemically Induced Aromatic Substitutions

Free radical substitution of aromatic compounds in dilute aqueous solution is a common process in biological systems. Hydroxylation is the chemically induced radical substitution in water about which the most information is available (214). An aqueous mixture of ferrous sulfate and hydrogen peroxide, Fenton's reagent, is the best known hydroxylating agent. Equations (22) and (95)–(98) describe the proposed (215,216) reactions in this system.

$$Fe^{2+} + H_2O_2 \longrightarrow Fe^{3+} + \cdot OH + OH^- \tag{95}$$

$$Fe^{2+} + \cdot OH \longrightarrow Fe^{3+} + OH^- \tag{96}$$

$$\cdot OH + H_2O_2 \longrightarrow H\dot{O}_2 + H_2O \tag{22}$$

$$Fe^{3+} + H_2O_2 \longrightarrow Fe^{2+} + H\dot{O}_2 + H^+ \tag{97}$$

$$Fe^{3+} + H\dot{O}_2 \longrightarrow Fe^{2+} + H^+ + O_2 \tag{98}$$

The available chemical and spectroscopic evidence (152,217–219) indicates the presence of free radicals in this system. Recent ESR studies strongly suggest that the hydroxylating agent is actually the hydroxyl radical complexed to a metal ion (220–223). Similarly in other chemical hydroxylating agents such as Hamilton's system (224) (aqueous H_2O_2 at pH 4 in the presence of catalytic amounts of ferric ion and catachol), Udenfriend's system (225) (aqueous ferrous ion, EDTA, ascorbic acid, and oxygen), and other systems of this type, it is very probable that the hydroxyl radical is complexed to the metal ion. There have been several studies involving the enzymatic hydroxylation (226–228) of aromatic compounds, but due to the complexities of these systems the nature of the hydroxylating agents has not been established. Conversely, radiation-induced hydroxylation is a relatively simpler system since in this case the generation of hydroxyl radicals *in situ* from water obviates the addition of chemical or biochemical reagents to the system. Indeed, in radiation-induced hydroxylation studies the importance of oxygen became apparent, a fact which was largely neglected in chemical hydroxylations.

Free-radical nitration by $\cdot NO_2$ or $\cdot NO_3$ has been studied by Russian workers (229), and the literature contains numerous references to free-radical halogenation and alkylation (230). However, the mechanisms of these reactions are inevitably complex and little progress has been made toward their elucidation. As in the case of hydroxylation, the radiation-induced substitutions, in principle, present a simpler system. It is certainly

possible to design experiments for radiation-induced nitration, halogenation, and indeed for other free-radical substitutions and to study the rates
of formation and decomposition of transients by pulse radiolysis and the
yield of the products, i.e., their G-values, by sensitive analytical techniques.
However, very little research has been conducted on the radiation-induced
nitration and chlorination of aromatic compounds and no work has been
done on the other substitutions.

2. Radiation-Induced Hydroxylation

Table V summarizes the available G-values for the formation of phenol
and substituted isomeric phenols in irradiated dilute aqueous solutions of
monosubstituted aromatic compounds. As previously mentioned, the lack
of sufficient knowledge on the radiation chemistry of water and the unavail-
ability of absolute rate constants for the primary species and of sensitive
analytical techniques prevented the early workers from optimizing their
conditions. In air or oxygen-saturated solutions containing a low concentration of an aromatic substrate, the hydrated electron preferentially reacts
with oxygen (eq. (9)) and thus the primary chemical species present is the
hydroxyl radical. If hydroxylation of an aromatic compound results in the
stoichiometric formation of monosubstituted phenols, their yield should,
of course, be 2.4, the G-value for the formation of hydroxyl radical in
irradiated water. A glance at Table V reveals that this situation exists only
for benzene in which case the yield of phenol in an air-saturated solution
is 2.3–2.7 (58,213,231,233,234). The yields of monosubstituted phenols in
irradiated aqueous solutions of all the substituted aromatic compounds,
C_6H_5X, listed in Table V are, however, considerably less than those ex-
pected if the reaction $C_6H_5X + \cdot OH \rightarrow X\dot{C}_6H_5OH \rightarrow XC_6H_4OH$ occurred
quantitatively. In addition, the yield of phenol in some cases depended
markedly on the absorbed *dose rate* (82,235,240) and post-radiolytic
processes were also noted (82). More significantly, oxygen must have roles
in addition to converting the hydrated electron into $\cdot O_2^-$ (eq. (9)) since the
total yields of the isomeric phenols and their distribution are considerably
different in the presence and in the absence of oxygen. (Compare the
G-values for the formation of monomeric phenols in N_2O saturated solutions with those in a mixture of N_2O and O_2.) Clearly the radiation-
induced hydroxylation of aromatic compounds in dilute aqueous solutions
involves more than a stoichiometric conversion of hydroxyl radical to
monosubstituted phenols. The presently available knowledge on the detailed mechanism of hydroxylation is rather scant; nevertheless, an attempt
is made here to discuss the present status in the hope that it will stimulate
further work in this field.

Whether oxygen is present or absent, the first step in the radiation-induced hydroxylation of monosubstituted aromatic compounds is the addition of hydroxyl radical to the aromatic ring forming a substituted hydroxycyclohexadienyl radical (eq. (29)). Pulse radiolytic measurements of the absorption spectra of substituted hydroxycyclohexadienyl radicals and their rates of formation (157) together with the lack of deuterium isotope effects in the reactions of $\cdot OH$ with C_6H_6 and C_6D_6 (82) and with $C_6H_5NO_2$ and $C_6D_5NO_2$ (158) amply substantiate this postulated mode of $\cdot OH$ attack. The subsequent fate of the substituted hydroxycyclohexadienyl radical is, however, considerably less clear and is dependent on the reaction conditions. For the sake of convenience the reactions of substituted hydroxycyclohexadienyl radicals in the presence and in the absence of oxygen are discussed separately.

When excess oxygen is present in the solution, it not only reacts with the hydrated electron (eq. (9)) but also with the substituted hydroxycyclohexadienyl radical (82,162) (eq. (30)). It is not unreasonable to expect that the disproportionation of the substituted hydroxycyclohexadienyl radical will be different from that of its oxygen adduct. This difference qualitatively accounts for the different yields of monosubstituted phenols and the different isomeric distributions in the presence and in the absence of oxygen. Using a high pulse intensity, ca. 36×10^{-18} eV per pulse, in the irradiation of a dilute aqueous air-saturated solution of benzene, the yield of phenol obtained was 0.19 (82). With decreasing dose rate the phenol yield increased. This dose-rate dependence for phenol formation is illustrated in Figure 11. Such dose-rate dependences of the yields in aqueous radiation

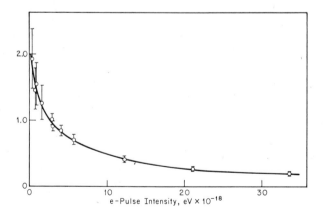

Fig. 11. Phenol yield as a function of pulse intensity. Taken from ref. 82.

TABLE V

Radiation-Induced Hydroxylation of Aromatic Compounds in Dilute Aqueous Solution

Compound	Conditions	G(Phenol)	G(o-Phenol)	G(m-Phenol)	G(p-Phenol)	G(Other products)	Ref.
C_6H_6	Deaerated, unbuffered	0.37				1.3, biphenyl	58
		0.36				1.2, biphenyl	231
		0.41				1.2, biphenyl	213
		0.30				0.58, biphenyl	232
	Air sat., unbuffered	2.1				0, biphenyl	58
		2.64				0, biphenyl	231
		2.3					213
		2.74					233
		2.63					234
		1.9[a]					82
	O_2 sat., pH 5.4	1.87[a]				0.84, mucondialdehyde	235
C_6H_5Cl	Air sat., pH 6	Yes[b]	15–20%[c]	20–25%[c]	50–60%[c]	Biphenyl, Cl^- [c]	179
	? ?					0.53, Cl^-	236
C_6H_5Br	? ?					0.90, Br^-	236
C_6H_5F	Air sat., pH 6.5	0.05	None	0.20	0.32	F^- [b]	180
	N_2O sat., pH 6.5	0.06	None	0.06	0.05	F^- [b]	180
	$N_2O + O_2$ sat., pH 6.5	0.10	None	0.38	0.60	F^- [b]	180
C_6H_5OH	Deaerated, pH 6		0.077	None	0.15		237
$C_6H_5OCH_3$	Argon sat., pH 6.5	0.46	0.07	None	None		238
	Air sat., pH 6.5	0.11	0.19	None	0.32		238
	N_2O sat., pH 6.5	0.28	0.54	None	0.14		238
	$N_2O + O_2$ sat., pH 6.5	0.23	0.38	None	0.59		238
$C_6H_5NO_2$	Air sat., pH 6	Yes[b]	35.5%[c]	29%[c]	35.5%[c]	NO_3^- [b]	239
	Air sat., pH 6.5	Yes[b]	0.48	0.48	0.64	0.5, NO_3^-	158

Argon sat., pH 6.5	Yes[b]	0.13	0.13	0.49		158
N$_2$O sat., pH 6.5	Yes[b]	0.37	0.15	0.67		158
O$_2$ sat., pH 6.5	Yes[b]	0.52	0.44	0.38		158
Air sat., pH 5.5		0.66[a,b]	0.42	0.42		240
N$_2$ sat., pH 5.5		0.12	0.08	0.22		240
130 atm O$_2$, pH 5.5		0.71				240
C$_6$H$_5$COOH/C$_6$H$_5$COO$^-$						
O$_2$ sat., pH 3		29.4%[c]	11.8%[c]	59%[c]		241
Air sat., unbuffered	0.05	0.74	0.42	0.33		242
Air sat., pH 6.5		0.67	0.37	0.37	0.73, CO$_2$	243
O$_2$ sat., pH 7.0		0.49	0.31	0.39	0.73, CO$_2$	244
N$_2$ sat., pH 7.0		0.03	0.04	0.13		244
N$_2$O sat., pH 7.0		0.11	0.12	0.22		244

[a] Yield depends on dose rate.
[b] Unspecified amount determined.
[c] Percentage of monohydroxylated product; total quantity unspecified.

chemistry are generally interpreted in terms of competing first- and second-order processes. The preferential formation of phenol at low dose rates was rationalized in terms of a predominance of a first-order reaction of the oxygen adduct of the hydroxycyclohexadienyl radical giving phenol (eq. (99)) over the second-order formation of peroxides and possibly hydroperoxides (82) (eqs. (100) and (101)).

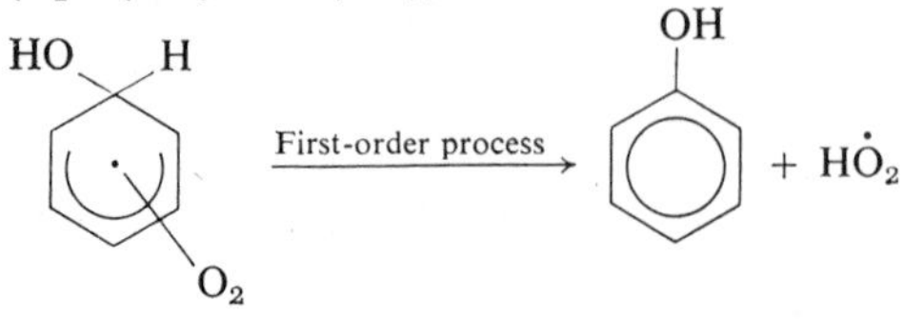

$$\text{First-order process} \longrightarrow \quad + \ \text{H}\overset{\cdot}{\text{O}}_2 \tag{99}$$

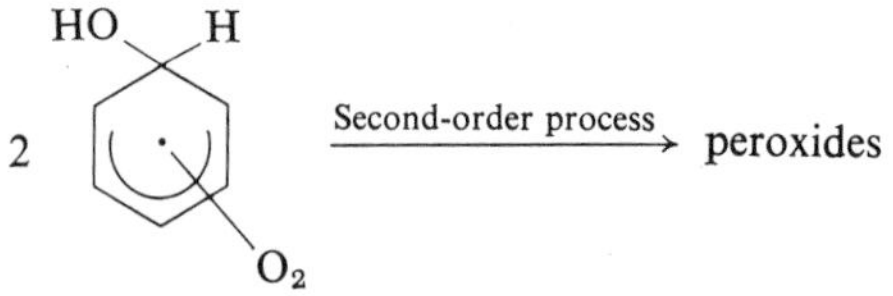

$$2 \quad \xrightarrow{\text{Second-order process}} \text{peroxides} \tag{100}$$

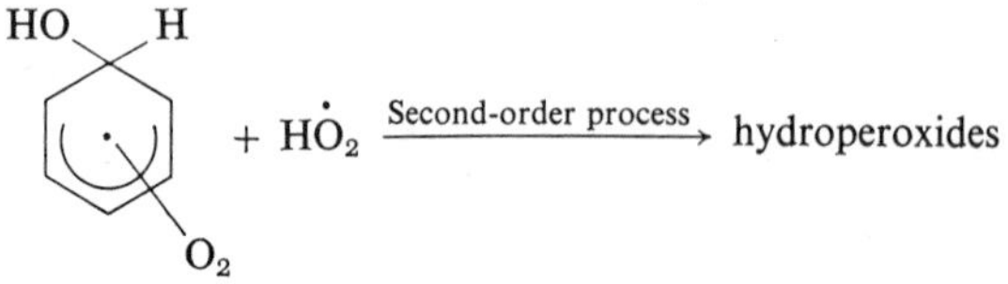

$$+ \ \text{H}\overset{\cdot}{\text{O}}_2 \xrightarrow{\text{Second-order process}} \text{hydroperoxides} \tag{101}$$

The formation of organic peroxides was indicated by chemical analysis (82). A similar dose-rate dependence was observed for the formation of o-nitrophenol in irradiated, air-saturated, aqueous solutions of nitrobenzene at pH 5.5 (240). In both cases it is clear that maximum formation of phenols occurs at low dose rates, i.e., eq. (99) or its analogs predominate; and therefore, in radiation-induced hydroxylation studies, it is preferable to use the smallest possible dose rate.

The formation of the oxygen adduct of the substituted hydroxycyclohexadienyl radical (eq. (30)) depends greatly on the aromatic substituent. For example, the rate constant for the reaction of oxygen with the nitro-substituted hydroxycyclohexadienyl radical is slower by a factor of 200 than that with the unsubstituted hydroxycyclohexadienyl radical (162). It is probable that not only the formation of this adduct but also its decomposition is dependent on the substituent present. Presumably in radiation-induced hydroxylation in air or oxygen-saturated solution, reaction (99) is more important than the second-order processes for unsubstituted aromatic compounds as compared to substituted ones. Indeed, the low yield of XC_6H_4OH in irradiated aqueous C_6H_5X compared to that of C_6H_5OH in irradiated aqueous C_6H_6 supports this conclusion. The use of considerably lower dose rates and complete pulse radiolytic analysis will undoubtedly throw more light on the intricacies of these processes.

The addition of $1 \times 10^{-4}M$ potassium dichromate to an aqueous, air-saturated solution of nitrobenzene increased the total yield of nitrophenols to a G-value of 2 (240). This effect could possibly be due to a rapid reaction of the dichromate with the nitro-substituted hydroxycyclohexadienyl radical forming a complex which, in turn, could produce nitrophenols by a simple disproportionation or by other reactions. The addition of $K_3Fe(CN)_6$ to nitrous oxide-saturated aqueous solutions containing benzoic acid was found to increase the total yield of the three isomeric hydroxybenzoic acids to 5.4 $\pm$ 0.5/100 eV (245). A mechanism in which the substituted hydroxycyclohexadienyl radical, formed via eqs. (1), (10), and (29), is oxidized by the ferricyanide ion was postulated (245). This mechanism was found to be in good agreement with the material balance equations, $G(\text{hydroxybenzoic acids}) + G(CO_2) = G_{OH} + G_{e_{aq}^-} + G_{H_2O_2} = 5.9$ and $G(-Fe(CN)_6{}^{3-}) = G_{OH} + G_{e_{aq}^-} + G_H = 5.8$ (245). The unexpected invariance in $G(\text{hydroxybenzoic acids})$ and in $G(CO_2)$ and the dramatic changes in the isomer distribution as a function of the $K_3Fe(CN)_6$ concentration was rationalized in terms of the differences in the composite rate constants for reaction (29) and for the oxidation of the substituted hydroxycyclohexadienyl radical by ferricyanide ion at the different positions of the aromatic ring. Pulse radiolytic data and product analyses under comparable conditions would allow a better insight into these types of hydroxylation.

In the absence of oxygen in N_2O-saturated solutions the hydrated electron is, in most cases, effectively scavenged forming an additional hydroxyl radical (eq. (10)). Under these conditions the yield of phenols should equal $\frac{1}{2}(G_{e_{aq}^-} + G_{OH})$. A yield of phenol approaching a G-value of 5 was found in irradiated, dilute, aqueous solutions of benzene saturated with nitrous oxide (180). However, the total yields of substituted isomeric monophenols are considerably less than this G-value. These different yields imply that substituted and unsubstituted hydroxycyclohexadienyl radicals disproportionate by different mechanisms. The formation of substituted phenols can be envisaged to occur by the disproportionation of two substituted hydroxycyclohexadienyl radicals (eq. (102)).

TABLE VI

Isomer Distribution of Nitrophenols and Methoxyphenols and Partial Rate Factors
for the Hydroxylation of Nitrobenzene and Anisole

	Isomer distribution, %			Partial rate factors[a]		
	o	m	p	o^R_f	m^R_f	p^R_f
Nitrobenzene						
Radiation induced[b]	51	31	18	1.80	1.10	1.91
Radiation induced[c]	40	21	39	1.42	0.71	2.76
Radiation induced[d]	50	31	19	1.77	1.10	1.34
Radiation induced[e]	30	30	60	1.06	1.06	4.25
Fenton's reagent[f]	24	30	46	0.85	1.06	3.26
Hamilton's reagent[f]	30	41	29	1.06	1.45	2.05
Anisole						
Radiation induced[g]	37	0	63	0.93	—	3.15
Radiation induced[h]	39	0	61	0.96	—	3.06
Radiation induced[i]	80	0	20	2.01	—	1.02
Fenton's reagent	86	0	14	2.16	—	0.70
Ferric-catechol system[f]	64	3	33	1.61	0.08	1.66
Ferric-hydroquinone system[f]	65	<5	35	1.64	0.12	1.76
Udenfriend's system[f]	88	0	12	2.20	—	0.60

^{a} $o^R_f = [k_{(C_6H_5X + \cdot OH)}/\frac{1}{6}k_{(C_6H_6 + \cdot OH)}] \cdot (\% \ o\text{-isomer}/200)$;
  $m^R_f = [k_{(C_6H_5X + \cdot OH)}/\frac{1}{6}k_{(C_6H_6 + \cdot OH)}] \cdot (\% \ m\text{-isomer}/200)$;
  $p^R_f = [k_{(C_6H_5X + \cdot OH)}/\frac{1}{6}k_{(C_6H_6 + \cdot OH)}] \cdot (\% \ p\text{-isomer}/100)$.
^{b} $[C_6H_5NO_2] = 5 \times 10^{-4}M$, $[N_2O] = 1.45 \times 10^{-2}M$, $[O_2] = 2.3 \times 10^{-4}M$; ref. 158.
^{c} $[C_6H_5NO_2] = 5 \times 10^{-4}M$, $[N_2O] = 2.0 \times 10^{-2}M$; ref. 158.
^{d} $[C_6H_5NO_2] = 5 \times 10^{-4}M$, air saturated; ref. 158.
^{e} $[C_6H_5NO_2] = 7 \times 10^{-3}M$, air saturated; ref. 158.
^{f} Refs. 214 and 246.
^{g} $[C_6H_5OCH_3] = 2.0 \times 10^{-4}M$, $[O_2] = 2.30 \times 10^{-4}M$; ref. 238.
^{h} $[C_6H_5OCH_3] = 2.0 \times 10^{-4}M$, $[N_2O] = 1.45 \times 10^{-2}M$, $[O_2] = 2.3 \times 10^{-4}M$; ref. 238.
^{i} $[C_6H_5OCH_3] = 2.0 \times 10^{-4}M$, $[N_2O] = 2.00 \times 10^{-2}M$; ref. 238.

Since two molecules of the substituted hydroxycyclohexadienyl radical
give only one molecule of the substituted phenol, its yield should be only
$\frac{1}{2}(G_{e_{aq}^-} + G_{OH})$. Furthermore, the formation of dimers (eq. (103)) competes
with the disproportionation (eq. (102)), thereby further reducing the yield
of phenols.

$$(103)$$

The total yield of isomeric nitrophenols, $G = 0.65$, in a nitrous oxide-saturated aqueous solution containing $5.0 \times 10^{-4}M$ nitrobenzene (158) in agreement with pulse radiolysis data (142) substantiates this argument.

Any assumption that the isomeric distribution of monosubstituted phenols reflects the extent of hydroxyl radical attack at the different ring positions is clearly a gross oversimplification in the light of the foregoing discussion. The partial rate factors, however, indicate some degree of selectivity (Table VI) which is a reflection of the partial electrophilic character of the hydroxyl radical.

A qualitative comparison of the partial rate factors for chemical and radiation-induced hydroxylations of nitrobenzene and anisole (Table VI) tends to indicate that they follow different mechanisms, and therefore the attacking species for the former is not likely to be a simple, uncomplexed hydroxyl radical. Alternatively it might indicate that in chemical hydroxylations oxygen may be rapidly depleted, and thus the hydroxylation mechanism resembles that observed for the radiation-induced hydroxylation in nitrous oxide-saturated solutions.

3. Radiation-Induced Nitration

The radiation-induced nitration of aromatic compounds has been observed and has been studied for benzene (247–250), benzoic (248) and salicylic (248) acids, and phenol (251). It usually involves the irradiation of the aromatic compound in a dilute aqueous solution containing 0.5–1.0M sodium or potassium nitrate or nitric acid. Depending on the conditions employed the products are phenol, nitroaromatics, and nitrophenols (see Table VII). In most cases, a fruitful mechanistic discussion of these results is grossly hampered by the high absorbed doses and conversions employed (247–251), the ambiguous analytical techniques used, and insufficient knowledge of the chemistry of the system. Indeed, the radiation chemistry of aqueous nitrates is incompletely understood (252–255), and at the concentration of nitrate ion ($> 0.5M$) employed in these radiation-induced reactions, concurrent direct and indirect effects of energy dissipation have to be considered. Pulse radiolytic studies have demonstrated the presence of nitrate radical, $\cdot NO_3$, in irradiated aqueous solutions containing $0.5M$ nitrate ion (254,256), and eqs. (104) and (105) were proposed for its formation (254).

$$NO_3^- \rightsquigarrow (\cdot NO_2 + O + e^-)_{\text{cage,spur}} \tag{104}$$

$$\cdot NO_2 + O \longrightarrow \cdot NO_3 \tag{105}$$

The role of the hydroxyl radical and the nature of the interactions between the species produced by direct and indirect processes, however, remain to be elucidated.

TABLE VII

Radiation-Induced Nitration of Aromatic Compounds

Compound	Conditions	G(Nitrobenzene)	G(Nitrophenol)			G(Phenol)	G(Other products)	Ref.
			o	p	$o + p$			
Benzene	0.5M NaNO$_3$, O$_2$ sat., pH 2	—	—	—	1.5	1.5	0.8, NO$_2^-$	250
Benzene	0.5M NaNO$_3$, O$_2$ sat., pH 5	—	—	—	—	1.0	1.2, NO$_2^-$	250
Benzene	0.5M NaNO$_3$, N$_2$ sat., pH 2	0.2	—	—	2.1	2.0	0.8, NO$_2^-$	250
Benzene	0.5M NaNO$_3$, N$_2$ sat., pH 5	—	—	—	0.7	1.4	1.2, NO$_2^-$	250
Benzene	0.2–1.7M HNO$_3$, air sat.	Yes[a]	—	0.2–1.0[b]	—	None	—	249
Nitrobenzene	HNO$_3$(M?), air sat.	—	—	None	—	—	—	249
p-Nitrophenol	0.5M NaNO$_3$, air sat. (?)	—	—	—	—	—	0.6, Dinitrophenol	250
Phenol	0.5M NaNO$_3$, air sat., pH 2	—	—	—	1.2	—	—	250
Benzene	1.0M NaNO$_3$, air sat.	None	0.41	—	—	—	—	251
Phenol	1.0M NaNO$_3$ + 0.05M H$_2$SO$_4$, air sat.	None	0.68	—	—	—	—	251
Phenol	1.0M NaNO$_3$ + 0.5M NaOH, air sat.	None	0.38	—	—	—	—	251

[a] The presence of an unspecified amount qualitatively determined.
[b] Dependent on dose rate and HNO$_3$ concentration.

The absence (251) or the small yield (250) of nitrobenzene in irradiated aqueous solutions of benzene containing nitrate ion and the fact that phenol could but nitrobenzene could not be nitrated radiolytically (Table VII) suggests that phenol is formed initially and is subsequently nitrated. Alternatively it is conceivable that the ease of nitration of phenol is considerably greater than that of benzene or of nitrobenzene. At higher doses secondary processes take place and result in the formation of dinitrobenzene, dinitrophenols, and picric acid (250).

When the radiation chemistry of concentrated aqueous nitrate solutions is completely understood, profitable research can certainly be pursued in this field by the judicious combination of pulse radiolysis with steady-state irradiation at low doses.

4. Radiation-Induced Carboxylation and Decarboxylation

The radiation-induced formation of amino acids in aqueous solutions of amines containing carbon dioxide has been reported (257). The suggested mechanism involved the abstraction of an amine hydrogen atom, usually from an α-carbon atom, concurrent reduction of carbon dioxide by e_{aq}^- and $\cdot H$ forming the carboxylate radical ion, $\cdot COO^-$, and the carboxylate radical, $\cdot COOH$, and subsequent carboxylation of the amine (257). In addition, the formation of cystine together with a host of other amino acids in aqueous solutions of ethylamine containing sodium bicarbonate and sodium sulfite has been reported (258). However, the complexity of these systems clearly does not allow a critical mechanistic discussion at the present time, but undoubtedly this potentially important field will receive considerable attention in the near future.

The irradiation of aromatic carboxylic acids in dilute aqueous solutions is an interesting case since the products formed indicate that both hydroxylation and decarboxylation occur. Furthermore, more than 80% of the G_{OH} resulted in the formation of isomeric hydroxybenzoic acids and carbon dioxide ($G(HOC_6H_4COOH) + G(CO_2) = 2.22$, Table V); and owing to the slow post-radiolytic increase in the carbon dioxide yield (259), these values represent an even greater stoichiometric conversion. Since the radiation-induced decarboxylation of benzoic (204) and salicylic (205) acids has been demonstrated to occur in a step subsequent to the addition of the hydroxyl radical to the aromatic ring (see Section III-C-2), a study of decarboxylation exclusively does not seem to be possible. Nevertheless, the effects of ring substituents on the rate of decarboxylation and the ratio of carbon dioxide to hydroxylated carboxylic acids could well be the subject of a worthwhile investigation. Using C-14 labelled benzoic acid as a hydroxyl radical scavenger, the rates of the reactions of the hydroxyl radical with a number of solutes have been determined competitively by measuring the $^{14}CO_2$ (259).

5. Scope for Research on Other Radiation-Induced Substitution Reactions

Radiation-induced halogenation, sulfation, sulfonation, and possibly amination and alkylation at an aromatic carbon atom undoubtedly will become the subject of profitable investigations. Obviously the necessary prerequisite for such research is a thorough understanding of the radiation chemistry of the system in which substitution occurs.

Pulse radiolytic work on aqueous solutions of sodium chloride has demonstrated the presence of the chlorine radical anion, $\cdot Cl_2^-$, in this system (260). This species, which has a broad absorption maximum in the 340–360 nm region and an extinction coefficient of $12,500 \pm 1000$ at 340 nm (260), is formed by the reaction of the hydroxyl radical with the chloride ion (eqs. (106) and (107)).

$$\cdot OH + Cl^- + H_3O^+ \longrightarrow \cdot Cl + 2H_2O \tag{106}$$

$$\cdot Cl + Cl^- \longrightarrow \cdot Cl_2^- \tag{107}$$

The rate constant for the third-order process, k_{106}, is $(1.5 \pm 0.3) \times 10^{10}$ $M^{-2} sec^{-1}$ and is invariant from pH 0.8–3.4 and from 3×10^{-4} to $1 \times 10^{-1} M$ sodium chloride (261). The reaction given in eq. (107) is very rapid compared to reaction (106) (260). When the hydrated electron was removed by saturating the solution with nitrous oxide (eq. (10)), the decay of $\cdot Cl_2^-$ (eq. (108)) at pH 3.0 was observed and was found to obey second-order kinetics.

$$\cdot Cl_2^- + \cdot Cl_2^- \longrightarrow 2Cl^- + Cl_2 \tag{108}$$

k_{108} was determined to be $(1.4 \pm 0.3) \times 10^{10} M^{-1} sec^{-1}$ (261). Therefore, the radiation-induced chlorination of aromatic compounds can be studied provided that the concentration of the aromatic solute is chosen so

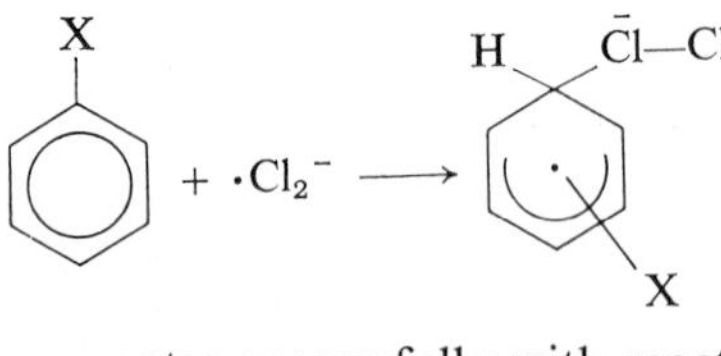

$$\tag{109}$$

that reaction (109) competes successfully with reaction (108) and so that the formation of $\cdot Cl_2^-$ is not prevented by removal of the hydroxyl radical (eq. (29)). Thus k_{109} cannot be less than ca. $5 \times 10^6 M^{-1} sec^{-1}$. By measuring the rate of disappearance of $\cdot Cl_2^-$ at its absorption maximum of 340 nm, $\varepsilon = 12,500 \pm 1000$ (260), in the presence of aromatic solutes in a manner analogous to that used for the measurement of the e_{aq}^- concentration decrease at 700 nm, pulse radiolytic rate constants for the reactions described by eq. (109) can be obtained easily. At the present time, such rate constant determinations have only been carried for thymine, uracil, cytosine, adenine, guanine, thymidylic acid, deoxycytidylic acid, deoxyadenylic acid, and deoxyguanylic acid (261). The rate constants for the reactions of $\cdot Cl_2^-$ with these compounds were found to be 20–200 times smaller than those for the corresponding hydroxyl radical reactions (261). Additional rate constants for the reactions of $\cdot Cl_2^-$, and the elucidation of the position of $\cdot Cl_2^-$ attack on substituted aromatic solutes and of the subsequent reactions of the transients formed would be extremely useful for a more complete understanding of the mechanism of radiation-induced chlorination.

The existence of the bromine radical anion, $\cdot Br_2^-$, has also been verified by pulse (262,263) and steady-state (264) radiolysis. The aqueous chemistry of this species, which has an absorption maximum at 360 nm, has been found to be similar to that of the chlorine radical anion in some respects. In addition, the iodine radical anion, $\cdot I_2^-$, has been observed by the flash photolysis of iodine solutions (265). Hence, a judicious selection of experimental conditions would allow pulse and steady-state radiolytic investigations of the radiation-induced bromination and iodination of aromatic compounds in dilute aqueous solutions. However, no such studies have been reported to date.

Investigations of radiation-induced halogenation and other substitutions of aromatic compounds in aqueous solution is limited only by the ingenuity of the chemist. Rapid progress in this field is almost a certainty and will allow detailed mechanistic investigations of aromatic substitutions for which no chemical analog exists.

E. Radiation-Induced Bioorganic Reactions

A considerable amount of work in the field of aqueous radiation chemistry has been motivated by a desire to gain an understanding of the mechanisms of radiation-induced chemical changes in living systems. Large organic molecules such as steroids, carbohydrates, amino acids, peptides, proteins, enzymes, vitamins, and hormones obviously play an important role in the basic functions of living organisms, and therefore, a great deal of attention has been devoted to the study of their radiation chemistry (18). The complexity of these large molecules inevitably necessitated an approach to their radiation chemistry which differs from that usually employed for simpler molecules (see Sections III-C-1 and 4 and III-D-2 and 3). The applicability of the theories of aqueous radiation chemistry in general and the role of the hydrated electron in particular to biological systems *in vivo* has been discussed (118). In these systems the attention has been largely limited to the rate at which some property or biological function changes as a function of the absorbed radiation. This type of approach is primarily the domain of radiation biology (11,266–272) and is clearly outside the scope of the present review. In recent years, the use of pulse radiolysis as an adjunct to carefully designed steady-state radiolysis experiments with sensitive analytical techniques resulted in considerable progress toward an understanding of the radiation chemistry of relatively simple bioorganic compounds in dilute aqueous solutions. Of these compounds, the simplest and the most completely understood are the amino acids and the pyrimidine bases; hence, their radiation chemistry is discussed in the following section to illustrate the potentialities of this approach for future research on more complex bioorganic systems.

1. Radiation Chemistry of Simple Amino Acids in Aqueous Solutions

Simple α-amino acids, $H_2NCH(R)COOH$, in aqueous solution are amphoteric; and depending on the conditions, reactions can occur at either the amino or the carboxyl group. The radiation chemistry of simple amino acids such as glycine and alanine has received a great deal of attention in the last 20 years (273,274). The observed products included ammonia arising from both oxidative and reductive deamination, hydrogen, carbon dioxide, aldehydes, ketones, and carboxylic acids. Their yields, however, were dependent to a very great extent on the concentration of the amino acid, the presence or absence of oxygen or other additives, the pH of the solution, the absorbed dose, and the dose rate (Table VIII). Unfortunately a comprehensive compilation of the G-values for the different products formed in irradiated aqueous amino acid solutions (see ref. (62) for a partial list) is somewhat meaningless since the experimental conditions are often incomparable or incompletely reported (62). Nevertheless, due to the accumulation of data on the radiation chemistry of simple α-amino acids, some mechanistic interpretation of the reactions involved is now possible. Since the radiation chemistry of amino acids (273) and of organic nitrogen compounds (274) has been reviewed, attention is focused in the ensuing discussion on the elucidated reaction paths and on research reported subsequent to these reviews.

Chart V summarizes the proposed reaction mechanisms. By following the rate of the hydrated electron disappearance pulse radiolytically at 700 nm in the presence of dissolved amino acids, the second-order rate constants for eq. (110) were found to be of the order of 10^6 to 10^8 M^{-1} sec^{-1}

TABLE VIII

G-Values for Products Formed in Irradiated Aqueous Glycine

Product	1.0M Glycine in oxygen-free solution at pH 6.5[a]	1.0M Glycine in oxygen-saturated solution at pH (?)[b]	0.3M Glycine in 0.03–0.04M Cu(II) oxygen-free solution at pH 3.0[c]	0.3M Glycine in 0.03–0.04M Cu(II) oxygen-free solution at pH 8.5[c]
NH_3	4.3	4.0	2.2 ± 0.2	5.0 ± 0.2
CO_2	0.9	—	0.5	3.7
H_2	2.18	0.45	—	0.42
CH_3NH_2	0.19	0.36	—	—
CH_3COOH	1.30	0.0	0.04	0.12
HCOOH	0.09	—	—	—
O ‖ HCCOOH	2.27	4.5	1.8	1.9 ± 0.1
HCHO	0.57	—	0.53	0.8 ± 0.1
CH_2—COOH | CH_2—COOH	—	—	0.07	0.27

[a] Ref. 275.
[b] Refs. 276 and 277.
[c] Ref. 278.

(279). Indeed, k_{110} for glycine at pH 6.4 equals $(8.3 \pm 0.3) \times 10^6 \ M^{-1}$ sec^{-1} and k_{110} for alanine at the same pH equals $5.9 \times 10^6 \ M^{-1} sec^{-1}$ (279). In the light of these low reactivities of the simple amino acids with e_{aq}^- (eq. (110)), the early observation that the product yields in irradiated aqueous amino acid solutions increase with increasing amino acid concentration (280) becomes explicable in terms of the competition between the amino acids (eq. (110)) and other primary chemical species (eqs. (3)–(6)) for the hydrated electron. To ensure that the hydrated electron is captured only by the dissolved amino acid, the concentration of the latter usually must exceed 0.3M for the doses commonly used in the steady-state irradiation of these systems (274). However, at higher amino acid concentrations, direct interactions between the energy and the solute may complicate the system. The fact that the hydrated electron and not the hydrogen atom is the reactive species in reaction (110) was demonstrated by the analysis of the products in irradiated aqueous solutions of alanine containing sodium formate (280,281). Since formate ion is relatively unreactive with the hydrated electron but is a good scavenger for the hydrogen atom and the hydroxyl radical (95) and since the yield of propionic acid was unaffected by the presence of formate ion (280,281), the hydrated electron is, therefore, the reactive species in reaction (110) and this reaction is followed by

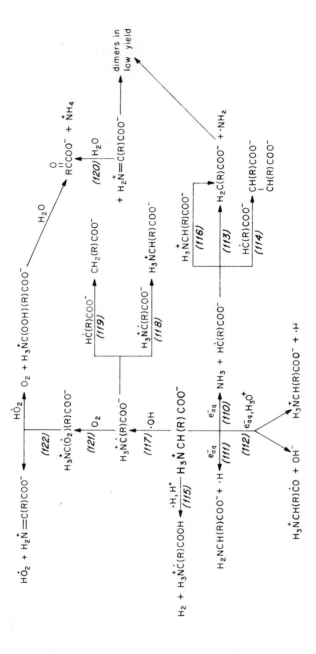

Chart V

reaction (113). However, the hydrogen yield was observed to be markedly higher in irradiated aqueous solutions of glycine than in alanine (275,280). The ability of the glycine zwitterion to act as a proton donor and converting e_{aq}^- to ·H (eq. (111)) in a manner analogous to eq. (2) was invoked to rationalize this discrepancy (274).

For simple amino acids, the attack of the hydrated electron generally occurs at the α-carbon atom forming the $\overset{\bullet}{C}H(R)COO^-$ anion radical (eq. (110)), but for more complex amino acids attack of the hydrated electron at other positions cannot be excluded. Such an attack at the carboxyl group is given in eq. (112). Indeed, the observed second-order rate constants, k_{110}, for glycine gradually decreased with increasing pH from pH 6.4 to 11.8 (279) which indicates that the glycine anion is the less reactive of these two forms. However, the plot of k_{110} versus pH gives a straight line (279) rather than the expected sigmoid curve with an abrupt change at the pK of glycine, 9.78. This type of pH-rate profile is apparently typical for the reaction of α-amino acids with e_{aq}^- (279) and could possibly be the result of a change in the mechanism or the position of e_{aq}^- attack as a function of pH. Clearly, further work is required to elucidate these anomalies.

The slow reaction of α-amino acids with e_{aq}^- and the magnitude of reactions (2)–(6) precludes the study of the reaction of the cationic forms of the simple amino acids with the hydrated electron at low pH values. Such studies, however, would be feasible and would certainly yield valuable information with the use of aromatic amino acids whose reactivity with e_{aq}^- is generally two orders of magnitude greater than that of its aliphatic analog (132).

In the light of the foregoing discussion, the significance of the linear free energy correlation between the rate constant for the reaction of e_{aq}^- with a number of amino acids and peptides and the dissociation constant of the amino group (132) (Fig. 6) must be reexamined. The good correlation in Figure 6 was interpreted to imply that the point of e_{aq}^- attack for these compounds is the protonated amino group. However, this mode of attack is only likely for amino acids when the driving force is proton transfer (eq. (111)). It has been shown that the linear correlation in Figure 6 is equally compatible with a mechanism involving initial attack of the hydrated electron at the carbonyl double bond (eq. (123)) followed by dissociation (eq. (124)) or hydrolysis (eq. (125)) since the R group induces a polarization of the carbonyl double bond which in turn influences the acidity of the amino group (282). The reductive C—N bond cleavage observed in the irradiation of the glycine–Cu^{2+} chelate supports this postulation (278).

Whether the hydrated electron reacts via eq. (110) or by eqs. (123)–

$$\overset{H}{\underset{R}{\overset{|}{\underset{|}{H_3\overset{+}{N}-C}}}}-\overset{O}{\overset{\parallel}{C}}\diagdown_{OH} + e_{aq}^- \longrightarrow \overset{H}{\underset{R}{\overset{|}{\underset{|}{H_3\overset{+}{N}-C}}}}-\overset{O^-}{\overset{}{C}}\cdot\diagdown_{OH} \qquad (123)$$

$$\overset{H}{\underset{R}{\overset{|}{\underset{|}{H_3\overset{+}{N}-C}}}}-\overset{O^-}{\overset{}{C}}\cdot\diagdown_{OH} \longrightarrow NH_3 + \cdot C-\overset{O}{\overset{\parallel}{C}}\diagdown_{OH} \qquad (124)$$

$$\overset{H}{\underset{R}{\overset{|}{\underset{|}{H_3\overset{+}{N}-C}}}}-\overset{O^-}{\overset{}{C}}\cdot\diagdown_{OH} + H_2O \longrightarrow NH_4^+ + OH^- + \cdot C-\overset{O}{\overset{\parallel}{C}}\diagdown_{OH} \qquad (125)$$

(125), the α-carbon anion radical, $\dot{C}H(R)COO^-$, produced yields primarily a carboxylic acid (eq. (113)). However, it can also dimerize to form dicarboxylic acids (eq. (114)). This reaction manifested itself in the formation of succinic acid in irradiated aqueous solutions of glycine containing Cu(II) ions (Table VIII). At low conversions dimer formation is insignificant (274). In addition, the $\dot{C}H(R)COO^-$ anion radical can react with a molecule of the amino acid forming a carboxylic acid and an amino acid radical (eq. (116)).

The observed $G(NH_3)$ value of 4.3 in irradiated oxygen-free solutions of glycine (Table VIII) and alanine (280) indicated that both oxidative (via reaction (117)) and reductive (via reaction (110)) deamination occur. By the addition of suitable e_{aq}^- and ·OH scavengers to the solutions, these processes have been separated. Thus, the $G(NH_3)$ in irradiated oxygen-free solutions of glycine and alanine decreased to values of 1.8 and 2.5, respectively, in the presence of $1.0M$ formate ion (282), a ·OH and ·H scavenger (95). Similarly a $G(NH_3)$ of 2.2 ± 0.2 was observed in irradiated, air-free solutions of $0.3M$ glycine in the presence of 0.03–$0.04M$ Cu(II) ion at pH 3.0 (278) (Table VIII). The hydrated electron is scavenged by the Cu^{2+} ion ($k_{(Cu^{2+} + e_{aq}^-)} = 3.0 \times 10^{10} M^{-1} sec^{-1}$) (283), or possibly by $[Cu(NH_2CH_2COO^-)]^+$, at pH values below 6 yielding Cu^+ without any apparent interaction between the hydrated electron and the glycine (278). Hence, this Cu^{2+} ion system represents the exclusive reaction of the hydroxyl radical with glycine.

The rate constants for reaction (117) have been determined pulse radiolytically by the use of the thiocyanate competition method (284). These observed rate constants showed considerably greater variation with pH than those observed for reaction (110) (279). Indeed, a sigmoid pH-rate profile was obtained for reaction (117) for glycine, the anionic form being more reactive by two orders of magnitude (284). As in the case of the hydrated electron, the proposed position of hydroxyl radical attack is at the α-carbon atom (284). This postulation has recently been verified by the analysis of the ESR spectra and the hyperfine coupling constants of the radicals formed when the hydroxyl radical, generated both from the $Ti^{3+}-H_2O_2$ system and from Fenton's reagent, reacted with amino acids in a flow system (285). Subsequently, the deamination of the amino acid radical, $H_3\overset{+}{N}\overset{\cdot}{C}(R)COO^-$, leads to a keto acid (eqs. (119) and (120)) unlike the reductive deamination which produces a carboxylic acid (eqs. (110), (113), and (116)). The absence of keto acids in irradiated aqueous solutions of amino acids when the hydroxyl radical has been scavenged by formate ion (280) and the absence of carboxylic acids when the hydrated electron has been scavenged by cupric ion (Table VIII) amply support this mechanism.

In the presence of a sufficient amount of oxygen, reactions (110), (111), and (112) cannot occur since the hydrated electron is scavenged according to eq. (9). The reactions of the hydroxyl radical, however, are not inhibited but the amino acid radical, $H_3\overset{+}{N}\overset{\cdot}{C}(R)COO^-$, which is formed can be removed by oxygen, at least partially, producing an unstable peroxy radical (eq. (121)). Conceivably the peroxy radical might dissociate into a labile imino acid and a hydroperoxy radical or react with the hydroperoxy radical, $H\overset{\cdot}{O}_2$, forming an unstable hydroperoxide and oxygen (eq. (122)). In any event the major product stoichiometry for glycine and alanine is best represented by $G(NH_3) \simeq G(R_2CO) \simeq G_{OH}$.

The reactions outlined in Chart V describe the radiation chemistry of simple amino acids in aqueous solution. However, the incursion of other reactions including different sites of attack of the primary chemical species will undoubtedly become more important for longer chained or aromatic amino acids. Consequently a mechanistic discussion of the radiation-induced processes in these more complex systems must await the accumulation of considerably more data.

2. Radiation-Induced Reactions of Aqueous Pyrimidine Bases

The radiation-induced reactions of uracil, thymine, cytosine, and other pyrimidine bases in aqueous solutions involve deamination, ring opening, and the formation of oxalic acid, urea, and ammonia (274,286–294). In irradiated aqueous solutions of uracil-2-^{14}C, dihydrouracil, 6-hydroxy-5-hydrouracil, isobarbituric acid, cis and trans glycols of uracil, and alloxan have been identified as products (295). At low doses, the

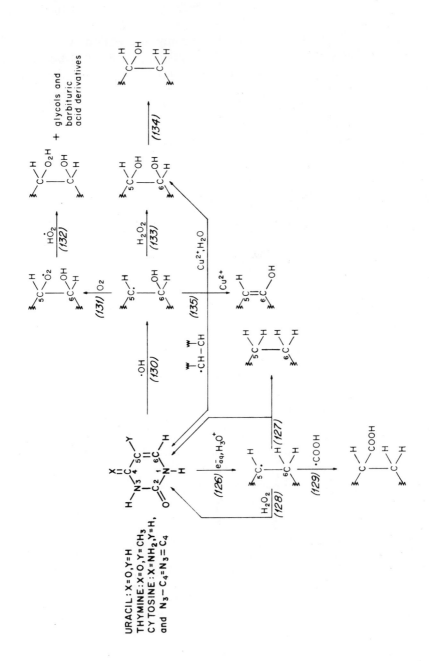

Chart VI

principal site of free radical attack was demonstrated to be the 5–6 double bond since identical initial G-values for the destruction of the pyrimidine bases were obtained by following the decrease in the ultraviolet absorption and by bromine titration of the 5–6 double bond (296,297). Quantum mechanical calculations have also indicated that the 5–6 double bond is the most reactive site in the pyrimidine molecule (298).

Absolute rate constants for reaction (126) (Chart VI) have been determined pulse radiolytically by following the disappearance of the hydrated electron at 700 nm (299,300). At pH 7, k_{126} for cytosine is 7.0×10^9 $M^{-1} \sec^{-1}$ (202), that for thymine is $1.7 \times 10^{10}\ M^{-1} \sec^{-1}$ (95), that for uracil is $1.5 \times 10^{10}\ M^{-1} \sec^{-1}$ (299), and that for dihydrouracil is 4.5×10^9 $M^{-1} \sec^{-1}$ (299). The magnitude of these rate constants for the reaction of e_{aq}^- with pyrimidines supports the proposed attack on the 5-6 double bond. The hydrated electron shows little sensitivity in its reaction with pyrimidine bases although the 4-amino group of cytosine is responsible for the slightly lesser reactivity of this compound. A decrease in the per cent of destruction of uracil with increasing pH of the solution has been observed in the steady-state radiolysis of aqueous solutions (301). Determinations of the absolute rate constants for reaction (126) as a function of pH confirms this observation; k_{126} at pH 11 is $9 \times 10^9\ M^{-1}$ $\sec^{-1}$ (299). Since the pK for uracil is 9.8, this decrease in the reactivity of the pyrimidine bases with e_{aq}^- as a function of increasing pH is explicable in terms of the lesser reactivity of the anionic (enolic) form resulting from the electrostatic repulsion between the two negative entities.

Equation (126) actually involves the formation of a pyrimidine radical anion which subsequently abstracts a proton from the solvent to form the hydropyrimidine radical. Evidence supporting this postulation has been obtained recently by analyzing the ESR spectra of the products formed from pyrimidine derivatives and photolytically generated electrons in H_2O and D_2O matrices at 80°K (302).

The hydrouracil radical disproportionates, forming dihydrouracil and uracil (eq. (127)), and, in addition, reforms uracil by its reaction with hydrogen peroxide (eq. (128)). Also, in the presence of formate ion which scavenges the hydroxyl radical forming the formic acid radical (95), ·COOH, the hydropyrimidyl radical combines with ·COOH giving a hydrocarboxypyrimidine (eq. (129)). Under these conditions, the formation of a hydrocarboxypyrimidine in high yield, $G \simeq 2.5 \simeq G_{OH}$, has been observed (303).

The transient spectra of irradiated aqueous pyrimidine bases in the presence of N_2O are rather complex (304); nevertheless, rate constants for reaction (130) have been determined pulse radiolytically both by the thiocyanate competition method and by following the rate of decrease in the absorption due to 5-6 double bond (299). Good agreement was found among the rate constants obtained by these methods and by other competition techniques (299,304). For uracil k_{130} at pH 7 is $(5 \pm 0.6) \times 10^9$ M^{-1} sec^{-1} (304). This high rate constant for reaction (130) is compatible with hydroxyl radical attack at the 5-6 double bond. In addition, ESR analyses of the structure of the radical formed by the interaction of thymine and the hydroxyl radical generated from Ti^{3+} and H_2O_2 confirmed that the C-5 position is the site of attack (305,306). With increasing pH there is a steady increase in k_{130} for uracil but no abrupt change occurs near the pK; this behavior indicates that the apparent reactivities of the keto and enolic forms of the pyrimidine base (299) are similar. Above pH 13, k_{130} is somewhat smaller, but this decrease might be due to the formation of $\cdot O^-$ by the reaction of the hydroxyl radical with hydroxide ion (307) (eq. (21)).

In the presence of oxygen the hydrated electron is scavenged (eq. (9)), and peroxy (eq. (131)) and hydroperoxy radicals are formed and subsequently decompose to give hydroperoxides (eq. (132)), glycols (eqs. (132) and (133)), and barbituric acid derivatives (eqs. (132) and (134)). A relatively stable hydroperoxide has been isolated by ion-exchange chromatography from an irradiated air-saturated solution of thymine and characterized by comparison with the independently synthesized cis and trans isomers of both 4-hydroxy-5-hydroxyperoxythymine and 4-hydroxyperoxy-5-hydroxythymine (296,308). The hydroperoxide formed in an irradiated, aqueous, air-saturated solution of thymine was found to be a mixture of cis and trans isomers in the ratio of $80:20\%$ (309). The G-values for the formation of hydroperoxides in irradiated aqueous solutions of thymine, uracil, and dimethyluracil were determined to be 1.05, 1.16, and 0.75, respectively (287). However, the total yields of oxidation products are considerably less than the G-values for the pyrimidine base destruction (287). The use of Cu^{2+} or Fe^{2+} (eq. (135)) instead of O_2 as the electron scavenger considerably simplifies the radiation chemistry of aqueous pyrimidine bases (310). In the presence of the metal ion, the hydroxypyrimidyl radical preferentially gives glycol ($G \simeq 2.3$) and isobarbituric acid ($G \simeq 0.5$) rather than the hydroperoxides formed in the presence of oxygen (310).

Pulse radiolytic data on the second-order rate constant for the decay of both the hydroxyl radical and the hydrated electron adducts of uracil and 5-bromouracil in the pH 1.3–11.0 range has been reported recently (300). However, the identification of the products formed in these processes must await further work.

3. Future Trends in the Elucidation of
Radiation-Induced Bioorganic Reactions

The present state of the art in radiation chemistry is such that the mechanisms of most of the radiation-induced reactions involving relatively simple organic molecules can be elucidated. The careful accumulation of data and its systematic correlation is now required in order to provide a solid foundation for research on the more complex systems. The subject of physical organic chemistry was essentially in a similar state of development in the years preceding the Second World War. The techniques were available and the utility of this physical approach was evident to a number of research groups who carefully systematized the mechanisms of a large number of organic reactions. This work in turn led to the birth and development of bioorganic mechanistic studies concerning ionic reactions (311). It is evident that the next few years will witness similar, if not more spectacular, progress toward an understanding of radiation-induced reactions and consequently allow an extension of the acquired knowledge to the more complex bioorganic molecules.

IV. RADIATION-INDUCED REACTIONS IN NONAQUEOUS SOLVENTS

In recent years, considerable progress has been made toward an understanding of the radiation chemistry of nonaqueous organic liquids (312). Most of the research work has centered around the radiation chemistry of hydrocarbons (41,57,313,314) and polar protic solvents such as alcohols (315–324). Important and novel data of mechanistic interest has been obtained by the application of radiation chemical techniques to nonaqueous solvent systems in spite of the less extensive research in these media as compared to that in water. The radiation-induced reactions in nonaqueous media which bear mechanistic significance for the organic chemist are presented in the following discussion. However, the emphasis is placed primarily on the pulse radiolytic studies of electron transfer processes. Since the radiation chemistry of organic liquids differs from that of water, the fundamental processes in nonaqueous systems are outlined initially.

A. Fundamentals of the Radiation Chemistry of Organic Liquids

The fundamental processes in the radiolysis of organic liquids, R, can be expressed by the following general equations:

$$R \xrightarrow{\text{(excitation)}} R^* \tag{136}$$

$$R \xrightarrow{\text{ionization}} R^+ + e^- \qquad (137)$$

$$R^* \text{ and } R^{**} \xrightarrow{\text{(dissociation)}} \begin{cases} \cdot R' + \cdot R'' \\ P + P' \end{cases} \qquad (138)$$

$$R^+ \xrightarrow{\text{(ion dissociation)}} \begin{cases} \cdot R'^+ + \cdot R'' \\ P^+ + P' \end{cases} \qquad (139)$$

where R^* and R^{**} represent excited states and P and P' are molecular products. A model in which the inhomogeneous formation of spurs along the path of the ionizing particle contain excited un-ionized molecules (R^*), positive ions (R^+), and electrons (e^-) can serve to describe the events which take place within 10^{-13} sec of the energy deposition. This and other models have been treated theoretically (47,87,325,326).

The excess energy of the electrons formed in eq. (137) is dissipated through excitation and ionization of the medium. Subsequently the electron may become solvated, generally attracting several solvent molecules to itself:

$$e^- + R \longrightarrow e_s^- \qquad (140)$$

Alternatively the electron may be neutralized by positive ions in the spur,

$$e^- + R^+ \longrightarrow R^{**} \qquad (141)$$

be captured by a solute, S, or the solvent,

$$e^- + R \text{ or } S \longrightarrow \cdot R^- \text{ or } \cdot S^- \qquad (142)$$

or undergo a dissociative capture,

$$e^- + SX \text{ or } RX \longrightarrow \cdot S \text{ or } \cdot R + X^- \qquad (143)$$

The positive ions formed in eq. (137) have been demonstrated to undergo fragmentation (327–332), proton transfer (333–337), molecular hydrogen transfer (338–340), hydrogen atom transfer (341,342), ion molecular (343,344), and neutralization reactions (314,344).

The radiation chemistry of organic liquids generally involves a greater number of excited states so that subsequent reactions are generally more complex than those in analogous photochemical cases, but many resemblances still remain. In the radiolysis of organic liquids both singlets and triplets can be formed initially by either excitation (eq. (136)) or electron neutralization (eq. (141)). Studies of excited molecules in the singlet state usually involve scintillators (345,346), with which the yields and lifetimes of the excited scintillator molecules, the lifetimes of the excited solvent molecules, and delayed fluorescence have been determined (345,347–350). The lifetimes of radiation-induced triplet states have been measured in kinetic (351,352) and pulse radiolytic (353) studies and studies in which chemical indicators such as 2-butene (352,354) and ferric chloride (355) were employed.

On the 10^{-11} to 10^{-8} sec time scale numerous *in spur* combinations and disproportionations occur between the various intermediates formed according to eqs. (136) to (139). The species which escape from the spur (for the most part within 10^{-8} sec after the energy deposition) and subsequently react homogeneously in or with the solvent are free ions, free radicals, or longer lived excited molecules. Indeed, transfer of the solvated electron to several solutes in numerous solvents has been investigated on this time scale and is the subject of the discussion in the following section.

The radiolysis of organic liquids differs considerably from the radiolysis of water since in many cases the primary species can react with the solvent. Therefore, the number of possible reactions which can occur in organic liquids far exceeds that which occurs in the radiolysis of water. In organic liquids, hydrogen atoms, for example, rarely combine to form molecular hydrogen, a common reaction in water (eq. (13)), but rather preferentially abstract hydrogen from or add to a solvent molecule. Similarly alkyl radicals also abstract hydrogen from the solvent in organic liquids. The relative rates of hydrogen abstraction in a number of systems have been determined recently (356,357) by utilizing the dissociative electron capture of organic halides in irradiated hydrocarbon solutions to produce alkyl (356) and fluoroalkyl (357) radicals. Using radioiodine scavenging techniques the ratios of the rate constants for hydrogen abstraction (eq. (144)) to radical scavenging (eq. (145)) have been determined.

$$\cdot CX_3 + RH \longrightarrow CX_3H + \cdot R \qquad (144)$$

$$\cdot CX_3 + I_2 \longrightarrow CX_3I + \cdot I \qquad (145)$$

The values of k_{144}/k_{145} at room temperature are 1.65×10^{-5} for $\cdot CF_3$ and 1.63×10^{-5} for $\cdot C_2F_5$ in cyclohexane (357) and 7×10^{-8} for $\cdot CH_3$ in 2,2,4-trimethylpentane (356). This increase in the ratios of the rates of abstraction to scavenging for the fluorinated radicals as compared to $\cdot CH_3$ is apparently primarily due to an increase in the rate of hydrogen abstraction.

Recently vigorous interest has been shown in the determination of free ion yields in different organic liquids. The value for the solvated electron free ion yield $(G_{e_s^-})^{fi}$ is ca. 1.0 in alcohols (81,170,315,317,319,358),

0.1 in hydrocarbons (52,58,60,359,360), 0.4 in ammonia (361), and 0.04 in dioxane (48). A correlation between the static dielectric constant of the solvent and the free ion yields has been proposed (47,325,326); however, subsequent work has demonstrated (362–364) that more complex, but so far incompletely understood, parameters govern the free ion yields in organic liquids. The elucidation of the radiation chemistry and free ion yields of dipolar aprotic solvents (365–368) is of considerable interest both to radiation chemists and to physical organic chemists. For this purpose, studies have been initiated in dioxane (368,369), formamide (370), and acetonitrile (371), but a great deal of additional work is warranted in this area.

B. Reactions of Aromatic Transients in Organic Solvents

When a molecule accepts an electron, a radical anion, $A^{\bar{\cdot}}$, is formed (372,373). Aromatic radical anions of some stability have been prepared by photochemical (374), electrolytic (375–377), alkali metal (378,379), and chemical (380–382) reduction of the aromatic compound usually dissolved in dipolar aprotic solvents such as 1,2-dimethoxyethane, tetrahydrofuran, and dioxane. The radical anions have been characterized by their ESR and absorption spectra (373,383–385), and their reactions have been investigated (372,386).

The loss of one electron from a neutral molecule results in the formation of a radical cation, $A^{\cdot+}$ (372,387). Although radical cations are less readily prepared than the corresponding anions, fairly stable polynuclear aromatic radical cations have been prepared and characterized by oxidation of the aromatic compound with persulfate ion in sulfuric acid (387–389). However, such a reactive medium also leads to simple protonation forming carbonium ions and radical–radical reactions which complicate the system. Carbonium ion formation can be achieved by electron abstraction from an electrically neutral free radical. Mass spectroscopic studies of this process have furnished useful information concerning the relative stabilities of carbonium ions in the gas phase (390,391). In solution, however, there are only a few chemical reactions which generate carbonium ions from free radicals (392–396).

By suitable adjustment of the experimental conditions, pulse radiolysis can serve as an eminently suitable tool not only for the identification of radical anions, excited species, radical cations, carbonium ions, and charge-transfer complexes but also for kinetic studies of the processes involved in the generation and reaction of these species. Hence, the ensuing discussion emphasizes this approach.

1. Radical Anions

To date, the most extensive studies of radical anions involve the electron transfer processes of aromatic compounds in alcohols. Chart VII summarizes the processes which have been identified.

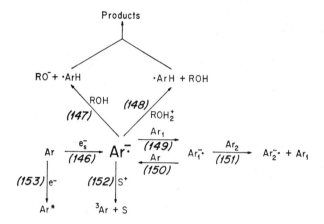

Chart VII

The absolute rate constants for the addition of the solvated electron to aromatic solutes (eq. (146)) in ethanol have been determined by following the disappearance of the solvated electron at 600 nm for all the compounds listed in Table IX with the exception of biphenyl (124). The overlapping absorption of the biphenylide ion and the solvated electron precluded direct measurement and, therefore, k_{146} for biphenyl was determined by competition kinetics (124). The rate constants for reaction (146) increase with increasing electron affinity (397) of the dissolved aromatic hydrocarbon. Since the available data for reaction (146) is rather limited

TABLE IX

Rate Constants for Reactions (146)–(148) in Ethanol[a]

Compound	$k_{146}, M^{-1} \sec^{-1}$	$k_{147}, M^{-1} \sec^{-1}$	$k_{148}, M^{-1} \sec^{-1}$
Biphenyl	$(4.3 \pm 0.7)10^9$	$(2.6 \pm 0.3)10^4$	$(3.3 \pm 0.5)10^{10}$
Naphthalene	$(5.4 \pm 0.5)10^9$	—	—
p-Terphenyl	$(7.2 \pm 0.6)10^9$	$(2.0 \pm 0.6)10^2$	$(1.9 \pm 0.3)10^{10}$
Naphthacene	$(10.2 \pm 0.8)10^9$	—	—
Anthracene	—	$(2.3 \pm 0.23)10^4$	$(3.7 \pm 0.6)10^{10}$

[a] Ref. 124.

and k_{146} is very nearly diffusion controlled, further work is required to ascertain whether a genuine correlation of this type exists and is meaningful.

The aromatic radical anion can react with the solvent alcohol (eq. (147)) or the counterion (eq. (148)), transfer its charge to another solute molecule (eq. (149)), or form a longer lived excited molecule (eq. (152)). A knowledge of the second-order rate constants for eq. (146), the electron yield, the dissociation constant of the alcohol, and hence the counterion concentration would allow the selection and study of any one of these reaction paths, i.e., reactions (147), (148), (149), or (152), by suitable adjustment of the experimental conditions.

When the concentration of the aromatic compound was increased so that reaction (146) was completed within 0.5 µsec of the pulse, the decay

TABLE X

Rate Constants and Arrhenius Parameters for the Proton Transfer from Aliphatic Alcohols to Biphenylide and Anthracenide Radical Anions[a]

	Methanol	Ethanol	Propanol	Isopropanol
k_{147},[b] $M^{-1} sec^{-1}$	$(6.9 \pm 1.2)10^4$	$(2.6 \pm 0.3)10^4$	$(3.2 \pm 0.3)10^4$	$(5.5 \pm 1.1)10^3$
E,[b] kcal mole^{-1}	2.7	3.1	—	5.8
$10^{-6}A$,[b] $M^{-1} sec^{-1}$	6.2	3.9	—	150
k_{147},[c] $M^{-1} sec^{-1}$	$(8.1 \pm 2.0)10^4$	$(2.3 \pm 0.23)10^4$	$(2.4 \pm 0.38)10^4$	$(3.6 \pm 0.6)10^3$
E,[c] kcal mole^{-1}	2.1	2.4	—	6.7
$10^{-6}A$,[c] $M^{-1} sec^{-1}$	3.7	1.2	—	320

[a] Refs. 128 and 398.
[b] Biphenylide radical anion.
[c] Anthracenide radical anion.

of the radical anion could be followed at the appropriate wavelength (383, 384). In the absence of added acid which would protonate the alcohol forming $ROH_2{}^+$, the radical anion followed a first-order decay according to eq. (147). Table X contains the second-order rate constants, k_{147}, for biphenylide and anthracenide radical anions in methanol, ethanol, propanol, and isopropanol which were obtained from the observed pseudo first-order rate constants by dividing by the alcohol concentration (124). The decrease in k_{147} with decreasing acidity of the alcohol (Table X) indicates that reaction (147) involves a proton transfer from the hydroxyl group of the alcohol to the radical anion. A correlation between the rate constants for eq. (147) and the dissociation constants of the alcohols is, once again, not straightforward. The differences in k_{147} for the various alcohols, and hence the differences in the free-energy requirements for this reaction, must be a delicately balanced composite involving the stability of the radical anion on the one hand and the acidity and solvating properties of the alcohol on the other. Good Arrhenius plots were obtained for reaction (147) by varying the temperature from $-70°$ to $80°$ (398), and the parameters are given in Table X. The relatively small rate constants obtained for reaction (147) are evidently a manifestation of a small preexponential factor rather than a high activation energy. Isopropanol, the weakest acid investigated, apparently has the least ability to transfer a proton to the radical anion, the highest activation energy, and the highest preexponential factor (Table X). Further work with other alcohols and solvents is clearly merited and is in progress (399).

Rate constants for reaction (148) have been determined by the addition of sufficient hydrochloric acid to the alcohol to obtain $(1.0–5.0) \times 10^{-5} M$ hydrogen ion concentrations. The neutralization of the radical anion by its alcohol counterion (eq. (148)) is considerably faster than the proton transfer (eq. (147)) from the alcohol (Table X), and, therefore, in most cases the contribution of reaction (147) to the rate of reaction (148) has been neglected (124).

The products of reactions (147) and (148) are identical and have been identified as the hydrogen adduct free radical, $\cdot$ArH (124). This radical decays by a second-order process probably involving radical–radical reactions.

Electron transfer reactions, eq. (149), have been studied successfully in alcohols by adjusting the experimental conditions in such a way that $k_{146}[Ar] \gg k_{146}[Ar_1]$ and that $[Ar_1]_0 \gg [Ar]_0$ where $[Ar]_0$ and $[Ar_1]_0$ are the initial concentrations (128,400). Under these selected conditions the electron rapidly (within less than 0.5 μsec) and exclusively reacts with Ar, hence the decay in the absorption due to $Ar^{\overline{\cdot}}$ and the increase in the absorption due to $Ar_1^{\overline{\cdot}}$ can be observed simultaneously provided that $Ar^{\overline{\cdot}}$ and $Ar_1^{\overline{\cdot}}$ do not exhibit extensive spectral overlap. For those cases in which $[Ar_1]_0 \gg [Ar]_0$, the observed kinetics are pseudo first order with respect to $Ar^{\overline{\cdot}}$. When the decay of $Ar^{\overline{\cdot}}$ is observed at sufficiently small pulse intensities, the counterion reaction (eq. (148)) can be neglected; and when the Ar_1 concentration is sufficiently high, k_{147} is negligible compared to the rate of the electron transfer (eq. (149)). The electron transfer is, of course, a reversible process and, consequently, when the equilibrium constant is small, the back reaction (eq. (150)) must be considered. The decay of $Ar^{\overline{\cdot}}$ is represented by:

$$\frac{-d[Ar^{\overline{\cdot}}]}{dt} = k_{149}[Ar^{\overline{\cdot}}][Ar_1] - k_{150}[Ar_1^{\overline{\cdot}}][Ar] \qquad (154)$$

On integrating and expressing the concentration of Ar_1^- in terms of absorbances, one obtains:

$$2.303 \log (OD_t - OD_e) = -(k_{149}[Ar_1] + k_{150}[Ar])t + C \quad (155)$$

where OD_t and OD_e are the absorbances of Ar_1^- at time t and at equilibrium, and C is a constant. Plots of the left-hand side of eq. (155) against t for different Ar_1 concentrations at a constant concentration of Ar give a family of straight lines whose slopes are $k_{149}[Ar_1] + k_{150}[Ar]$. A plot of these slopes against the Ar_1 concentration results in a straight line whose slope is k_{149} and intercept k_{150}. Such a plot is illustrated in Figure 12 for the electron transfer equilibrium of 9.10-dimethylanthracene and pyrene (128). Rate equations similar to eq. (154) can be derived for cases in which the appearance of Ar_1^- is followed pulse radiolytically and in which k_{147} is not negligible. The subsequent treatment of the data is analogous to that described above. The available rate constants for these aromatic electron transfer processes (eq. (149)) in alcohols are collected in Table XI.

The direction of electron transfer between two aromatic compounds indicates, at least qualitatively, their relative electron affinities. Indeed,

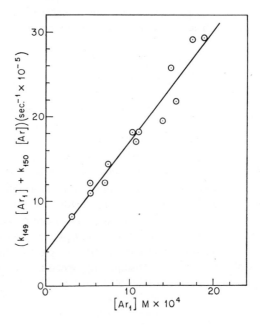

Fig. 12. Plot of $k_{149}[Ar_1] + k_{150}[Ar]$ vs. $[Ar_1]$. Ar = pyrene, Ar_1 = 9,10-dimethylanthracene. Taken from ref. 401.

TABLE XI

Electron-Transfer Reactions in Isopropanol at 25°

| Donor radical anion | Acceptor molecule | k_{149}, M^{-1} sec^{-1} | | $\Delta F^{\circ\prime}$ [b] |
		Experimental[a]	Calculated[a]	
Anthracenide[c]	Pyrene	$(2.1 \pm 0.9)10^7$	1.2×10^7	$+2.61$
9,10-Dimethyl-anthracenide[c]	Pyrene	$(3.7 \pm 1.7)10^7$	2.2×10^7	$+2.01$
Biphenylide	Naphthalene	$(2.6 \pm 0.8)10^8$	3.0×10^8	-0.99
Biphenylide	Phenanthrene	$(6 \pm 3)10^8$	1.6×10^9	-3.28
Pyrenide	9,10-Dimethyl-anthracene	$(1.3 \pm 0.3)10^9$	6.5×10^8	-2.01
Pyrenide	Anthracene	$(1.8 \pm 0.3)10^9$	1.0×10^9	-2.61
m-Terphenylide	p-Terphenyl	$(2.3 \pm 0.4)10^9$	—	—
Biphenylide	p-Terphenyl	$(3.2 \pm 0.7)10^9$	—	—
Biphenylide	Pyrene	$(5.0 \pm 1.8)10^9$	7.2×10^{10}	-12.2
Biphenylide	Anthracene	$(6.4 \pm 2.0)10^9$	9.8×10^{10}	-14.8
p-Terphenylide	Pyrene	$(3.6 \pm 1.1)10^9$	—	—
p-Terphenylide	Anthracene	$(5.5 \pm 0.9)10^9$	—	—
m-Terphenylide	Pyrene	$(3.5 \pm 1.2)10^9$	—	—
o-Terphenylide	Pyrene	$(4.0 \pm 1.8)10^9$	—	—
$(CH_3)_2\overset{\cdot}{C}O^-$ [d]	Fluorenone	2×10^9	—	—

[a] Refs. 400 and 401.
[b] Standard free energy of reaction = difference in reduction potential.
[c] Determined as the intercept of a plot of $(k_{148}[Ar_1] + k_{149}[Ar])$ vs. $[Ar_1]$.
[d] In ethanol, ref. 402.

combined polarographic and ESR techniques demonstrated that the formation of radical anions in dimethylformamide paralleled their reduction potentials (403). Since pulse radiolysis affords a direct determination of electron transfer rate constants in polar solvents, it consequently allows a comparison between kinetically and thermodynamically obtained equilibrium constants and offers an experimental test for the theories of homogeneous electron transfer processes in polar liquids (404–411). From the data in Table XI the equilibrium constant, $K = k_{149}/k_{150}$, was calculated for the pyrene–anthracene pair to be ca. 86 and for the pyrene-9,10-dimethylanthracene pair to be ca. 35 (401). Using the reduction potential differences, $\Delta F^{\circ\prime}$, for these compounds determined potentiometrically in tetrahydrofuran (385,412–414), the equilibrium constants for these two pairs were calculated from eq. (156) to be, respectively, 81 and 30.

$$\Delta F^{\circ\prime} = RT \ln K \tag{156}$$

In view of the different solvents used in these two studies, i.e., isopropanol (401) and tetrahydrofuran (385,412–414), such good agreement between the equilibrium constants may well be fortuitous. Further research on electron transfer reactions, especially on the effect of solvents, is currently being conducted (399).

Taking a value of 5 Å for the encounter radii in calculating the reorganization parameter of Marcus' electron transfer theory in conjunction with the appropriate solvent dielectric properties and the reduction potentials for the donor–acceptor pairs, the free energy changes for reaction (146) have been calculated (401). The introduction of a collision number of $10^{11} M^{-1} \sec^{-1}$ for the electron transfer reaction (412–414) allowed the calculation of k_{149} (Table XI). Reasonable agreement was obtained between the calculated and experimental rate constants for those electron transfer reactions whose rate constants were at least two orders of magnitude smaller than diffusion controlled.

The spectra and kinetics of "cascade" electron transfers have been observed pulse radiolytically in aqueous solution for acetone$^-$ $\rightarrow$ acetophenone$^-$ $\rightarrow$ benzophenone$^-$ and for isopropanol$^-$ $\rightarrow$ acetophenone$^-$ $\rightarrow$ benzophenone$^-$ $\rightarrow$ ferricyanide$^-$ (415). The inherent interest in this type of study will undoubtedly promote further work.

2. Excited States of Aromatic Molecules

If a sufficiently high concentration of an organic solute is dissolved in a *nonpolar* solvent to scavenge the free electrons, its radical anion will be formed according to eq. (146). The neutralization of the radical anion, however, will lead to a neutral excited aromatic molecule (eq. (152)). The cation S^+ in eq. (152) may be the solvent or the aromatic solute. Excited solvent molecules in nonpolar systems are, of course, also formed by excitation, reaction (136), in which case R^* may be in a singlet or a triplet state. The excited solvent molecules may then transfer energy to the solute:

$$^1R + Ar \longrightarrow R + {}^1Ar \tag{157}$$

$$^3R + Ar \longrightarrow R + {}^3Ar \tag{158}$$

Intersystem crossing may also occur for both the solvent and the solute,

$$^1R \longrightarrow {}^3R \tag{159}$$

$$^1Ar \longrightarrow {}^3Ar \tag{160}$$

and subexcitation electrons, e^-, might lead directly to either a singlet or a triplet-state arene (eq. (153)). Studies of the radiation and the photo-induced *cis–trans* isomerization of several organic compounds, particularly stilbenes, have contributed significantly to the elucidation of the mechanisms involved in triplet-state formation and subsequent reactions (416–420).

Pulse-radiolytic observations of excited aromatic molecules have been aided considerably by the reported absorption spectra and, in some cases, extinction coefficients of the triplet states observed in the photolysis of organic glasses (421) and in the flash photolysis of liquids (422). The absorption spectra and G-values of the triplet states of anthracene, phenanthrene, naphthalene, and acetylanthracene have been determined in pulse-irradiated paraffin oil solutions (423,424). The yield of anthracene triplet in liquid paraffin was estimated to be 0.5 molecule/100 eV (423). The anthracene triplet has also been observed in benzene with an estimated G-value of 0.7 (425). In addition, the delayed fluorescence emission at 430 nm observed in this study was attributed to arise from triplet–triplet quenching (425). The anthracene, naphthalene, and benzophenone triplet yields in cyclohexane and in benzene were found to depend on the solute concentration (426–428).

The rate constants for the formation of the triplet states of anthracene, 1,2-benzanthracene, naphthalene, and biphenyl in acetone have been determined pulse radiolytically from the rate of absorption increase due to the triplet formation (128,429). The precursor of the solute triplet is most likely the excited solvent triplet (eq. (158)). The aromatic radical anion has been ruled out as a precursor (eq. (152)) on the grounds of the observed differences in the kinetics when the decay of the radical anion (at 730 nm for anthracene) and the increase due to the triplet (at 424 nm for anthracene) were followed simultaneously (429). The calculated half-life of the excited acetone molecule in the absence of solutes, > 5 μsec, excludes the possibility that a much shorter lived singlet acetone molecule is the precursor in the formation of the aromatic triplet (eq. (157)). Indeed, in calculating k_{158}, corrections have been made for the spontaneous de-excitation of the solvent molecules, i.e., for ^{3}acetone $\rightarrow$ acetone. Table XII gives the rate constants and the yields for the formation of aromatic triplets in acetone. The large uncertainties in the G-values in Table XII are due to the uncertainties in the molecular extinction coefficients (422). The decay of the aromatic triplets in acetone is slower than their formation and exhibits rather complex kinetic behavior (429) which indicates the incursion of a large number of concurrent and/or consecutive reactions.

As mentioned previously, the fate of the electron and of the aromatic solute depends on the solvent. Triplet states were detected, for example, in the pulse radiolysis of naphthalene in benzene and in cyclohexane but not in ether, tetrahydrofuran, or methanol (430). Similarly, benzophenone

TABLE XII

Rate Constants (k_{158}) and Yields of Aromatic Triplets in Organic Solvents

Solute	k_{158}, M^{-1} sec^{-1} in acetone[a]	G-Value in acetone[a]	G-Value in cyclohexane[b]	G-Value in benzene[b]
Anthracene	$(6.2 \pm 0.6)10^9$	1.1 ± 0.7	$0.2(10^{-4}M)$[c] $1.05(10^{-2}M)$[c]	$0.15(10^{-4}M)$[c] $1.5(5 \times 10^{-2}M)$[c]
Naphthalene	$(4.5 \pm 0.9)10^9$	<3	$0.25(10^{-4}M)$[c] $2.4(9 \times 10^{-1}M)$[c]	$0.4(10^{-3}M)$[c] $2.3(1.0M)$[c]
1,2-Benzan-thracene	—	1.9 ± 1.0	—	—
Biphenyl	$\sim 5 \times 10^8$	—	—	—

[a] Ref. 128.
[b] Ref. 427.
[c] Solute concentration given in parentheses.

formed triplets in benzene, but the triplet yield was negligible in cyclo-hexane and in tetrahydrofuran (430). The triplet yield of naphthalene in cyclohexane was found to increase with increasing naphthalene concentration until it reached a G-value of 4 (426). Two distinct kinetic processes were identified (426,431). One occurred within less than 1 nsec after the pulse while the other took place considerably more slowly. The faster process was identified as reaction (153) and the slower one with reaction (152) (426,431). A great deal of work on both sides of the Atlantic is being pursued currently in this interesting and important area of research (432).

3. Radical Cations and Carbonium Ions

Although radical cations have been observed in irradiated low-temperature glasses (433,434), their unambiguous identification in organic liquids is less straightforward. Radical anions and triplet states may also be present in the irradiated system, and these species often have spectra similar to those of the radical cations (128,435,436). The addition of specific scavengers whose spectra do not interfere with those of the transients and kinetic analyses of the processes have been used as criteria for the assignment of the observed absorption to a particular species. Nitrous oxide and sulfur hexafluoride are powerful electron scavengers and normally do not affect radical cations or triplet states; however, the addition of nitrous oxide was observed to reduce the yields of ^{3}TPA and ^{3}DMPD and to eliminate ^{3}TMPD* in irradiated solutions of the respective parent amines in benzene (437). Such complications may not occur with

* TPA = triphenylamine;   DMPD = N,N-dimethyl-p-phenylenediamine;   and TMPD = N,N,N',N'-tetramethyl-p-phenylenediamine.

sulfur hexafluoride as the electron scavenger. Methanol does not affect triplet states and radical anions but is an effective cation radical scavenger. Oxygen scavenges radical anions and triplet states but does not affect radical cations. Naphthalene is a good energy acceptor but, since it absorbs strongly at 400–420 nm, its use in pulse radiolytic studies is somewhat limited. Ferric acetylacetonate is an excellent quencher of triplet states (438).

TMPD readily loses an electron forming a stable radical known as Wurster's blue whose photochemistry in various solvents has been examined extensively (435,436,439,440). Pulse-radiolytic studies revealed some of the complexities involved in the radiation-induced reaction of TMPD in different organic liquids (437,441,442). Figure 13 allows a comparison of the absorption spectra of the pulse radiolytically generated transients in cyclohexane and 2,2,4-trimethylpentane in the presence of different scavengers with that of Wurster's blue generated chemically (441). The transient spectra in these solvents have been resolved into that of a triplet state of TMPD and that of the radical cation, TMPD$^{\ddot{+}}$ (441). The main part of the absorption was assigned to the triplet state of TMPD since the difference spectrum of TMPD obtained from the spectra observed in the presence and absence of N_2O (curve 3 in Fig. 13) showed a hypsochromic shift of 10 nm and matched the assigned ^{3}TMPD spectrum (441). The decay of ^{3}TMPD, curve 3 in Figure 13 observed as a function of time, obeyed first-order kinetics at sufficiently high concentrations of TMPD (441). Similarly, first-order kinetics have been observed for the decay of ^{3}TMPD in benzene, liquid paraffin, and dioxane (437). A comparison of the rate constants for the triplet decay of TMPD in different solvents is, however, not warranted because of the observed dose-rate dependencies (437). Nevertheless, the lifetimes of these triplets correspond to those observed by flash photolysis (443).

In the presence of nitrous oxide, the triplet formation in cyclohexane and in 2,2,4-trimethylpentane is completely inhibited (notice the good match between curves 2 and 5 in Fig. 13), and the TMPD radical cation decays by second-order kinetics (441). The rate constant for this process was calculated to be $7.7 \times 10^{11} M^{-1} \sec^{-1}$ in cyclohexane and $7.9 \times 10^{11} M^{-1} \sec^{-1}$ in 2,2,4-trimethylpentane by using the extinction coefficient of TMPD$^{\ddot{+}}$ determined from curve 5 in Figure 13.

The yield of ^{3}TMPD, expressed as $G\varepsilon$, was found to increase with the square root of the TMPD concentration, indicating that the solute reacts with the positive ion before initial recombination occurs. The yield of TMPD$^{\ddot{+}}$, on the other hand, was found to be independent of the TMPD concentration (441). Using the value for ε determined from curve 5 in

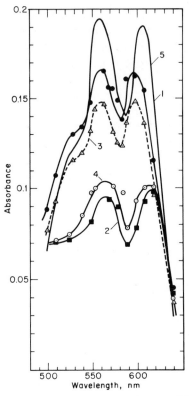

Fig. 13. Curves *1–4*, transient spectra from solutions of TMPD as measured at the end of a 5-μsec pulse of 1.9-MeV electrons: curve *1*, Ar-bubbled solution in cyclohexane, normalized to a dose of 1.4×10^{17} eV/ml; curve *2*, N_2O-bubbled solution in cyclohexane, normalized to 6.9×10^{17} eV/ml; curve *3*, difference between curves *2* and *1*, normalized to 1.4×10^{17} eV/ml; curve *4*, N_2O-bubbled solution in 2,2,4-trimethylpentane, normalized to 7.35×10^{17} eV/ml; curve *5*, absorption spectrum (obtained with a Cary Model 14 recording spectrophotometer) of $1.8 \times 10^{-5}M$ aqueous Wurster's blue solution, prepared by oxidation of TMPD with N_2O_4. Taken from ref. 441.

Figure 13, $G(TMPD^{\ddot{+}})$ was calculated to be 0.16 in cyclohexane and 0.21 in 2,2,4-trimethylpentane (441). The *G*-value in cyclohexane, 0.16, agrees very well with that for the free ions in cyclohexane, 0.15, determined by the clearing field method (363,364), but the yield in 2,2,4-trimethylpentane, 0.21, is somewhat lower than the free ion yield in 2,2,4-trimethylpentane, 0.33 (363,364). Apparently, all of the ions formed in cyclohexane but only some of those formed in 2,2,4-trimethylpentane transfer their charges to TMPD.

Pulse radiolysis of TMPD in dimethylsulfoxide revealed the predominate formation TMPD^{+} (437). This assignment was based on the observed second-order decay of the transient and the lack of an effect of nitrous oxide, naphthalene, ferric acetylacetonate, and oxygen on the initial absorption spectra of pulse-irradiated TMPD solutions (437). Reactions (161)–(164) have been suggested to occur in this system.

$$DMSO \rightsquigarrow DMSO^{+} + e^{-} \qquad (161)$$

$$DMSO^{+} + TMPD \longrightarrow DMSO + TMPD^{+} \qquad (162)$$

$$e^{-} + DMSO \longrightarrow DMSO^{-} \qquad (163)$$

$$TMPD^{+} + DMSO^{-} \longrightarrow TMPD + DMSO \qquad (164)$$

The observed second-order decay of TMPD^{+} corresponds to eq. (164).

Reactions similar to (161)–(164) have been observed recently in the nanosecond pulse radiolysis of CCl$_4$ (442). CCl$_4{}^{+}$ was found to have an absorption maximum at 475 nm which was not affected by oxygen or nitrous oxide but was removed by the addition of methanol; thereby, the presence of radical cations was confirmed and the possibility of ^{3}CCl$_4$ being the transient was excluded. CCl$_4{}^{+}$ decayed in two consecutive steps. A rapid first-order process with a half-life of 15 ± 2 nsec was followed by a much slower process, $t_{1/2}$ = several μsec (442). When CCl$_4$ was pulse irradiated in the presence of aniline or TMPD (442), new absorptions characteristic of aniline^{+} (444) and TMPD^{+} (441) appeared at the expense of the absorption at 475 nm due to CCl$_4{}^{+}$. The spectrum of CCl$_4{}^{+}$ was also observed in the presence of biphenyl, toluene, benzene, anthracene, and pyrene (442).

Radical cations of biphenyl, p-terphenyl, $trans$-stilbene, and anthracene have been observed by pulse radiolysis in 1,1-dichloroethane, 1,2-dichloroethane, 1,1,2-trichloroethane, 1,1,2,2-tetrachloroethane, and n-butyl chloride (445). The chlorinated solvent, RCl, produces secondary electrons (eq. (165)) which are preferentially scavenged by the solvent rather than by the solute (eq. (146)).

$$e^{-} + RCl \longrightarrow RCl^{+} + 2e^{-} \qquad (165)$$

The positive charge center can migrate between the solvent molecules (eq. (166)) and can transfer its charge to the aromatic solute (eq. (167)) in competition with ion recombination (eqs. (168) and (169)).

$$RCl_2^+ + RCl \longrightarrow RCl + RCl_2^+ \tag{166}$$

$$RCl_2^+ + Ar \longrightarrow Ar_2^+ + RCl \tag{167}$$

$$RCl_2^+ + Cl^- \longrightarrow RCl + \cdot Cl \tag{168}$$

$$Ar_2^+ + Cl^- \longrightarrow Ar + \cdot Cl \tag{169}$$

In chlorinated solvents the favorable competition of reaction (167) with reactions (168) and (169) allowed the pulse radiolytic observation of radical cations, which typically were formed within 0.2 μsec and had lifetimes exceeding a few μsec (445). Since the transient absorption spectra of these radical cations were found to be very similar to those of anion radicals, specific scavengers were used to substantiate their proposed structures. Neither nitrous oxide nor oxygen, efficient electron and anion radical scavengers, decreased the absorbance or the lifetime of the transient species produced in these systems; however, aniline and dimethylaniline, which do not scavenge radical anions in ethanol (124), were found to decrease the lifetimes of the radical cations. The charge transfer equilibrium (eq. (170)), a process analogous to eqs. (149) and (150), has also been observed for a number of donor–acceptor pairs.

$$Ar_1^+ + Ar_2 \rightleftharpoons Ar_1 + Ar_2^+ \tag{170}$$

The direction of this equilibrium was found to be dependent on the relative ionization potential of the aromatic solute (445).

The formation of radical cations rather than triplet states in dimethylsulfoxide (437), CCl_4 (442), and other chlorinated solvents (445) emphasizes the importance of the solvent in the generation of a particular species. Further work will certainly shed a considerable amount of light on the role of solvents in these systems.

The formation of triphenylmethyl cation, $(C_6H_5)_3C^+$, and triphenylmethyl radical, $(C_6H_5)_3C\cdot$, in pulse-irradiated $10^{-3}M$ solutions of triphenylcarbinol in cyclohexane has been reported recently (446). The assignment of the observed spectra for these species was based on the absorption spectrum of the stable carbonium ion which is easily obtained by dissolving triphenylcarbinol in concentrated sulfuric acid and on the known absorption spectrum of the neutral triphenylmethyl radical (81,447). The decay of the carbonium ion and the neutral radical can be followed without the addition of scavengers since the absorption spectra of these species do not overlap (Fig. 14). Both the carbonium ion and the radical decayed by clean second-order kinetics; the half-life for the former was 3.0×10^{-5} sec, however the latter was somewhat longer lived (446). The carbonium ion yield was found to be independent of the triphenylcarbinol concentration from 10^{-1} to $10^{-5}M$, but the radical yield showed a square

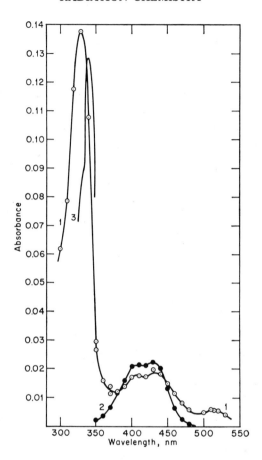

Fig. 14. Curve *1*, spectrum obtained on pulse radiolysis of $10^{-3}M$ $(C_6H_5)_3COH$ in cyclohexane. Curve *2*, spectrum of $(C_6H_5)_3COH$ dissolved in concentrated sulfuric acid. Curve *3*, spectrum of neutral $(C_6H_5)_3C\cdot$ in toluene. Curves *1* and *2* are taken from ref. 446 and curve *3* from ref. 447.

root dependence (446). The mechanism for the carbonium ion formation was interpreted in terms of a proton transfer to the carbinol from the free solvent ions which escape initial recombinations (446). The pulse radiolytic observation of triphenylmethyl carbonium ion opened the door to kinetic investigations of rapid carbonium ion reactions in the presence of suitable solute and scavengers in different solvents. Physical organic chemists can anticipate the rapid accumulation of interesting and important information in this area.

4. Charge Transfer Complexes

The transient absorption spectra of several charge-transfer complexes have been observed in pulse-irradiated solutions of aromatic compounds in carbon tetrachloride (448,449). Figure 15 illustrates the absorption spectra of pure benzene, pure carbon tetrachloride, and their mixtures at the end of ca. 2 μsec pulses. The transient spectra of the pulse-irradiated benzene in carbon tetrachloride consist of a band in the ultraviolet region and another centered at 490 nm, neither of which could be observed in the transient spectra of either component at the time resolution used (449). The absorption at 490 nm has been assigned to a charge transfer complex in which the chlorine atom functions as an electron acceptor and the benzene molecule as an electron donor (eq. (171)).

$$\cdot Cl + C_6H_6 \underset{\longleftarrow}{\overset{K_I}{\longrightarrow}} Cl \leftarrow C_6H_6 \underset{\longleftarrow}{\overset{K_{II}}{\longrightarrow}} Cl^- + C_6H_6^+ \tag{171}$$

The transient absorption cannot be attributed to excited species since carbon tetrachloride is an efficient scavenger of these species and since

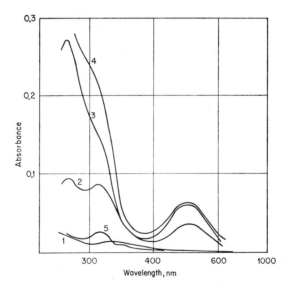

Fig. 15. Absorption spectra obtained on pulse radiolysis of carbon tetrachloride, benzene, and several concentrations of benzene in CCl₄: curve 1, CCl₄; curve 2, 2 × 10⁻⁴M benzene in CCl₄; curve 3, 10⁻²M benzene in CCl₄; curve 4, 3M benzene in CCl₄; curve 5, benzene. Taken from ref. 449.

identical spectra were obtained in the presence and in the absence of oxygen (448,449). Analogous results have been reported for the flash photolytic observation of iodine atom charge transfer complexes with different electron donors (450). The possibility of the formation of the complex with trichloromethyl radical, $\cdot CCl_3$, and of its absorption being at 490 nm was eliminated by the observation of identical absorptions at 490 nm from pulse-irradiated pure chlorobenzene and from a chlorobenzene solution of carbon tetrachloride (448). The reported linear dependence of the absorption band energy, $h\nu_{max}$, of the chlorine atom charge transfer complexes on the ionization potentials of such donors as o-dichlorobenzene, t-butylbenzene, benzene, cumene, chlorobenzene, bromobenzene, toluene, o-xylene, biphenyl, naphthalene, 1-methylnaphthalene, anthracene, and naphthacene (448,449) has, however, substantiated the structure of the proposed charge-transfer complexes. Such linear correlations between $h\nu_{max}$ and the ionization potential of the donor have been used as evidence for charge-transfer complexes of the radical type (450) as well as of the molecular type (451).

The chlorine atom acceptor–aromatic donor complexes generally were formed in carbon tetrachloride in less than 0.3 μsec and consequently using μsec resolution the rate constants for complex formation could not be measured. Depending on the concentration of the donor and on that of the acceptor, i.e., the absorbed dose, the transient charge transfer complexes were sufficiently long lived so that the kinetics of their decay could be followed. At the lower doses (less than 1000 rads) the decay obeyed first-order kinetics, but at higher doses mixed second- and first-order processes were observed. A typical half-life of 24 μsec has been reported for the chlorine atom–benzene charge transfer complex decomposition at 210 rads and $10^{-2}M$ benzene in carbon tetrachloride (449). The Benesi-Hildebrand equation (452),

$$\frac{[\cdot Cl]L}{OD_{490}} = \frac{1}{\varepsilon} + \frac{1}{\varepsilon K_I[C_6H_6]} \tag{172}$$

where L is the path length, allowed the calculation of the equilibrium constant for the charge transfer complex formation in eq. (171). The reported value of ε for the chlorine atom–benzene charge transfer complex in CCl_4 was estimated to be 1600 M^{-1} cm^{-1} and the lower limit for K_I has been calculated to be 1000 M^{-1} (449). Uncertainties in these values rise from the inaccuracies inherent in the Benesi-Hildebrand plot and from the difficulties in measuring *initial* absorbances of the charge-transfer complexes. The magnitude of the equilibrium constant K_I nevertheless indicates that the species in eq. (171) are overwhelmingly in the form of the complex. The rates and hence the equilibrium constants for the transient charge-transfer complexes depend markedly on the nature of the electron donor and on the solvent. No charge-transfer complexes have been detected in pulse-irradiated solutions of phenol, aniline, and N,N-dimethylaniline in carbon tetrachloride (448,449). The transient spectra in these cases resembled those of phenoxyl (168,169,453) and anilino radicals (168,169). The μsec resolution of the spectra, however, does not permit exclusion of the initial formation of charge-transfer complexes followed by their rapid decomposition. The stability of the transient charge-transfer complexes was found to decrease when carbon tetrachloride was replaced by methylene chloride and by chloroform (449). The G-values for the chlorine–aromatic donor charge transfer formation corresponded closely to that for the chlorine atom, $G(\cdot Cl) = 2.34$ (454), indicating that complex formation is a main part of the radiation-induced reactions in these systems.

The determination of the transient charge-transfer spectra for different donor–acceptor systems and a systematic study of the parameters affecting the complex stabilities should be of considerable interest to researchers in several areas of chemistry. At the present time, the mechanism of halogen atom formation in pulse-irradiated haloaliphatic organic liquids, the occurrence of side reactions both in the formation and in the decomposition of transient charge-transfer complexes, and the products formed in these reactions have not been elucidated. The application of nanosecond pulse radiolysis to these and related systems can provide the solutions to such mechanistic problems (432,444).

5. Valence Isomers

A great deal of interest has been shown in the preparation and chemistry of Dewar benzenes (455), benzvalenes (456), and prismanes— the "valence isomers" of benzene (457). Photochemical, gas–liquid partition chromatographic, and absorption, mass, and nuclear magnetic resonance spectroscopic techniques have been used for the preparation, isolation, and characterization of these compounds (455–457).

Radiation-induced valence isomerizations of hexafluorobenzene (458), octafluorotoluene (459), and o-decafluoroxylene (459) have been reported recently. The parent liquids were irradiated in a cobalt-60 source, and the products were separated by gas–liquid partition chromatography (458,459). Comparison of the isolated products with authentic samples prepared photochemically and analyses of their ^{19}F nuclear magnetic resonance

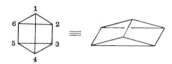

Dewar benzene
Bicyclo [2.2.0] hexa-2,5-diene

Benzvalene
Tricyclo [3.1.0.0^{2,6}] hex-3-ene

Prismane
Tetracyclo [2.2.0.0^{2,6}0^{3,5}] hexane

spectra (460,461) revealed that the products were substituted Dewar benzenes, e.g., hexafluorobicyclo[2.2.0]hexa-2,5-diene (458,459).

Radiolysis of Dewar hexafluorobenzene resulted primarily in reformation of the Kekule hexafluorobenzene, with a G-value of ca. 10, and in polymer formation (459). In addition, a new product was observed gas chromatographically which decomposed to Dewar and Kekule hexafluorobenzenes. This new product has been tentatively ascribed to be hexafluoroprismane (459) by analogy with the behavior of hexamethylprismane (457, 462).

The mechanism for the radiation-induced formation of these substituted Dewar benzenes is not clear. Direct excitation or ion–electron recombinations followed by collisional deactivation may be responsible for their formation (459). It is very improbable that triplet-excited intermediates are involved since oxygen did not affect the photochemical valence isomerization of hexafluorobenzene (463).

It is possible that radiation-induced valence isomerizations also occur in hydrocarbons, in which case the interconvertible isomers may act as an "energy sink" to store a part of the energy absorbed by the solvent (459). The isolation and characterization of other radiation-induced valence isomers and possibly the observation of their reactions pulse radiolytically will help to clarify a number of points interesting to both radiation and physical organic chemists.

V. CONCLUSION

The application of radiation chemistry to the study and elucidation of organic reaction mechanisms has been outlined in this review. Repeatedly we suggested possibilities for further work and pointed the way toward new avenues of research. If physical organic chemists accept radiation chemistry as a powerful tool in their armory and apply it to well designed and executed experiments, our efforts in writing this review will be more than adequately rewarded. The results of such interdisciplinary studies will, in turn, lead to new and novel information on the nature and properties of the primary species involved in radiation chemistry.

Acknowledgments

The authors wish to express their sincere gratitude to their colleagues both in the U.S. and abroad for their invaluable comments and constructive criticism and for their willingness to provide information and manuscripts prior to publication. The permission which was kindly granted by the copyright holders to reproduce Figures 2, 3, and 5–15 is gratefully acknowledged.

References

1. H. Eyring, C. J. Christensen, and H. S. Johnston, *Ann. Rev. Phys. Chem.*, Annual Reviews, Palo Alto, Calif., 1950 to present.
2. E. Segré, G. Friedlander, and H. P. Hoyes, *Ann. Rev. Nucl. Sci.*, Annual Reviews, Palo Alto, Calif., 1952 to present.
3. M. Haïssinsky, Ed., *Actions Chim. Biol. Radiations*, Masson et Cie, Paris, 1955 to present.
4. M. Ebert and A. Howard, *Current Topics in Radiation Research*, Interscience, New York, 1965 to present.
5. L. G. Augenstein, R. Mason, and H. Quastler, *Advan. Radiation Biol.*, Academic Press, New York, 1964 to present.
6. M. Burton and J. L. Magee, *Advan. Radiation Chem.*, *1*, Wiley, New York, 1969.
7. G. Porter, *Progr. Reaction Kinetics*, Pergamon Press, New York, 1961 to present.
8. G. O. Phillips, R. B. Cundall, and F. S. Dainton, *Radiation Res. Rev.*, *1*, Elsevier, Amsterdam, 1968 to present.
9. Proceedings of the 1962 Tihany Symposium on *Radiation Chemistry*, J. Dobó, Ed., Akadémiai Kiadó, Budapest, 1964.
10. *Pulse Radiolysis*, Proceedings of the International Symposium held at Manchester, April, 1965, M. Ebert, J. P. Keene, A. J. Swallow, and J. H. Baxendale, Eds., Academic Press, New York, 1965.
11. *Energy Transfer in Radiation Processes*, Proceedings of the International Symposium held at Cardiff, January 1965, G. O. Phillips, Ed., Elsevier, New York, 1966.
12. *Solvated Electron*, *Advan. Chem. Ser.*, *50*, 1965.

13. Proceedings of the Second Tihany Symposium on *Radiation Chemistry*, J. Dobó and P. Hedvig, Eds., Akadémiai Kiadó, Budapest, 1967.
14. *The Chemistry of Ionization and Excitation*, Proceedings of a Conference on Radiation Chemistry and Photochemistry, University of Newcastle upon Tyne, September 21–23, 1966, G. R. A. Johnson and G. Scholes, Eds., Taylor and Francis Ltd., London, 1967.
15. Symposium on Photochemistry and Radiation Chemistry, U.S. Army Natick Laboratories, April 22–24, 1968; *J. Phys. Chem.*, *72*, 3709–3935 (1968).
16. *Radiation Chem.*, Vols. I and II, *Advan. Chem. Ser.*, *81, 82*, 1968.
17. *Radiation Chemistry of Aqueous Systems*, Proceedings of the 19th Farkas Symposium, December 27–29, 1967, Israel, Wiley, New York, 1969.
18. A. J. Swallow, *Radiation Chemistry of Organic Compounds*, Pergamon Press, New York, 1960.
19. M. Anbar, in *Fundamental Processes in Radiation Chemistry*, P. Ausloos, Ed., Wiley, New York, 1968, p. 651.
20. J. W. T. Spinks and R. J. Woods, *An Introduction to Radiation Chemistry*, Wiley, New York, 1964.
21. I. V. Vereshchinskii and A. K. Pikaev, *Introduction to Radiation Chemistry*, Daniel Davey, New York, 1964.
22. A. O. Allen, *The Radiation Chemistry of Water and Aqueous Solutions*, Van Nostrand, Princeton, N.J., 1961.
23. E. J. Hart and R. L. Platzman, in *Mechanisms and Radiobiology*, Vol. 1, M. Errera and A. Forssberg, Eds., Academic Press, New York, 1961, p. 93.
24. H. Fricke, E. J. Hart, and H. P. Smith, *Radiation Res.*, *17*, 262 (1962).
25. J. H. O'Donnell and D. F. Sangster, *Principles of Radiation Chemistry*, Edward Arnold, London, 1969.
26. A. O. Allen, *J. Chem. Ed.*, *45*, 290 (1968).
27. R. W. Hummel, G. R. Freeman, A. B. Van Cleave, and J. W. T. Spinks, *Science*, *119*, 159 (1954).
28. G. R. Freeman, A. B. Van Cleave, and J. W. T. Spinks, *Can. J. Chem.*, *32*, 322 (1954).
29. R. F. Platford and J. W. T. Spinks, *Can. J. Chem.*, *37*, 1022 (1959).
30. K. Schmidt, *Z. Naturforsch.*, *16b*, 206 (1961).
31. F. S. Dainton and J. Rowbottom, *Trans. Faraday Soc.*, *49*, 1160 (1953).
32. K. N. Jha and G. R. Freeman, *J. Chem. Phys.*, *48*, 5480 (1968).
33. J. C. Russell and G. R. Freeman, *J. Phys. Chem.*, *72*, 816 (1968).
34. A. Weissberger, *Technique of Organic Chemistry*, Vol. 3, Interscience, New York, 1956.
35. Rept. of the Intern. Comm. on Radiological Units and Measurements (ICRU), Nat. Bureau Stand. (U.S.), Handbook 78, 1961.
36. G. N. Whyte, *Principles Radiation Dosimetry*, 2nd ed., Wiley, New York, 1959.
37. F. H. Attix and W. C. Roesch, Eds., *Radiation Dosimetry*, Academic Press, New York, 1968.
38. J. Weiss, A. O. Allen, and H. A. Schwarz, *Proc. Intern. Conf. Peaceful Uses Atomic Energy*, United Nations, New York, *14*, 179 (1956).
39. L. M. Dorfman and M. S. Matheson, *Progr. Reaction Kinetics*, *3*, 237 (1965).
40. R. W. Fessenden and R. H. Schuler, *Advan. Radiation Chem.*, *2* (1969).
41. R. A. Holroyd, in *Aspects of Hydrocarbon Radiolysis*, T. Gäumann and J. Hoigné, Eds., Academic Press, New York, 1968.

42. R. H. Schuler, *J. Phys. Chem.*, *62*, 37 (1958).
43. I. Mani and R. J. Hanrahan, *J. Phys. Chem.*, *70*, 2233 (1966).
44. D. Perner and R. H. Schuler, *J. Phys. Chem.*, *70*, 2224 (1966).
45. J. L. Magee, *Ann. Rev. Phys. Chem.*, *12*, 389 (1961).
46. F. Williams, *J. Am. Chem. Soc.*, *86*, 3954 (1964).
47. G. R. Freeman and J. M. Fayadh, *J. Chem. Phys.*, *43*, 86 (1965).
48. A. Hummel, A. O. Allen, and F. H. Watson, Jr., *J. Chem. Phys.*, *44*, 3431 (1966).
49. A. Mozumder and J. L. Magee, *J. Chem. Phys.*, *47*, 939 (1967).
50. A. Mozumder, *J. Chem. Phys.*, *48*, 1659 (1968).
51. A. Hummel, *J. Chem. Phys.*, *48*, 3268 (1968).
52. S. J. Rzad and J. M. Warman, *J. Chem. Phys.*, *49*, 2861 (1968).
53. K.-D. Asmus and J. H. Fendler, *J. Phys. Chem.*, *72*, 4285 (1968).
54. K.-D. Asmus and A. Henglein, *Ber. Bunsenges. Phys. Chem.*, *68*, 348 (1964).
55. F. S. Dainton and D. B. Peterson, *Proc. Roy. Soc. (London)*, Ser. A, *267*, 443 (1962).
56. G. Scholes and M. Simic, *Nature*, *202*, 895 (1964).
57. J. M. Warman, K.-D. Asmus, and R. H. Schuler, in *Radiation Chemistry—II*, *Advan. Chem. Ser. 82*, 25 (1968).
58. H. C. Christensen, *Nukleonik*, 7, 1 (1965) and references cited therein.
59. J. Goodman and J. Steigman, *J. Phys. Chem.*, *62*, 1020 (1958).
60. G. Czapski, in *Radiation Chemistry—I*, *Advan. Chem. Ser.*, *81*, 106 (1968).
61. B. H. J. Bielski and A. O. Allen, *Intern. J. Radiation Phys. Chem.*, *Haïssinsky Birthday Issue*, 1969, in press.
62. M. Haïssinsky and M. Magat, *Selected Constants, Radiolytic Yields*, Pergamon Press, New York, 1961.
63. A. K. Pikaev, *Pulse Radiolysis of Water and Aqueous Solutions*, Indiana Univ. Press, Bloomington, Ind., 1967.
64. L. M. Dorfman, *Science*, *141*, 493 (1963).
65. J. P. Keene, in *Pulse Radiolysis*, M. Ebert, J. P. Keene, A. J. Swallow, and J. H. Baxendale, Eds., Academic Press, New York, 1965, p. 1.
66. J. W. Boag, in *Chimiques et Biologiques des Radiations*, M. Haïssinsky, Ed., Masson et Cⁱᵉ, Paris, 1963, p. 1.
67. G. Porter, in *Technique of Organic Chemistry*, 2nd ed., Vol. 8, Part 2, Interscience, New York, 1963, p. 1055.
68. J. W. Boag, *Am. J. Roentgenol. Radium Therapy, Nucl. Med.*, *90*, 896 (1963).
69. K.-D. Asmus, G. Beck, A. Henglein, and A. Wigger, *Ber. Bunsenges. Phys. Chem.*, *70*, 869 (1966).
70. K. Schmidt and W. L. Buck, *Science*, *151*, 70 (1966).
71. Z. P. Zagorski and K. Shested, in *Pulse Radiolysis*, M. Ebert, J. P. Keene, A. J. Swallow, and J. H. Baxendale, Eds., Academic Press, New York, 1965, p. 29.
72. M. A. Dillon and M. Burton, in *Pulse Radiolysis*, M. Ebert, J. P. Keene, A. J. Swallow, and J. H. Baxendale, Eds., Academic Press, New York, 1965, p.259.
73. R. W. Fessenden and R. H. Schuler, *J. Chem. Phys.*, *39*, 2147 (1963).
74. R. W. Fessenden, *J. Phys. Chem.*, *68*, 1508 (1964).
75. E. C. Avery, J. R. Remko, and B. Smaller, *J. Chem. Phys.*, *49*, 951 (1968).
76. L. Kevan, *J. Am. Chem. Soc.*, *87*, 1481 (1965) and references to earlier work cited therein.
77. J. P. Keene, *J. Sci. Instr.*, *41*, 493 (1964).

78. R. W. Fessenden and I. A. Taub, unpublished results.
78a. M. J. Bronskill, R. K. Wolff, and J. W. Hunt, *J. Phys. Chem.*, *73*, 1175 (1969)
79. A. A. Frost and R. G. Pearson, *Kinetics and Mechanism*, 2nd ed., Wiley, New York, 1961.
80. S. W. Benson, *The Foundations of Chemical Kinetics*, McGraw-Hill, New York, 1960.
81. I. A. Taub, D. A. Harter, M. C. Sauer, Jr., and L. M. Dorfman, *J. Chem. Phys.*, *41*, 979 (1964).
82. L. M. Dorfman, I. A. Taub, and R. E. Bühler, *J. Chem. Phys.*, *36*, 3051 (1962).
83. K. Schmidt, Argonne Natl. Lab. Rept. ANL-7400 (1968).
84. E. J. Hart, E. M. Fielden, and M. Anbar, *J. Phys. Chem.*, *71*, 3993 (1967).
85. M. Anbar and P. Neta, *Intern. J. Appl. Radiation Isotopes*, *18*, 493 (1967).
86. M. Anbar, *Selected Specific Rates of Reactions in Aqueous Solution. I. Hydrated Electron. II. Hydrogen Atom. III. Hydroxyl Radical. IV. Perhydroxyl Radical*, to be published, summer 1969 (for information write Radiation Chemistry Data Center, Radiation Laboratory, Univ. of Notre Dame, Notre Dame, Ind. 46556).
87. I. Santar and J. Bednar, in *Radiation Chemistry—I, Advan. Chem. Ser.*, *81*, 523 (1968) and references cited therein.
88. G. V. Buxton, *Radiation Res. Rev.*, *1*, 209 (1968).
89. M. Haïssinsky, *Actions Chimiques et Biologiques des Radiations*, *11*, Masson et Cie, Paris, 1967, p. 133.
90. J. P. Keene, *Radiation Res.*, *22*, 1 (1964).
91. S. Gordon, E. J. Hart, M. Matheson, J. Rabani, and J. K. Thomas, *J. Am. Chem. Soc.*, *85*, 1375 (1963).
92. L. M. Dorfman and I. A. Taub, *J. Am. Chem. Soc.*, *85*, 2370 (1963).
93. S. Gordon, E. J. Hart, M. S. Matheson, J. Rabani, and J. K. Thomas, *Discussions Faraday Soc.*, *36*, 193 (1963).
94. M. S. Matheson and J. Rabani, *J. Phys. Chem.*, *69*, 1324 (1965).
95. E. J. Hart, J. K. Thomas, and S. Gordon, *Radiation Res. Suppl.*, *4*, 74 (1964).
96. E. J. Hart, S. Gordon, and E. M. Fielden, *J. Phys. Chem.*, *70*, 150 (1966).
97. S. Gordon, E. J. Hart, and J. K. Thomas, *J. Phys. Chem.*, *68*, 1262 (1964).
98. H. A. Schwarz, *J. Phys. Chem.*, *67*, 2827 (1963).
99. H. Fricke and J. K. Thomas, *Radiation Res. Suppl.*, *4*, 35 (1964).
100. J. P. Sweet and J. K. Thomas, *J. Phys. Chem.*, *68*, 1363 (1964).
101. S. O. Nielsen, P. Pagsberg, J. Rabani, H. Christensen, and G. Nilsson, *Chem. Commun.*, *1968*, 1523; P. Pagsberg, H. Christensen, J. Rabani, G. Nilsson, and S. O. Nielsen, *J. Phys. Chem.*, *73*, 1029 (1969).
102. J. K. Thomas, J. Rabani, M. S. Matheson, E. J. Hart, and S. Gordon, *J. Phys. Chem.*, *70*, 2409 (1966).
103. F. S. Dainton and S. A. Sills, *Proc. Chem. Soc.*, *1962*, 223.
104. H. A. Schwarz, *J. Phys. Chem.*, *66*, 255 (1962).
105. G. E. Adams, J. W. Boag, J. Currant, and B. D. Michael, in *Pulse Radiolysis*, M. Ebert, J. P. Keene, A. J. Swallow, and J. H. Baxendale, Eds., Academic Press, New York, 1965, p. 131.
106. M. Eigen and L. De Meyer, *Z. Elektrochem.*, *59*, 986 (1955).
107. G. Ertl and H. Gerischer, *Z. Elektrochem.*, *66*, 560 (1962).
108. L. P. Hammett, *Physical Organic Chemistry*, McGraw-Hill, New York, 1940, chapts. 3, 4, and 7.
109. S. Ehrenson, *Progr. Phys. Org. Chem.*, *2*, 195 (1964).

110. C. D. Ritchie and W. F. Sager, *Progr. Phys. Org. Chem.*, *2*, 323 (1964).
111. P. R. Wells, *Chem. Rev.*, *63*, 171 (1963).
112. J. E. Leffler and E. Grunwald, *Rates and Equilibria of Organic Reactions*, Wiley, New York, 1963.
113. H. H. Jaffé, *Chem. Rev.*, *53*, 191 (1953).
114. D. C. Walker, *Quart. Rev.*, *21*, 79 (1967) and references cited therein.
115. J. W. Boag and E. J. Hart, *Nature*, *197*, 45 (1963).
116. E. J. Hart and J. W. Boag, *J. Am. Chem. Soc.*, *84*, 4090 (1962).
117. J. P. Keene, *Nature*, *197*, 47 (1963).
118. M. Anbar, *Advan. Phys. Org. Chem.*, *7*, 115 (1969).
119. E. J. Hart and M. Anbar, to be published.
120. M. Anbar and E. J. Hart, *J. Am. Chem. Soc.*, *86*, 5633 (1964).
121. L. M. Stock and H. C. Brown, *Advan. Phys. Org. Chem.*, *1*, 35 (1963).
122. E. Berliner, *Progr. Phys. Org. Chem.*, *2*, 253 (1964).
123. S. D. Ross, *Progr. Phys. Org. Chem.*, *1*, 31 (1963).
124. S. Arai and L. M. Dorfman, *J. Chem. Phys.*, *41*, 2190 (1964).
125. R. S. Alger, *Electron Paramagnetic Resonance: Techniques and Applications*, Interscience, New York, 1968.
126. M. H. Studier and E. J. Hart, *J. Phys. Chem.*, in press.
127. S. Gordon, *Radiation Res. Suppl.*, *4*, 21 (1964).
128. L. M. Dorfman, N. E. Shank, and S. Arai, in *Radiation Chemistry—II*, *Advan. Chem. Ser.*, *82*, 58 (1968).
129. I. A. Taub, M. C. Sauer, and L. M. Dorfman, *Discussions Faraday Soc.*, *36*, 206 (1963).
130. M. Anbar and E. J. Hart, *J. Phys. Chem.*, *69*, 271 (1965).
131. C. A. Bunton, *Nucleophilic Substitution at a Saturated Carbon Atom*, Elsevier, New York, 1963.
132. R. Braams, in *Pulse Radiolysis*, M. Ebert, J. P. Keene, A. J. Swallow, and J. H. Baxendale, Eds., Academic Press, New York, 1965, p. 171.
133. M. Anbar, Z. B. Alfassi, and H. Bregman-Reisler, *J. Am. Chem. Soc.*, *89*, 1263 (1967) and references cited therein.
134. M. Anbar and E. J. Hart, *J. Phys. Chem.*, *71*, 3700 (1967).
135. B. Cercek and M. Ebert, *J. Phys. Chem.*, *72*, 766 (1968).
135a. S. R. Logan, *J. Phys. Chem.*, *73*, 227 (1969).
136. P. Neta and L. M. Dorfman, *J. Phys. Chem.*, *73*, 413 (1969).
137. M. Anbar and D. Meyerstein, *J. Phys. Chem.*, *68*, 3184 (1964).
138. M. Anbar and P. Neta, *J. Chem. Soc. (A)*, *1967*, 834.
139. M. Anbar, D. Meyerstein, and P. Neta, *Nature*, *209*, 1348 (1966).
140. M. C. Sauer, Jr., and B. Ward, *J. Phys. Chem.*, *71*, 3971 (1967).
141. E. J. Land and M. Ebert, *Trans. Faraday Soc.*, *63*, 1181 (1967).
142. K.-D. Asmus, B. Cercek, M. Ebert, A. Henglein, and A. Wigger, *Trans. Faraday Soc.*, *63*, 2435 (1967).
143. G. Stein, in *Energetics and Mechanism in Radiation Biology*, G. O. Phillips, Ed., Academic Press, New York, 1968, p. 467.
144. G. Navon and G. Stein, *J. Phys. Chem.*, *69*, 1384 (1965).
145. G. Navon and G. Stein, *J. Phys. Chem.*, *69*, 1389 (1965).
146. G. Navon and G. Stein, *J. Phys. Chem.*, *70*, 1390 (1966).
147. G. Navon and G. Stein, *Israel J. Chem.*, *2*, 151 (1964).
148. G. Czapski, N. Frohwirth, and G. Stein, *Nature*, *207*, 1191 (1965).

149. L. K. Mee, G. Navon, and G. Stein, *Nature*, *204*, 1056 (1964).
150. L. K. Mee, G. Navon, and G. Stein, *Biochim. Biophys. Acta*, *104*, 151 (1965).
151. B. E. Holmes, G. Navon, and G. Stein, *Nature*, *213*, 1087 (1967).
152. R. O. C. Norman and G. K. Radda, *Proc. Chem. Soc.*, *1962*, 138.
153. W. T. Dixon, R. O. C. Norman, and A. L. Buley, *J. Chem. Soc.*, *1964*, 3625.
154. I. Kraljic and C. N. Trumbore, *J. Am. Chem. Soc.*, *87*, 2547 (1965).
155. M. Anbar, D. Meyerstein, and P. Neta, *J. Phys. Chem.*, *70*, 2660 (1966).
156. F. S. Dainton and B. Wiseall, *Trans. Faraday Soc.*, *64*, 694 (1968).
157. P. Neta and L. M. Dorfman, in *Radiation Chemistry—I, Advan. Chem. Ser.*, *81*, 222 (1968).
158. J. H. Fendler and G. L. Gasowski, *J. Org. Chem.*, *33*, 1865 (1968).
159. M. Anbar, D. Meyerstein, and D. Neta, *J. Chem. Soc. (B)*, *1966*, 742.
160. C. H. Hochanadel, *Radiation Res.*, *17*, 286 (1962).
161. A. Hummell and A. O. Allen, *Radiation Res.*, *17*, 302 (1962).
162. B. Cercek, *J. Phys. Chem.*, *72*, 3832 (1968).
163. A. Streitwieser, *Molecular Orbital Theory for Organic Chemists*, Wiley, New York, 1961.
164. S. W. Charles, J. T. Pearson, and E. Whittle, *Trans. Faraday Soc.*, *59*, 1156 (1963).
165. K.-D. Asmus, A. Wigger, and A. Henglein, *Ber. Bunsenges. Phys. Chem.*, *70*, 862 (1966).
166. K.-D. Asmus, A. Henglein, A. Wigger, and G. Beck, *Ber. Bunsenges. Phys. Chem.*, *70*, 756 (1966); M. Simic, P. Neta, and E. Hayon, *J. Phys. Chem.*, to be published.
167. J. Lilie and A. Henglein, *Ber. Bunsenges. Phys. Chem.*, in press.
168. E. J. Land, in *Progr. Reaction Kinetics*, *3*, 369 (1965).
169. A. Habersbergerová, I. Janovsky, and J. Teplý, *Radiation Res. Rev.*, *1*, 109 (1968).
170. S. Arai and M. C. Sauer, *J. Chem. Phys.*, *44*, 2297 (1966).
171. M. Anbar and E. J. Hart, *J. Phys. Chem.*, *69*, 1244 (1965).
172. W. C. Gottschall and E. J. Hart, *J. Phys. Chem.*, *71*, 2102 (1967).
173. J. H. Baxendale, *Current Topics Radiation Res.*, *3*, 1 (1967).
174. B. Chutny, *Nature*, *213*, 593 (1967) and references cited therein.
175. B. Cercek, *J. Phys. Chem.*, *71*, 2354 (1967) and unpublished results.
176. A. Wigger, A. Henglein, and K.-D. Asmus, *Ber. Bunsenges. Phys. Chem.*, *71*, 513 (1967).
176a. K.-D. Asmus, J. H. Fendler, W. Grünbein, and A. Henglein, unpublished results.
177. E. J. Land, G. Porter, and E. Strachan, *Trans. Faraday Soc.*, *57*, 1885 (1961).
178. W. T. Dixon and R. O. C. Norman, *Proc. Chem. Soc.*, *1963*, 97.
179. G. R. A. Johnson, G. Stein, and J. Weiss, *J. Chem. Soc.*, *1951*, 3275.
180. J. H. Fendler, unpublished results.
181. B. Cercek and M. Ebert, in *Radiation Chemistry—I, Advan. Chem. Ser.*, *81*, 210 (1968).
181a. W. Grünbein and A. Henglein, *Ber. Bunsenges. Phys. Chem.*, in press.
182. C. Corvaja, G. Farna, and E. Vianello, *Electrochim. Acta*, *11*, 919 (1966).
182a. A. Henglein and W. Grünbein, unpublished results.
183. O. Volkert, G. Fermens, and D. Schulte-Frohlinde, *Z. Phys. Chem. (Frankfurt)*, *56*, 261 (1967).

184. N. Waterman and H. Limbourg, *Biochem. Z.*, *263*, 400 (1933).
185. J. H. Baxendale and D. Smithies, *Z. Phys. Chem. (Frankfurt)*, 7, 242 (1956).
186. C. Vermeil and L. Salomon, *Compt. Rend.*, *249*, 268 (1959).
187. J. H. Baxendale and G. Hughes, *Z. Phys. Chem. (Frankfurt)*, *14*, 306 (1958).
188. H. Diebler, M. Eigen, and P. Mathies, *Z. Elektrochem.*, *65*, 634 (1961).
189. H. Diebler, M. Eigen, and P. Mathies, *Z. Naturforsch.*, *16b*, 629 (1961).
190. L. Michaelis, *Ann. N.Y. Acad. Sci.*, *11*, 37 (1940).
191. L. Michaelis, *Chem. Rev.*, *16*, 243 (1935).
192. L. Michaelis and M. P. Schubert, *Chem. Rev.*, *22*, 437 (1938).
193. G. M. Coppinger, *J. Am. Chem. Soc.*, *79*, 501 (1957).
194. T. H. James, J. M. Snell, and A. Weissberger, *J. Am. Chem. Soc.*, *60*, 2085 (1938).
195. L. Loisleur and R. Latarjet, *Compt. Rend. Soc. Biol.*, *135*, 1534 (1941).
196. C. Vermeil, G. Roquet, and L. Salomon, *J. Chim. Phys.*, *60*, 659 (1963).
197. B. H. J. Przybielski-Bielski and R. R. Becker, *J. Am. Chem. Soc.*, *82*, 2164 (1960).
198. G. W. Black and B. H. J. Bielski, *J. Phys. Chem.*, *66*, 1203 (1962).
199. B. H. J. Bielski and J. M. Gebicki, *Advan. Radiation Chem.*, to be published.
200. G. Czapski and L. M. Dorfman, *J. Phys. Chem.*, *68*, 1169 (1964).
201. G. E. Adams and B. D. Michael, *Trans. Faraday Soc.*, *63*, 1171 (1967).
202. E. J. Hart, S. Gordon, and J. K. Thomas, *J. Phys. Chem.*, *68*, 1271 (1964).
203. R. O. C. Norman and R. J. Pritchett, *J. Chem. Soc. (B)*, *1967*, 926.
204. D. F. Sangster, *J. Phys. Chem.*, *70*, 1712 (1966).
205. C. B. Amphlett, G. E. Adams, and B. D. Michael, in *Radiation Chemistry—I*, *Advan. Chem. Ser.*, *81*, 231 (1968).
206. D. Turnbull and S. H. Maron, *J. Am. Chem. Soc.*, *65*, 212 (1943).
207. G. W. Wheland, *Advanced Organic Chemistry*, Wiley, New York, 1960, p. 719.
208. K.-D. Asmus, A. Henglein, and G. Beck, *Ber. Bunsenges. Phys. Chem.*, *70*, 459 (1966).
209. K. Eiben and R. W. Fessenden, *J. Phys. Chem.*, *72*, 3387 (1968).
210. K.-D. Asmus and I. A. Taub, *J. Phys. Chem.*, *72*, 3882 (1967).
210a. K. Eiben and R. W. Fessenden, unpublished results.
211. M. J. Day and G. Stein, *Nature*, *164*, 671 (1949).
212. M. E. J. Carr, *Nature*, *167*, 363 (1951).
213. F. T. Farmer, G. Stein, and J. Weiss, *J. Chem. Soc.*, *1949*, 3241 and subsequent papers.
214. R. O. C. Norman and J. R. Lindsay Smith, *Oxidases and Related Reactions*, T. E. King, H. S. Mason, and M. Morrison, Eds., Wiley, New York, 1965, p. 131, and references cited therein.
215. F. Haber and J. Weiss, *Proc. Roy. Soc. (London)*, *Ser. A*, *147*, 332 (1934).
216. W. G. Barb, J. H. Baxendale, P. George, and K. R. Hargrave, *Trans. Faraday Soc.*, *47*, 462 (1951).
217. J. H. Merz and W. A. Waters, *J. Chem. Soc.*, *1949*, 2427.
218. J. R. Lindsay Smith and R. O. C. Norman, *J. Chem. Soc.*, *1963*, 2897.
219. J. H. Baxendale, M. G. Evans, and G. S. Park, *Trans. Faraday Soc.*, *42*, 155 (1946).
220. L. H. Piette, G. A. Bulow, and K. O. Loeffler, Symposium on Use of ESR in Elucidation of Reaction Mechanism, Vol. 9, No. 2-C, Amer. Chem. Soc., Washington, 1964.

221. T. Shiga, *J. Phys. Chem.*, *69*, 3805 (1965).
222. R. O. C. Norman and B. C. Gilbert, *Advan. Phys. Org. Chem.*, 5, 53 (1967).
223. H. Fischer, *Adv. Polymer Sci.*, *5*, 463 (1968).
224. G. A. Hamilton and J. P. Friedman, *J. Am. Chem. Soc.*, *85*, 1008 (1963).
225. S. Udenfriend, C. T. Clark, J. Axelrod, and B. B. Brodie, *J. Biol. Chem.*, *208*, 731 (1954).
226. C. Mitoma, H. S. Posner, H. C. Reitz, and S. Udenfriend, *Arch. Biochem.*, *61*, 431 (1956).
227. O. Hayaishi, *Ann. Rev. Biochem.*, *31*, 25 (1962).
228. H. Kampffmeyer and M. Kiese, *Naunyn-Schmiedebergs Arch. Exp. Path. Pharmak.*, *250*, 1 (1965) and references cited therein.
229. A. I. Titov, *Tetrahedron*, *19*, 557 (1963) and references cited therein.
230. G. H. Williams, *Homolytic Aromatic Substitution*, Pergamon Press, New York, 1960.
231. P. V. Phung and M. Burton, *Radiation Res.*, *7*, 199 (1957).
232. J. H. Baxendale and D. H. Smithies, *J. Chem. Soc.*, *1959*, 779.
233. G. R. Freeman, A. B. Van Cleave, and J. W. T. Spinks, *Can. J. Chem.*, *31*, 448 (1953).
234. H. Hotta and A. Terakawa, *Bull. Chem. Soc. Japan*, *33*, 335 (1960).
235. I. Loeff and G. Stein, *J. Chem. Soc.*, *1963*, 2623.
236. H. Minder, W. Minder, and A. Liechti, *Radiol. Clin.*, *18*, 108 (1949).
237. G. Stein and J. Weiss, *J. Chem. Soc.*, *1951*, 3265.
238. J. H. Fendler and G. L. Gasowski, *J. Org. Chem.*, *33*, 2755 (1968).
239. H. Loebel, G. Stein, and J. Weiss, *J. Chem. Soc.*, *1950*, 2704.
240. R. W. Matthews and D. F. Sangster, *J. Phys. Chem.*, *71*, 4056 (1967).
241. H. Loebl, G. Stein, and J. Weiss, *J. Chem. Soc.*, *1951*, 405.
242. A. M. Downes, *Australian J. Chem.*, *11*, 154 (1958).
243. W. A. Armstrong, B. A. Black, and D. W. Grant, *J. Phys. Chem.*, *64*, 1415 (1960).
244. K. F. Nakken, T. Brustad, and A. K. Hansen, in *Radiation Chemistry—I*, *Advan. Chem. Ser.*, *81*, 251 (1968).
245. O. Volkert and D. Schulte-Frohlinde, *Tetrahedron Letters*, *1968*, 2151.
246. G. A. Hamilton, J. P. Friedman, and P. M. Campbell, *J. Am. Chem. Soc.*, *88*, 5266 (1966); G. A. Hamilton, J. W. Hanifin, Jr., and J. P. Friedman, *ibid.*, 5269.
247. R. Broszkiewicz, S. Minc, and Z. P. Zagorski, *Bull. Acad. Pol. Sci.*, *8*, 103 (1960).
248. A. I. Chernova, V. D. Orekhov, and M. A. Proskurnin, *Trans. Second All Union Conf. Radiation Chemistry*, Moskow, 1962, p. 233.
249. K. Sugimoto, W. Ando, and S. Oae, *Bull. Chem. Soc. Japan*, *36*, 124 (1963).
250. A. I. Chernova and V. D. Orekhov, *Kinetics Catalysis*, *7*, 41 (1966).
251. R. Broszkiewicz, *Nature*, *209*, 1235 (1966).
252. A. K. Pikaev, *Russ. Chem. Rev.*, *29*, 235 (1960).
253. V. A. Sharpatyi, *Russ. Chem. Rev.*, *30*, 279 (1961).
254. M. Daniels, in *Radiation Chemistry—I*, *Advan. Chem. Ser.*, *81*, 153 (1968) and references cited therein.
255. T. J. Sworski, R. W. Matthews, and H. A. Mahlman, in *Radiation Chemistry—I*, *Advan. Chem. Ser.*, *81*, 164 (1968) and references cited therein.
256. M. Daniels, *J. Phys. Chem.*, *70*, 3022 (1966).
257. N. Getoff and G. O. Schenck, *Radiation Res.*, *31*, 486 (1967) and references cited therein.

258. N. Getoff and G. O. Schenck, in *Radiation Chemistry—I, Advan. Chem. Ser.*, *81*, 337 (1968).

259. R. W. Matthews and D. F. Sangster, *J. Phys. Chem.*, *69*, 1938 (1965).

260. M. Anbar and J. K. Thomas, *J. Phys. Chem.*, *68*, 3829 (1964).

261. J. F. Ward and I. Kuo, in *Radiation Chemistry—I, Advan. Chem. Ser.*, *81*, 368 (1968).

262. H. C. Sutton, G. E. Adams, J. W. Boag, and B. D. Michael, in *Pulse Radiolysis*, M. Ebert, J. P. Keene, A. J. Swallow, and J. H. Baxendale, Eds., Academic Press, New York, 1965, p. 61.

263. B. Cercek, M. Ebert, C. W. Gilbert, and A. J. Swallow, in *pulse Radiolysis*, M. Ebert, J. P. Keene, A. J. Swallow, and J. H. Baxendale, Eds., Academic Press, New York, 1965, p. 83.

264. A. Rafi and H. C. Sutton, *Trans. Faraday Soc.*, *61*, 877 (1965).

265. L. I. Grossweiner and M. S. Matheson, *J. Phys. Chem.*, *61*, 1089 (1957).

266. Z. M. Bacq and P. Alexander, *Fundamentals of Radiobiology*, Butterworths, London, 1955.

267. D. E. Lea, *Actions of Radiations on Living Cells*, 2nd ed., Cambridge University Press, Cambridge, 1956.

268. W. D. Claus, Ed., *Radiation Biology and Medicine*, Addison-Wesley, Reading, Mass., 1958.

269. M. Burton, J. S. Kirby-Smith, and J. L. Magee, Eds., *Comparative Effects of Radiation*, Wiley, New York, 1960.

270. A. Hollaender, Ed., *Radiation Protection and Recovery*, Pergamon Press, New York, 1960.

271. L. Augenstein, R. Mason, and B. Rosenberg, Eds., *Physical Processes in Radiation Biology*, Academic Press, New York, 1964.

272. M. Errera and A. Forssberg, Eds., *Mechanisms in Radiobiology*, Academic Press, New York, 1961 to present.

273. J. Liebster and J. Kopoldová, *Advan. Radiation Biol.*, *1*, 157 (1964).

274. W. M. Garrison, *Current Topics Radiation Res.*, *4*, 43 (1968).

275. C. R. Maxwell, D. C. Peterson, and N. E. Sharpless, *Radiation Res.*, *1*, 530 (1954).

276. B. M. Weeks and W. M. Garrison, *Radiation Res.*, *9*, 291 (1958).

277. B. M. Weeks, *U.S. Atomic Energy Commission Rept.* UCRL-3071 (1955).

278. R. L. S. Willix and W. M. Garrison, *J. Phys. Chem.*, *67*, 1579 (1965).

279. J. V. Davies, M. Ebert, and A. J. Swallow, in *Pulse Radiolysis*, M. Ebert, J. P. Keene, A. J. Swallow, and J. H. Baxendale, Eds., Academic Press, New York, 1965, p. 165.

280. B. M. Weeks, S. A. Cole, and W. M. Garrison, *J. Phys. Chem.*, *69*, 4131 (1965).

281. W. M. Garrison, *Radiation Res. Suppl.*, *4*, 158 (1964).

282. R. L. S. Willix and W. M. Garrison, *Radiation Res.*, *32*, 452 (1967).

283. J. H. Baxendale, E. M. Fielden, and J. P. Keene, *Proc. Chem. Soc.*, *1963*, 242 (1963).

284. G. Scholes, P. Shaw, R. L. Willson, and M. Ebert, in *Pulse Radiolysis*, M. Ebert, J. P. Keene, A. J. Swallow, and J. H. Baxendale, Eds., Academic Press, New York, 1965, p. 151.

285. H. Taniguchi, K. Fukui, S. Ohnishi, H. Hatano, H. Hasegawa, and T. Maruyama, *J. Phys. Chem.*, *72*, 1926 (1968).

286. J. J. Weiss, *Progr. Nucleic Acid Res. Mol. Biol.*, *3*, 103 (1964).

287. G. Scholes, *Progr. Biophys. Mol. Biol.*, *13*, 59 (1963).
288. G. Scholes, J. F. Ward, and J. Weiss, *J. Mol. Biol.*, *2*, 379 (1960).
289. B. Ekert and R. Monier, *Nature*, *188*, 309 (1960).
290. R. Latarjet, B. Ekert, and P. Demerseman, *Radiation Res. Suppl.*, *3*, 247 (1963).
291. J. Holian and W. M. Garrison, *J. Phys. Chem.*, *71*, 462 (1967).
292. H. Reuschl, *Z. Naturforsch.*, *21b*, 643 (1966).
293. E. Gilbert, O. Volkert, and D. Schulte-Frohlinde, *Z. Naturforsch.*, *22b*, 477 (1967).
294. O. Volkert, W. Bors, and D. Schulte-Frohlinde, *Z. Naturforsch.*, *22b*, 480 (1967).
295. K. C. Smith and J. E. Hays, *Radiation Res.*, *33*, 129 (1968).
296. B. Ekert and R. Monier, *Ann. Inst. Pasteur*, *92*, 556 (1957).
297. G. Scholes and J. Weiss, *Nature*, *185*, 305 (1960).
298. B. Pullman and A. Pullman, *Quantum Biochemistry*, Interscience, New York, 1963.
299. C. L. Greenstock, M. Ng, and J. W. Hunt, in *Radiation Chemistry—I, Advan. Chem. Ser.*, *81*, 397 (1968).
300. R. M. Danziger, E. Hayon, and M. E. Langmuir, *J. Phys. Chem.*, *72*, 3842 (1968).
301. N. S. Ranadive, K. S. Korgaonkar, and M. B. Sakasrabudhe, *Proc. Intern. Conf. Peaceful Uses Atomic Energy*, United Nations, New York, *11*, 299 (1956).
302. R. A. Holroyd and J. W. Glass, *Intern. J. Radiation Biol.*, *14*, 445 (1968).
303. A. Kamal and W. M. Garrison, *Nature*, *206*, 1315 (1965).
304. L. S. Myers, Jr., M. L. Hollis, and L. M. Theard, in *Radiation Chemistry—I, Advan. Chem. Ser.*, *81*, 345 (1968).
305. J. F. Ward, A. A. Al-Thannon, and L. S. Meyers, Jr., unpublished work cited in ref. 243.
306. M. G. Ormerod and B. B. Singh, *Intern. J. Radiation Biol.*, *10*, 533 (1966).
307. J. Rabani, in *Radiation Chemistry—I, Advan. Chem. Ser.*, *81*, 131 (1968).
308. B. Ekert and R. Monier, *Nature*, *184*, BA 58 (1959).
309. R. Latarjet, B. Ekert, S. Apelgot, and N. Reybeyrotte, *J. Chim. Phys.*, *58*, 1046 (1961).
310. J. Holian and W. Garrison, *Nature*, *212*, 394 (1966).
311. T. C. Bruice and S. J. Benkovic, *Bioorganic Mechanisms*, Vols. 1 and 2, Benjamin, New York, 1966.
312. R. A. Holroyd, in *Fundamental Processes in Radiation Chemistry*, P. Ausloos Ed., Wiley, New York, 1968.
313. A. O. Allen, in *Current Topics Radiation Res.*, *4*, 1 (1968).
314. P. J. Dyne, *Can. J. Chem.*, *43*, 1080 (1965).
315. M. C. Sauer, Jr., S. Arai, and L. M. Dorfman, *J. Chem. Phys.*, *42*, 708 (1965).
316. L. M. Dorfman, in *Solvated Electron, Advan. Chem. Ser.*, *50*, 36 (1965).
317. J. C. Russell and G. R. Freeman, *J. Phys. Chem.*, *71*, 755 (1967).
318. L. R. Dalton, J. L. Dye, F. M. Fielden, and E. J. Hart, *J. Phys. Chem.*, *70*, 3358 (1966).
319. R. A. Basson, *Nature*, *211*, 629 (1966).
320. H. Seki and M. Imamura, *J. Phys. Chem.*, *71*, 870 (1967).
321. A. Habersbergerová, I. Janovsky, and J. Teplý, *Collection Czech. Chem. Commun.*, *32*, 1860 (1967).

322. J. Teplý and A. Habersbergerová, *Collection Czech. Chem. Commun.*, *32*, 1608 (1967).

323. J. Telpý and A. Habersbergerová, *Collection Czech. Chem. Commun.*, *32*, 1350 (1967).

324. R. A. Basson and H. J. van der Linde, *J. Chem. Soc.* (*A*), *1967*, 28.

325. G. R. Freeman, *J. Chem. Phys.*, *46*, 2822 (1967).

326. G. R. Freeman, in *Radiation Chemistry—II, Advan. Chem. Ser.*, *82*, 339 (1968).

327. C. E. McCauley and R. H. Schuler, *J. Am. Chem. Soc.*, *79*, 4008 (1957).

328. R. A. Holroyd and G. W. Klein, *J. Am. Chem. Soc.*, *87*, 4983 (1965).

329. T. F. Williams, *Trans. Faraday Soc.*, *57*, 755 (1961).

330. K. Tanno, T. Miyazaki, K. Shinsaka, and S. Shida, *J. Phys. Chem.*, *71*, 4290 (1967).

331. R. D. Koob and L. Kevan, *J. Phys. Chem.*, *70*, 1336 (1966).

332. F. Williams, *Nature*, *194*, 348 (1962).

333. J. J. J. Myron and G. R. Freeman, *Can. J. Chem.*, *43*, 1484 (1965).

334. D. R. Howton and G. S. Wu, *J. Am. Chem. Soc.*, *89*, 516 (1967).

335. W. R. Busler, D. H. Martin, and F. Williams, *Discussions Faraday Soc.*, *36*, 102 (1963).

336. D. P. Stevenson and D. O. Schissler, *J. Chem. Phys.*, *23*, 1353 (1955).

337. A. Henglein and G. A. Muccini, *Z. Naturforsch.*, *17a*, 452 (1966).

338. A. A. Scala, S. G. Lias, and P. Ausloos, *J. Am. Chem. Soc.*, *88*, 5701 (1966).

339. P. Ausloos, A. A. Scala, and S. G. Lias, *J. Am. Chem. Soc.*, *88*, 1583 (1966).

340. P. Ausloos, A. A. Scala, and S. G. Lias, *J. Am. Chem. Soc.*, *89*, 3677 (1967).

341. R. D. Doepker and P. Ausloos, *J. Chem. Phys.*, *42*, 3746 (1965).

342. F. P. Abramson and J. H. Futrell, *J. Phys. Chem.*, *71*, 1233 (1967).

343. P. C. Chang, N. C. Yang, and C. D. Wagner, *J. Am. Chem. Soc.*, *81*, 2060 (1959).

344. C. D. Wagner, *J. Phys. Chem.*, *66*, 1158 (1962).

345. J. B. Burks, *Theory and Practice of Scintillation Counting*, Macmillan, New York, 1964.

346. S. Lipski, in *Physical Processes in Radiation Biology*, Academic Press, New York, 1964, p. 215.

347. P. K. Ludwig and M. Burton, in *Radiation Chemistry—II, Advan. Chem. Ser.*, *82*, 542 (1968).

348. H. Dreeskamp and M. Burton, *Z. Elektrochem.*, *64*, 165 (1960).

349. P. C. Sjölin, *Nucl. Instrum. Methods*, *37*, 45 (1965).

350. M. Burton, P. K. Ludwig, M. S. Kenward, and R. J. Povinelli, *J. Chem. Phys.*, *41*, 2563 (1964).

351. J. Nosworthy, *Trans. Faraday Soc.*, *61*, 1138 (1965).

352. R. B. Cundall and P. A. Griffiths, *Trans. Faraday Soc.*, *61*, 1968 (1965).

353. R. B. Cundall and P. A. Griffiths, *Chem. Commun.*, *1966*, 194.

354. R. B. Cundall and W. Tippett, in *Radiation Chemistry—II, Advan. Chem. Ser.*, *82*, 387 (1968).

355. E. A. Cherniak, E. Collinson, and F. S. Dainton, *Trans. Faraday Soc.*, *60*, 1408 (1964).

356. R. H. Schuler and R. R. Kuntz, *J. Phys. Chem.*, *67*, 1004 (1963).

357. P. P. Infelta and R. H. Schuler, *J. Phys. Chem.*, in press.

358. E. Hayon and M. Moreau, *J. Phys. Chem.*, *69*, 4053 (1965).

359. P. H. Tewari and G. R. Freeman, *J. Chem. Phys.*, *49*, 4394 (1968).

360. S. J. Rzad and R. H. Schuler, *J. Phys. Chem.*, **72**, 228 (1968).
361. D. M. J. Compton, J. F. Bryant, R. A. Cesena, and B. L. Gehman, in *Pulse Radiolysis*, M. Ebert, J. P. Keene, A. J. Swallow, and J. H. Baxendale, Eds., Academic Press, New York, 1965, p. 43.
362. W. G. Burns, *J. Chem. Phys.*, **48**, 1876 (1968).
363. W. F. Schmidt and A. O. Allen, *Science*, **160**, 301 (1968).
364. W. F. Schmidt and A. O. Allen, *J. Phys. Chem.*, **72**, 3730 (1968).
365. A. J. Parker, *Quart. Rev.*, 163 (1962).
366. A. J. Parker, *Advan. Org. Chem.*, **5**, 1 (1965).
367. A. J. Parker, *Advan. Phys. Org. Chem.*, **5**, 173 (1967).
368. R. R. Hentz and W. V. Sherman, *J. Phys. Chem.*, **72**, 2635 (1968).
369. J. H. Baxendale and M. A. J. Rodgers, *J. Phys. Chem.*, **72**, 3849 (1968).
370. A. Matsumoto and N. N. Lichtin, in *Radiation Chemistry—II, Advan. Chem. Ser.*, **82**, 547 (1968).
371. A. Singh, H. D. Gesser, and A. R. Scott, *Chem. Phys. Letters*, **2**, 271 (1968).
372. M. Szwarc, *Progr. Phys. Org. Chem.*, **6**, 323 (1968).
373. E. T. Kaiser and L. Kevan, Eds., *Radical Ions*, Interscience, New York, 1968.
374. P. B. Ayscough and F. P. Sargent, *Proc. Chem. Soc.*, *1963*, 94.
375. P. H. Rieger, I. Bernal, W. H. Reinmuth, and G. K. Fraenkel, *J. Am. Chem. Soc.*, **85**, 683 (1963).
376. D. H. Geske and A. H. Maki, *J. Am. Chem. Soc.*, **82**, 2671 (1960).
377. G. J. W. Gutch and W. A. Waters, *Chem. Commun.*, *1966*, 39.
378. W. T. Dixon and R. O. C. Norman, *Nature*, **196**, 891 (1962).
379. T. R. Tuttle and S. I. Weismann, *J. Am. Chem. Soc.*, **80**, 5342 (1958).
380. P. B. Ayscough and R. Wilson, *J. Chem. Soc.*, *1963*, 5412.
381. P. B. Ayscough, R. P. Sargent, and R. Wilson, *J. Chem. Soc.*, *1963*, 5418.
382. P. L. Kolker and W. A. Waters, *Proc. Chem. Soc.*, *1963*, 55.
383. P. Balk, G. J. Hoijtink, and J. W. H. Schreurs, *Rec. Trav. Chim.*, **76**, 813 (1957)
384. E. DeBoer and S. I. Weissman, *Rec. Trav. Chim.*, **76**, 824 (1957).
385. D. Gill, J. Jagur-Grodzinski, and M. Szwarc, *Trans. Faraday Soc.*, **60**, 1424 (1964).
386. G. A. Russell, E. G. Janzen, and E. T. Strom, *J. Am. Chem. Soc.*, **86**, 1807 (1964).
387. M. C. R. Symons, *Advan. Phys. Org. Chem.*, **1**, 284 (1963).
388. A. Carrington, F. Dravnieks, and M. C. R. Symons, *J. Chem. Soc.*, *1959*, 947.
389. J. R. Bolton and A. Carrington, *Mol. Phys.*, **5**, 161 (1962).
390. D. Bethell and V. Gold, *Carbonium Ions, An Introduction*, Academic Press, New York, 1967.
391. M. M. Bursey and F. W. McLafferty, in *Carbonium Ions*, Vol. 1, G. A. Olah and P. v. R. Schleyer, Eds., Interscience, New York, 1968, p. 257.
392. E. J. Corey, N. L. Bauld, R. T. LaLonde, J. Casanova, and E. T. Kaiser, *J. Am. Chem. Soc.*, **82**, 2645 (1960).
393. W. J. Koehl, *J. Am. Chem. Soc.*, **86**, 4686 (1964).
394. E. J. Corey and J. Casanova, *J. Am. Chem. Soc.*, **85**, 165 (1963).
395. J. K. Kochi, *J. Am. Chem. Soc.*, **84**, 3271 (1962).
396. J. K. Kochi and A. Bemis, *J. Am. Chem. Soc.*, **90**, 4038 (1968).
397. G. Brieglieb, *Angew. Chem.*, **76**, 326 (1964).
398. S. Arai, E. L. Tremba, J. R. Brandon, and L. M. Dorfman, *Can. J. Chem.*, **45**, 1119 (1967).

334   E. J. FENDLER AND J. H. FENDLER

399. L. M. Dorfman, private communication.
400. S. Arai, D. A. Grev, and L. M. Dorfman, *J. Chem. Phys.*, *46*, 2572 (1967).
401. S. Arai and L. M. Dorfman, in *Radiation Chemistry—II, Advan. Chem. Ser.*, *82*, 378 (1968).
402. G. E. Adams, B. D. Michael, and J. T. Richards, *Nature*, *215*, 1248 (1967).
403. J. M. Fritsch, T. P. Layloff, and R. N. Adams, *J. Am. Chem. Soc.*, *87*, 1724 (1965).
404. R. A. Marcus, *J. Chem. Phys.*, *24*, 966 (1956).
405. R. A. Marcus, *J. Chem. Phys.*, *26*, 867 (1957).
406. R. A. Marcus, *J. Chem. Phys.*, *26*, 872 (1957).
407. R. A. Marcus, *Discussions Faraday Soc.*, *29*, 21 (1960).
408. R. A. Marcus, *J. Phys. Chem.*, *67*, 853 (1963).
409. R. A. Marcus, *J. Chem. Phys.*, *43*, 679 (1965).
410. R. A. Marcus, *J. Chem. Phys.*, *43*, 2654 (1965).
411. R. A. Marcus, *J. Chem. Phys.*, *43*, 3477 (1965).
412. J. Chaudhuri, J. Jagur-Grodzinski, and M. Szwarc, *J. Phys. Chem.*, *71*, 3063 (1967).
413. J. Jagur-Grodzinski, M. Feld, S. L. Yang, and M. Szwarc, *J. Phys. Chem.*, *69*, 628 (1965).
414. R. V. Slates and M. Szwarc, *J. Phys. Chem.*, *69*, 4124 (1965).
415. G. E. Adams, B. D. Michael, and R. L. Willson, in *Radiation Chemistry—I, Advan. Chem. Ser.*, *81*, 289 (1968).
416. G. S. Hammond, J. Saltiel, A. A. Lamola, N. J. Turro, J. S. Bradshaw, D. O. Cowan, R. C. Counsell, V. Vogt, and C. Dalton, *J. Am. Chem. Soc.*, *86*, 3197 (1964).
417. G. Stein, in *The Chemistry of Ionization and Excitation*, G. R. A. Johnson and G. Scholes, Eds., Taylor and Francis Ltd., London, 1957, p. 25.
418. R. R. Hentz, D. B. Peterson, S. B. Srivastava, H. F. Barzynski, and M. Burton, *J. Phys. Chem.*, *70*, 2362 (1966).
419. E. Fischer, H. P. Lehmann, and G. Stein, *J. Chem. Phys.*, *45*, 3905 (1966).
420. E. Fischer, G. Fischer, and G. Stein, *J. Chem. Phys.*, *46*, 3680 (1967).
421. D. S. McClure, *J. Chem. Phys.*, *19*, 670 (1951).
422. G. Porter and M. W. Windsor, *Proc. Royal Soc. (London)*, *Ser. A*, *245*, 238 (1958).
423. J. D. McCollum and W. A. Wilson, *U.S. Tech. Report ASD-61-170* (1961).
424. J. D. McCollum and T. D. Nevitt, *U.S. Tech. Report ASD-TDR-63-616* (1963).
425. J. M. Nosworthy and J. P. Keene, *Proc. Chem. Soc.*, *1964*, 114.
426. J. W. Hunt and J. K. Thomas, *J. Chem. Phys.*, *46*, 2954 (1967).
427. E. J. Land and A. J. Swallow, *Trans. Faraday Soc.*, *64*, 1247 (1968).
428. J. K. Thomas, K. Johnson, T. Klippert, and R. Lowers, *J. Chem. Phys.*, *48*, 1608 (1968).
429. S. Arai and L. M. Dorfman, *J. Phys. Chem.*, *69*, 2239 (1965).
430. F. S. Dainton, T. J. Kemp, G. A. Salmon, and J. P. Keene, *Nature*, *203*, 1050 (1964).
431. J. K. Thomas, *Preprints*, ACS Division of Petroleum Chemistry, *13*, D18 (1968).
432. L. M. Dorfman, E. J. Land, and J. K. Thomas, private communications.
433. J. P. Guarino and W. H. Hamill, *J. Am. Chem. Soc.*, *86*, 777 (1964).
434. T. Shida and W. H. Hamill, *J. Chem. Phys.*, *44*, 2375 (1966).
435. K. D. Cadogan and A. C. Albrecht, *J. Chem. Phys.*, *43*, 2550 (1965).

436. N. Yamamoto, Y. Nakato, and H. Tsubomura, *Bull. Chem. Soc. Japan, 39,* 2603 (1966).
437. T. J. Kemp, J. P. Roberts, G. A. Salmon, and G. F. Thompson, *J. Phys. Chem., 72,* 1464 (1968).
438. A. J. Fry, R. S. H. Liu, and G. S. Hammond, *J. Am. Chem. Soc., 88,* 4781 (1966).
439. G. N. Lewis and D. Lepkin, *J. Am. Chem. Soc., 64,* 2801 (1942).
440. G. N. Lewis and J. Bigeleisen, *J. Am. Chem. Soc., 65,* 520 (1943).
441. C. Capellos and A. O. Allen, *J. Phys. Chem., 72,* 4265 (1968).
442. R. Cooper and J. K. Thomas, in *Radiation Chemistry—II, Advan. Chem. Ser., 82,* 351 (1968).
443. G. Porter and M. W. Windsor, *Discussions Faraday Soc., 17,* 178 (1954).
444. E. J. Land and G. Porter, *Trans. Faraday Soc., 59,* 2027 (1963).
445. S. Arai, H. Ueda, R. F. Firestone, and L. M. Dorfman, *J. Chem. Phys., 50,* 1072 (1969).
446. C. Capellos and A. O. Allen, *Science, 160,* 302 (1968).
447. T. L. Chu and S. I. Weissman, *J. Chem. Phys., 22,* 21 (1954).
448. R. E. Bühler and M. Ebert, *Nature, 214,* 1220 (1967).
449. R. E. Bühler, *Helv. Chim. Acta, 51,* 1558 (1968).
450. T. A. Gover and G. Porter, *Proc. Roy. Soc. (London), Ser. A, 262,* 476 (1961).
451. G. Briegleb, *Elektronen-Donator-Acceptor-Komplexe,* Springer, Berlin, 1961.
452. H. A. Benesi and J. H. Hildebrand, *J. Am. Chem. Soc., 71,* 2703 (1949).
453. G. Dobson and L. I. Grossweiner, *Trans. Faraday Soc., 61,* 708 (1965).
454. E. Collinson, F. S. Dainton, and H. Gillis, *J. Phys. Chem., 65,* 695 (1961).
455. E. E. van Tamelen and S. P. Pappas, *J. Am. Chem. Soc., 85,* 3297 (1963).
456. K. E. Wilzbach, J. S. Ritscher, and L. Kaplan, *J. Am. Chem. Soc., 89,* 1031 (1967).
457. W. Schäfer and H. Hellman, *Angew. Chem., Intern. Ed., 6,* 518 (1967).
458. J. Fajer and D. K. MacKenzie, *J. Phys. Chem., 71,* 784 (1967).
459. J. Fajer and D. R. MacKenzie, in *Radiation Chemistry—II, Advan. Chem. Ser., 82,* 469 (1968).
460. G. Gamaggi, F. Gozzo, and G. Cevidalli, *Chem. Commun., 1966,* 313.
461. I. Haller, *J. Am. Chem. Soc., 88,* 2070 (1966).
462. D. M. Lemal and J. P. Lokensgard, *J. Am. Chem. Soc., 88,* 5934 (1966).
463. I. Haller, *J. Chem. Phys., 47,* 1117 (1967).

Author Index

Numbers in parentheses are reference numbers and indicate that the author's work is referred to although his name is not mentioned in the text. Numbers in *italics* show the pages on which the complete references are listed.

Subject Index

Progress in Physical Organic Chemistry

CUMULATIVE INDEX, VOLUMES 1–7